Matthew Cooke

Injurious insects of the Orchard, Vineyard,

Field, Garden, Conservatory, Household, Storehouse, Domestic Animals, etc. ...

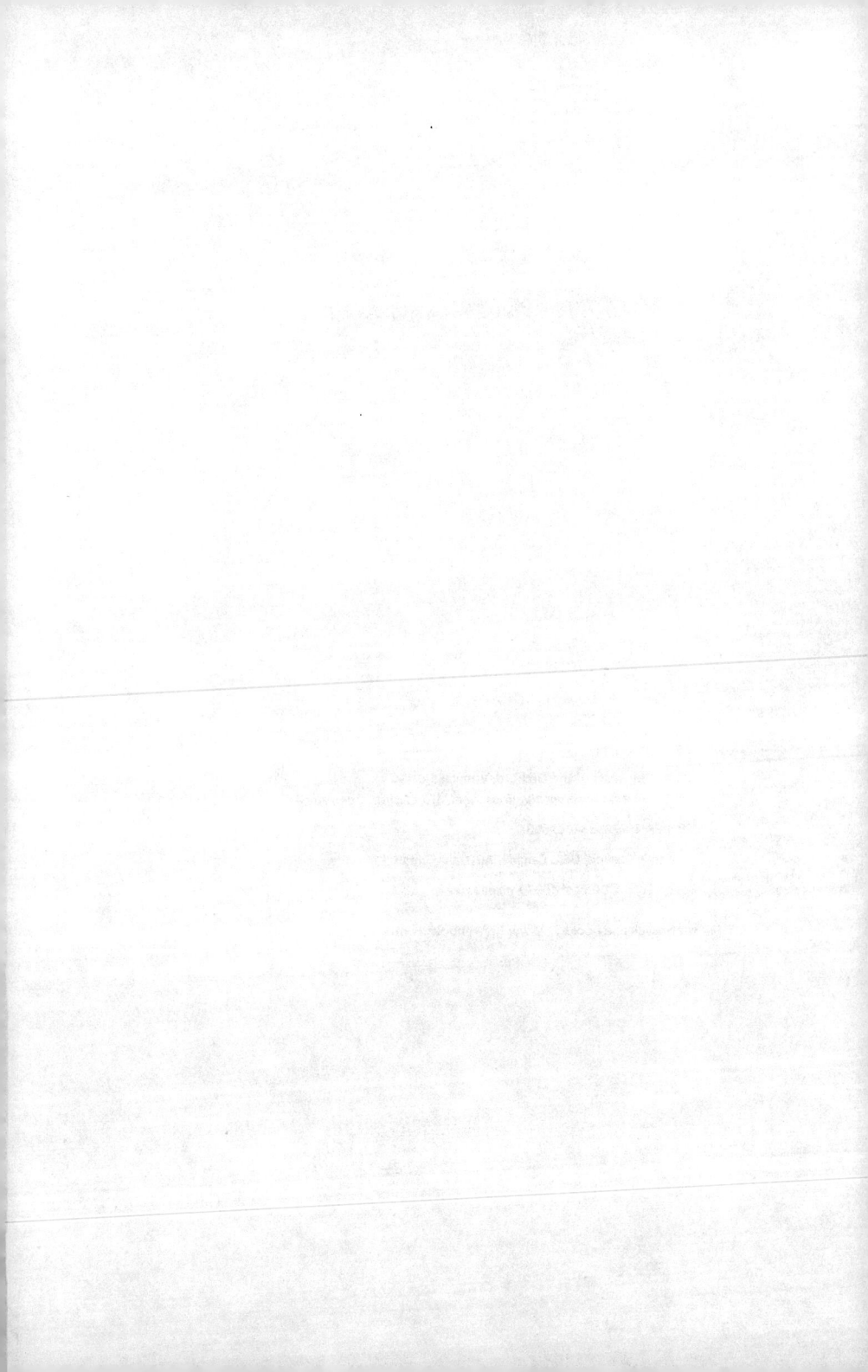

Matthew Cooke

Injurious insects of the Orchard, Vineyard,
Field, Garden, Conservatory, Household, Storehouse, Domestic Animals, etc. ...

ISBN/EAN: 9783337082666

Printed in Europe, USA, Canada, Australia, Japan

Cover: Foto ©berggeist007 / pixelio.de

More available books at **www.hansebooks.com**

OF THE

ORCHARD, VINEYARD,

FIELD, GARDEN, CONSERVATORY,

HOUSEHOLD, STOREHOUSE, DOMESTIC ANIMALS, ETC.,

WITH

REMEDIES FOR THEIR EXTERMINATION

BY MATTHEW COOKE,

LATE CHIEF EXECUTIVE HORTICULTURAL OFFICER OF CALIFORNIA

SACRAMENTO:

H. S. CROCKER & CO., PRINTERS AND STATIONERS.

1883.

PREFACE.

This book is designed for the use of orchardists, vineyardists, farmers, and others interested in the subjects treated. It
is designed to convey practical information concerning some
of the species of insects injurious to the industries of cultivators of the soil, and those interested in earth produce generally.

It has been my aim to free the volume, so far as was possible,
from technical terms, and I have retained the technical or scientific names of insects only to aid the reader in reference to
scientific works.

It must not be assumed that there has been any intention to
present this work as a scientific treatise, and I may be par-
doned for supplementing the statement by the information, that
I have never laid claim whatever to scientific education. My
advantages were limited in youth to a common school system,
and since that period I have, from time to time, pursued the
study of economic entomology, as opportunity allowed, up to
the year 1874. Since that date I have been enabled to engage
in extended practical investigations into the realm which before
I had knowledge of only by reading.

I was led to the field of experiment and investigation through
my business. In 1874 I engaged in the manufacture, in the
City of Sacramento, of fruit boxes. The next year the codlin
moth appeared in some orchards, and the fruit of the country
was threatened. Naturally, its injury would affect my business, and thus I was drawn to a consideration of means to protect it. I therefore entered upon the field of investigation, as
I found the text-books and treatises did not afford the desired
information—at least such as I was enabled to find in the lit-
erature of the subject.

In 1878 I began to give the result of my inquiries to the
public through the columns of the daily and the weekly press.
Very few persons in this State, especially of those who should
have manifested the deepest interest at that time, paid any
attention to economic entomology. In time, however, through
the assistance of the Sacramento RECORD-UNION and the PACIFIC
RURAL PRESS, and some other journals, and the discussions

engaged in at the meetings of the State Horticultural Society, the public was brought to a degree of appreciation of the dangers threatening the fruit interests, and as a result, protective legislation was had in March, 1881.

In connection with the investigations made, the results of which appear in this volume, I deem it not out of place to state that I have been engaged in mechanical pursuits nearly forty years. I became interested in the fruit box business, as a manufacturer, through financial assistance extended me by Mrs. MARY E. GREGORY, and the late HENRY MILLER, both of Sacramento. To the former, as a token of sincere acknowledgment, this volume is respectfully dedicated.

It is due that I should acknowledge my special indebtedness to Professor CHARLES V. RILEY, Entomologist of the United States Department of Agriculture, for his kindness in affording me the use of his valuable reports and works, and also for his identification of specimens sent to him for that purpose from time to time.

To Professor CYRUS THOMAS, late State Entomologist of Illinois, and at present connected with the Smithsonian Institution at Washington, D. C., for the use of his valuable volumes of reports, I am much indebted.

To Professor J. H. COMSTOCK, late Entomologist of the United States Department of Agriculture, and at present Professor of Entomology in Cornell University, at Ithaca, New York, my thanks are due for favors extended.

To my respected friend, Professor C. H. DWINELLE, of the State University, at Berkeley, California, my sincere thanks are due for his kindly assistance and his many favors to me during the period I have been engaged upon this work.

I desire also to acknowledge the kind favors of R. B. BLOWERS, Esq., of Woodland, California, extended during my investigation of grapevine and other insect pests.

I have freely consulted, in the preparation of the volume now presented, the reports of Drs. LE BARON and WALSH, Illinois; Miss ELEANOR A. ORMEROD, Consulting Entomologist to the Royal Agricultural Society of England; of Drs. FITCH and EMMONS, of New York; Professor A. S. PACKARD's "GUIDE TO THE STUDY OF INSECTS;" Dr. HARRIS' "INSECTS IN-

JURIOUS TO VEGETATION;" Professor SAUNDERS' recent work, "INSECTS INJURIOUS TO FRUITS;" FIGUIER's "INSECT WORLD;" "THE CANADIAN ENTOMOLOGIST," and the "BULLETINS OF THE UNITED STATES ENTOMOLOGICAL COMMISSION."

In the preparation of this book I have been placed under special obligations to D. W. COQUILLET, Esq., late Assistant State Entomologist of Illinois, who has been assiduous in his labors for the last four months, in preparing manuscript, reading proof, and assisting me in my labors connected with the issuance of the volume.

In the matter of illustrations, some of those presented have been copied from previous works of this character, and due credit has been given in another place.

I should be doing an injustice to my own sense of propriety, if I submitted these pages to the criticism of the reader without acknowledging the courtesy of my publishers, H. S. CROCKER & Co.. of Sacramento, at whose establishment the typographic work has been done; and also if I failed to add that the binding in which the volume appears, was the work of a fellow mechanic of Sacramento, FRANK FOSTER, Esq., the pioneer bookbinder of the city. Very much is due to the skill, patience, and care of O. H. TUBBS. foreman for CROCKER & Co., for the freedom of the book from errors, and for its neatness of appearance.

This volume presents to the reader the results of my investigations during the eight years last past. Two years of that time I had the honor to fill the position of Chief Executive Horticultural Officer of this State, a fact that it is deemed proper to state, that the reader may be fully informed of the opportunities enjoyed for the preparation of the work now offered him.

I am much indebted to the fruit-growers of California, and to the cultivators of the soil generally throughout the State, for their cordial assistance in my work; and I feel it no more than right to close this page by saying that to the press of the State which first enabled me to awaken the people to the danger threatening them from insect pests, and has since given me free use of its columns, not only MY thanks are due, but those of all the citizens of California.   M. C.

SACRAMENTO, CAL., September, 1883.

ACKNOWLEDGMENTS.

The electrotypes of the following Figures were purchased from Prof. C. V. Riley : 14, 25, 29, 30, 33-39, 43-45, 52-60, 63, 68, 72, 77, 85, 86, 90, 91, 103, 105, 109, 120, 121, 146, 148-153, 159-178, 180, 181, 183, 187-193, 195$\frac{1}{2}$, 196, 198, 200, 205-213, 215, 216, 218, 219, 232, 264-266, 269, 271-281, 285-288, 290-293, 298-309, 312-333, 338, 339, 359, 360, 365 and 368.

The following were purchased from Lippincott & Co., of Philadelphia: 31, 69-71, 88 89, 104, 147, 197, 199, 201 and 211.

Electrotypes of Figures 11-13, 51, 96, 97, 138 and 155, were purchased of Dr. A. S. Packard, Jr.

The following were copied from Dr. Fitch's Reports : 73, 74, 114, 204, 259-263, 268, 273-275, 289, 336, 337, 356 and 357.

Figures 26-28, 116-119, were copied from Riley's Reports.

Figures 156, 202, 203, 233, 254, 347, 351, 363, 364 and 369, were copied from Packard's "Guide to the Study of Insects."

The following are from Harris' "Insects Injurious to Vegetation :" 67, 108, 113, 157, 227, 229, 234, 244, 245, 255 and 294.

The following were copied from Figuier's "Insect World :" 87, 124-127, 194, 195, 348, 349, 361 and 362.

Figures 256, 257, 267, 282, 284, 334 and 340, were copied from Miss Ormerod's "Manual of Injurious Insects."

The following are from the U. S. Agr. Reports : 40-42, 82-84, 277-280, 297, 341-343, 366, 367, 370.

Figures 350 and 352 were copied from the report of the Ontario Entomological Society.

Figure 283 is from Dr. LeBaron's Fourth Illinois Report, and Figure 8 was copied from the Mass. Agr. Report.

Figure 253 is from the "Scientific American Supplement."

The following were kindly loaned me by the "Pacific Rural Press :" 80, 122, 128, 132-137, 140, 141, 154, 186, 226, 235, 237$\frac{1}{4}$, 237$\frac{1}{2}$, 241, 242, 248-252.

To Mr. C. Muller, 135 Montgomery St., San Francisco, I am indebted for the use of the electrotypes of Figures 9 and 10.

The remaining cuts are original, and were made expressly for this work. Of these, Figures 32, 40, 41, 42, 51a, 64, 65, 66, 75, 76, 78, 82, 83, 84, 92, 93, 94, 95, 106, 107, 110, 111, 112, 115, 127, 146, 179, 182, 184, 185, 220, 221, 230, 231, 246, 277, 278, 279, 280, 297, 310, 341, 342, 343, 353, 354, 355, 358, 370, 371 and 372, were drawn and engraved by Mr. Buchi, of Messrs. M. Schmidt & Co., of San Francisco. These were received too late to be inserted in the text, and so are given on Plates 1-4. Several of the remaining figures were drawn and engraved by Mr. R. Philip, of this city, and by Mr. V. Craig and Mr. Patten, 659 Clay St., San Francisco.

In order to assist fruit-growers and others in readily identifying insects, twenty-five pages of classified figures are given.

CONTENTS.

INTRODUCTORY.

CHAPTER I.

The Sentiments of an Enterprising Fruit Grower.

The following is quoted from an address by A. T. Hatch, Esq., of Cordelia, Solano County, California, before the Committee of Agriculture of the Senate, at the State Capitol, January 13, 1883:

"Our watchword must ever be onward and upward, and falter not, although difficulties apparently insurmountable arise; he who will, may overcome them. The enterprising fruit growers of California are filled with a spirit that no power on earth can curb. It falters not at misfortune's door, or any obstacle to success, but boldly advances and removes them all; at least, it *has* been so, and must *ever* be. The time was when our glorious climate, fruitful soil, and exemption from all diseases and pests, made our Golden State the wonder of all who were conversant of its fruits and flowers. Now, alas! the spoiler's hand is felt—a change has come over the spirit of our dream. It seems as though all that is detrimental to the fruit interest is here, or coming, making eternal vigilance the price of success in this, *the* industry of the State. The time has come when every one who by this occupation would thrive, will find ceaseless use for head and hand; even then, *the fittest only can survive.* Who will supinely sit and see misfortune spoil the result of years of toil, while others gird on their armor with energies stimulated by the presence of the forces arrayed against them on every hand?"

CHAPTER II.

History of Legislation to Prevent the Spread of Injurious Insects.

As California is the first of the United States, and it may be said of any part of the world, that has attempted to prevent

2

the spread of injurious insects of the orchard, etc., by legislation; and as fruit growers in distant parts of the world are watching our success, it may be well to place on the record what has been accomplished in the space of thirty months, or, from October, 1880, to April 1. 1883.

During 1879 and 1880, the subject of the spread of the codlin moth and other injurious insects of the orchard was discussed in the newspapers of this State, and at the meetings of the State Horticultural Society; and these discussions resulted in the above society, at its October meeting, 1880, appointing a committee, consisting of Professor C. H. Dwinelle, of Alameda County; Dr. Behr, of San Francisco; A. T. Hatch, of Solano County; W. H. Jessup, of Alameda; and Matthew Cooke, of Sacramento, to prepare a bill for the protection of the horticultural industries of the State, to be presented to the Legislature, which would meet in the month of December following.

The committee met at the office of the Hon. J. N. Young, at Sacramento, early in the month of November, and a bill was prepared, which, in due time, was presented in the Assembly by Mr. Young. About the same time a bill was presented in the Senate by Senator Baker, of Santa Clara. These bills were referred to a joint committee of the Senate and Assembly. The chairman of this committee called a meeting, and the fruit growers were represented by A. T. Hatch, W. H. Jessup, and the late James B. Saul, and others ; as the bills presented for the protection of horticulture conflicted somewhat with each other, and also with another bill (viticultural) before the Legislature, the committee recommended that a new bill be presented. A new bill was prepared, and on its being brought before the Senate was passed to a second reading, but amended to such an extent that it was thought best to have it withdrawn. Later in the session, the Hon. William Johnston, of Sacramento County, then Vice-President of the Senate, had the viticultural bill, which had passed the Assembly, amended in the Senate by adding Section 8. The viticultural law, as enacted, reads as follows :

AN ACT

To define and enlarge the duties and powers of the Board of State Viticultural Commissioners, and to authorize the appointment of certain officers, and to protect the interests of horticulture and viticulture.

[Approved March 4, 1881.]

The People of the State of California, represented in Senate and Assembly, do enact as follows:

SECTION 1. The Board of State Viticultural Commissioners, in addition to the duties and powers provided for by the Act entitled " An Act for the promotion of viticultural industries of the State," approved April 15, 1880, shall, in respect to diseases of grape vines and vine pests, constitute a Board of Health. It shall, in addition to laboratory work, cause practical experiments to be made to determine or demonstrate the utility of known and new remedies against such diseases and pests.

SEC. 2. The Board shall elect of their own number, or appoint from without their number, a competent person to serve as Chief Executive Viticultural Officer, who shall perform also the duties of Viticultural Health Officer, under direction of said Board, and subject to removal from such office at any time by the Board.

SEC. 3. The Viticultural Health Officer shall have power, subject to the approval of the Board, to prevent the spread of vine diseases and vine pests, by declaring and enforcing rules and regulations in the nature of quarantine, to govern the manner of, restrain, or prohibit the importation into the State, and the distribution and disposal within the State, of all vines, vine cuttings, debris of vineyards, empty fruit boxes, or other material, on, or by which the contagion of vine diseases and germs of vine pests may be introduced into the State, or transported from place to place within the State ; to declare and enforce regulations approved by the Board for the disinfection of vines, vine cuttings, vineyard debris, empty fruit boxes, and other suspected material dangerous to vineyards, while in transit, or about to be distributed, or transported into, or within the State ; to classify the vineyards and viticultural regions of the State, according to the degree of health, or vine disease prevailing therein, and to change the same as circumstances may require to be done, subjecting each class to such varying rules and regulations, respecting the introduction or transportation of vines, vine cuttings, and other material liable to spread contagion of disease among vines, as may, in the opinion of the Board, become necessary and expedient for the preservation of vineyards. Such rules and regulations shall be circulated in printed form by the Board among the vine growers and fruit dealers of the State, shall be published at least thirty days in two daily newspapers of general circulation in the State, not of the same city or county, and shall be posted in a conspicuous place at the county seat of each county affected by their provisions.

SEC. 4. The Viticultural Health Officer may appoint local resident Inspectors in any and all of the viticultural regions of the State, whose duties shall be to report to him concerning the health of grapevines, the progress of vine diseases and pests, and all violations of the rules and regulations of the Board; to certify to the proper disinfection of vines, vine cuttings, empty fruit boxes, and other transportable articles required by the Board to be disinfected before transportation, or while in transit, or after delivery at any point of destination: the methods of disinfection to be determined and approved by the Health Officer and the Board; to seize upon and destroy all vines, vine cuttings, debris of vineyards, empty fruit boxes, and other material liable to spread contagion, which may be found in transit, or delivered after transportation, not certified to as required by the Board; *provided*, that the same may be exempt from such destruction if the cost of disinfection by such Inspector shall be provided for by the owner or agent in charge thereof, as may be prescribed for such cases of negligence, carelessness, or violation of quarantine rules, and to keep a record of all proceedings as such Inspectors; *provided*, that there shall be no compensation for such services of inspection, excepting a fee, not to exceed one dollar for each certificate of disinfection, in case of compliance with quarantine regulations, and not to exceed five dollars for each certificate of disinfection after seizure for non-compliance; *provided, however*, such inspection may be employed at the option of the owners of property requiring disinfection to disinfect the same. All vines, or other articles absolutely prohibited of importation or transportation, may be promptly destroyed by any Inspector discovering the same transported or in transit, in violation of regulations, and the cost of such seizure, together with a fee of ten dollars, shall be paid to such Inspector out of any fine that may be collected from the party or parties guilty of such violation. Willful violation of the quarantine regulations of the Board shall be considered a misdemeanor, and punishable by a fine of not less than twenty-five nor more than one hundred dollars. Whenever required for the convenience of vine or fruit growers, or fruit dealers, a resident Inspector shall be appointed upon petition of any three neighboring vine or fruit growers, or dealers in grapes, to reside in their vicinity, if not already provided for; and there shall be not less than two Inspectors appointed for each county which is subjected to such quarantine regulations, and they shall each be subject to removal at the will of the Viticultural Health Officer, if incompetent, or if they fail to perform their duties, or are unreasonably distasteful to vine growers and grape dealers.

SEC. 5. It shall be also the duty of the Chief Executive Viticultural Officer to personally visit, examine, and report upon the several viticultural regions of the State; to prepare documents for publication, as required by the Board, relating to any and all branches of viticultural industry, including treatises for the instruction of the public; to supervise the preparation of reports for publication, and especially report upon the practibility and means of eradicating diseases from vineyards, and to superintend experiments with known and new remedies.

SEC. 6. All printing heretofore ordered by the Board shall be paid for out of the appropriations heretofore made for its use. All printing required hereafter shall be done by the State Printer.

SEC. 7. The salary of the Chief Executive Viticultural Officer shall be fixed by the Board, not to exceed one hundred and fifty dollars per month, for services while engaged as such officer, and his actual traveling expenses shall be allowed, not to exceed five hundred dollars per annum.

SEC. 8. The Board of State Viticultural Commissioners shall also appoint an officer, who shall be especially qualified, by practical experience in horticulture, for the duties of his office, to perform similar duties respecting the protection of fruit and fruit trees as are provided for in this Act in reference to grapevines, with like powers; and the salary and traveling expenses of such officer shall be fixed by the said Board, at the same amounts provided for in the case of the Chief Executive Viticultural Officer; and the said Board shall have power to establish such quarantine rules and regulations as are required for the protection of fruit and fruit trees from the spread of insect pests.

SEC. 9. There is hereby appropriated for the uses of the Board of State Viticultural Commissioners, as set forth in this Act, and in the Act providing for its organization, out of any moneys in the State treasury not otherwise appropriated, the sum of ten thousand dollars for the year commencing July first, eighteen hundred and eighty-one, and ten thousand dol-

lars for the year commencing July first, eighteen hundred and
eighty-two; and the State Controller will draw his warrants
upon the State Treasurer in favor of the Treasurer of the said
Board for the said sums, or any part thereof, when they become
available, upon proper demand being made for the same by
said Board; *provided*, that no claim shall be paid out of said
appropriation until the same shall have been presented to and
approved by the State Board of Examiners.

SEC. 10. This Act shall take effect and be in force from
and after its passage.

Assemblyman Reynolds, of Santa Clara, introduced the
following bill, which passed both houses and was approved by
the Governor March 14, 1881 :

AN ACT

To protect and promote the horticultural interests of the State.

*The People of the State of California represented in Senate
and Assembly, do enact as follows:*

SECTION 1. Whenever a petition is presented to the Board of
Supervisors of any county, and signed by five or more persons
who are resident freeholders and possessors of an orchard, or
both, stating that certain or all orchards, or nurseries, or trees
of any variety, are infested with scale bug, codlin moth, or other
insects that are destructive to trees, and praying that a commission be appointed by them, whose duty it shall be to supervise their destruction, as hereinafter provided, the Board of
Supervisors shall, within twenty days thereafter, select three
commissioners for the county, to be known as a County Board
of Horticultural Commissioners. The Board of Supervisors
may fill any vacancy that may occur in said commission by
death, resignation, or otherwise, and appoint one Commissioner each year, one month or thereabouts previous to the
expiration of the term of office of any member of said commission. The said Commissioners shall serve for a period of three
years from the date of their appointment, except the Commissioners first appointed, one of whom shall serve for one year,
one of whom shall serve for two years, and one of whom shall
serve for three years, from the date of appointment. The
Commissioners first appointed shall themselves decide, by lot,
or otherwise, who shall serve for one year, who two years, and
who three years, and shall notify the Board of Supervisors of
the result of their choice.

SEC. 2. It shall be the duty of the County Board of Horticultural Commissioners in each county, whenever they shall be informed by complaint of any person residing in such county, that an orchard, or nursery, or trees, or any fruit packing house, storeroom, saleroom, or any other place in their jurisdiction, is infested with scale bug, codlin moth, red spider, or other noxious insects liable to spread contagion dangerous to the trees or fruit of complainant, or their eggs or larvæ, injurious to fruit or fruit trees, they shall cause an inspection to be made of the said premises, and if found infected they shall notify the owner or owners, or the person or persons in charge or possession of said trees or places, as aforesaid, that the same are infested with said insects, or any of them, or their eggs or larvæ, and shall require such person or persons to disinfect the same within a certain time to be specified. If, within such specified time, such disinfection has not been accomplished, the said person or persons shall be required to make application of such treatment for the purpose of destroying them as said Commissioners shall prescribe. Said notices may be served upon the person or persons owning or having charge or possession of such infested trees, or places, or articles as aforesaid, by any Commissioner, or by any person deputed by the said Commissioners for that purpose, or they may be served in the same manner as a summons in a civil action. If the owner or owners, or the person or persons in charge or possession of any orchard, or nursery, or trees, or places, or articles, infested with said insects, or any of them, or their larvæ or eggs, after having been notified as above to make application of treatment as directed, shall fail, neglect, or refuse so to do, he or they shall be deemed guilty of maintaining a public nuisance, and any such orchards, nurseries, trees, or places, or articles thus infested, shall be adjudged and the same is hereby declared a public nuisance, and may be proceeded against as such. If found guilty, the Court shall direct the aforesaid County Board of Horticultural Commissioners to abate the nuisance. The expenses thus incurred shall be a lien upon the real property of the defendant.

SEC. 3. Said County Board of Horticultural Commissioners shall have power to divide the county into districts, and to appoint a local Inspector for each of said districts. The duties of such local Inspectors shall be prescribed by said County Board.

SEC. 4.· It shall be the duty of said County Board of Commissioners to keep a record of their official doings, and to make a report to the Board of State Viticultural Commissioners on or before the first day of November of each year, who shall incorporate the same in their annual reports.

SEC. 5. It shall be the duty of the Commissioners at large,

appointed by the Board of State Viticultural Commissioners
for such purpose, to recommend, consult, and act with the
County Board of Commissioners in their respective counties,
as to the most efficacious treatment to be adopted for the
extermination of the aforesaid insects, or larvæ, or eggs thereof,
and to attend to such other duties as may be necessary to
accomplish or carry out the full intent and meaning of this
Act.

SEC. 6.   Each County Commissioner and local Inspector
may be paid five dollars for each day actually engaged in the
performance of his duties under this Act, payable out of the
county treasury of his county; *provided*, that no more shall be
paid for such services than shall be determined by resolution
of the Board of Supervisors of the county for services actually
and necessarily rendered.

SEC. 7.   Each of said Commissioners may select one or more
persons, without pay, to assist him in the discharge of his
duties, as he may deem necessary.

SEC. 8.   If any County Board of Commissioners, after having received complaint in writing, as provided for in section
two of this Act, shall fail to perform the duties of their office,
as required by this Act, they may be removed from office by
the Board of Supervisors, and the vacancy thus formed shall
be filled in the same manner as provided for in this Act.

SEC. 9.   Nothing in this Act shall be construed so as to
affect vinyards or their products.

SEC. 10.   This Act shall take effect immediately.

The appointment of a Chief Horticultural Officer and the
enforcement of quarantine rules, etc., being given to the State
Board of Viticultural Commissioners, they appointed a State
Board of Horticultural Commissioners to recommend a competent person for Horticultural Officer, and also such regulations, etc., as were necessary for the protection of horticultural
industries.

The Board consisted of C. H. Dwinelle, Alameda County;
Elwood Cooper, Santa Barbara County; Albert S. White, San
Bernardino County; Dr. Chapin, Santa Clara County; A. Cad-
well, Sonoma County; W. W. Smith, Solano County; Felix
Gillett, Nevada County; W. B. West, San Joaquin County; E.
J. Wickson, Jr.,* Alameda County; M. T. Brewer, Sacramento
County; Matthew Cooke, Sacramento County.

---

* Mr. Charles H. Shinn was a member of the Board for one year, but
resigned on account of his leaving the State, and Mr. Wickson was
appointed his successor.

The State Board of Horticulture met quarterly. At first there was a laxity of action by the fruit growers in regard to petitioning for County Commissioners; however, in eight months sixteen counties had appointed Boards, and in twelve months twenty-one counties, in all, had appointed Boards, namely:

Sacramento, Yolo, Solano, Santa Barbara, El Dorado, San Bernardino, Santa Clara, Santa Cruz, San Joaquin, Amador, Contra Costa, Nevada, Placer, San Diego, Alameda, Los Angeles, San Benito, Fresno, Marin, Kern, and Butte (leaving only six of what may be termed fruit growing counties that had not asked for Commissioners).

The State Board of Horticulture did not think it advisable to enforce quarantine regulations in 1881, They therefore had prepared, by the Chief Horticultural Officer, "*A Treatise on the Injurious Insects of California, and Remedies for their Extermination.*" Ten thousand copies of this book were distributed. A State Convention of Fruit Growers was held at Sacramento, in December, 1881. Quarantine regulations were issued, to be enforced after January 1, 1882, as follows:

HORTICULTURAL QUARANTINE RULES.

To all whom it may concern: Be it known that I, Matthew Cooke, Chief Executive Horticultural and Health Officer of the Board of State Viticultural Commissioners, being duly authorized and instructed by said Board, do declare the following quarantine rules and regulations for the protection of the horticultural interests of the State, and due notice thereof is hereby given as provided by law, to wit, thirty days of publication in two daily newspapers of general circulation in the State, and by posting notices in all counties to be affected by these rules. All parties concerned therein are required to conform thereto, subject to penalties provided for by law, for any infraction or evasion of said rules and regulations:

Quarantine Rules and Regulations for the protection of fruit and fruit trees

From insect pests, namely, insects injurious to fruit and fruit trees, authorized and approved by the State Board of Viticultural Commissioners of California. In pursuance of an Act entitled "An Act to define and enlarge the duties and powers of the Board of State Viticultural Commissioners, and to authorize the appointment of certain officers, and to protect the interests of Horticulture and Viticulture." approved March 4, 1881, the Chief Executive Horticultural and Health Officer may appoint local resident Inspectors in any and all of the fruit-growing regions of the State, whose duties shall be as provided in Section IV of an Act entitled "An Act to define and enlarge the duties and powers of the Board of State Viticultural Commissioners, and to authorize the appointment of certain officers, and to protect the interests of Horticulture and Viticulture," provided that there shall be no compensation for such services of inspection excepting a fee, not to exceed one dollar for each certificate of disinfection, in case of compliance with quarantine regulations, and not to exceed five dollars for each certificate of disinfection after seizure for non-compliance; provided, however, such inspection may be employed at the option of the owners of property requiring disinfection, to disinfect the same. And also said local resident Inspectors will be entitled to such other fees as are provided for in cases of conviction and seizures.

1. All tree or plant cuttings, grafts or scions, plants or trees of any kind, infested by any insect or insects, or the germs thereof, namely their eggs, larvæ, or pupæ, that are known to be injurious to fruit or fruit trees, and liable to spread contagion; or any tree or plant cuttings, grafts, scions, plants, or trees of any kind, grown or planted in any county or district within the State of California, in which trees or plants, in orchards, nurseries, or places, are known to be infested by any insect or insects, or the germs thereof, namely, their eggs, larvæ, or pupæ, known to be injurious to fruit or fruit trees, and liable to spread contagion, are hereby required to be disinfected before removal for distribution or transportation from any orchard, nursery, or place where said tree or plant, cuttings, grafts or scions, plants, or trees of any kind are grown, or offered for sale or gift, as hereinafter provided.

2. All tree or plant cuttings, grafts, or scions, plants, or trees of any kind, imported or brought into this State from any foreign country, or from any of the United States or Territories, are hereby required to be disinfected immediately after their arrival in this State, and before being offered for sale or removed for distribution or transportation, as hereinafter described; provided, that if on examination of any such importations by a local resident Inspector, or the Chief Executive Horticultural Officer, a bill of health is certified to by such examining officer, then disinfection will be unnecessary.

3. Fruit of any kind, infested by any species of scale insect or scale insects, or the germs thereof, namely, their eggs, larvæ

or pupæ, known to be injurious to fruit and fruit trees, and liable to spread contagion, is hereby required to be disinfected, as hereinafter provided, before removal off premises where grown, for the purpose of sale, gift, distribution, or transportation.

4. Fruit of any kind, infested by any insect or insects, or the germs thereof, namely, their eggs, larvæ, or pupæ, known to be injurious to fruit or fruit trees, and liable to spread contagion, imported or brought into this State from any foreign country, or from any of the United States or Territories, are hereby prohibited from being offered for sale, gift, distribution, or transportation.

5. Fruit of any kind infested by the insect known as codlin moth, or its larvæ or pupæ, is hereby prohibited from being kept in bulk, or in packages or boxes of any kind, in any orchard, storeroom, salesroom, or place, or being dried for food, or any other purposes, or being removed for sale, gift, distribution or transportation.

6. Fruit boxes, packages, or baskets, used for shipping fruit to any destination, are hereby required to be disinfected, as hereinafter provided, previous to their being returned to any orchard, storeroom, salesroom, or place to be used for storage, shipping or any other purpose.

7. Transportable material of any kind, infested by any insect or insects, or the germs thereof, namely, their eggs, larvæ, or pupæ, known to be injurious to fruit or fruit trees, and liable to spread contagion, is hereby prohibited from being offered for sale, gift, distribution, or transportation.

8. Tree or plant cuttings, grafts, scions, plants, or trees of any kind, may be disinfected by dipping in a solution composed of not less than one pound (1 lb.) of commercial concentrated lye to each and every two (2) gallons of water used as such disinfectant, or in any other manner satisfactory to the Chief Executive Horticultural and Health Officer.

9. Empty fruit boxes, packages, or baskets, may be disinfected by dipping in boiling water, and allowed to remain in said boiling water not less than two minutes; said boiling water used as such disinfectant to contain, in solution, not less than one pound (1 lb.) of commercial potash, or three-fourths (¾) of one pound (1 lb.) of concentrated lye, to each and every twenty gallons of water, or in any other manner satisfactory to the Chief Executive Horticultural and Health Officer.

10. Fruit on deciduous and citrus trees infested by any species of scale insect or scale insects, or the germs thereof, namely, their eggs, larvæ, or pupæ, may be disinfected before removal from the tree, or from the premises where grown, by washing or thoroughly spraying said fruit with a solution

composed of one pound (1 lb.) of whale oil soap and one-fourth of one pound of flour of sulphur to each and every one and one-quarter (1¼) gallons of water used as such disinfectant, or in any other manner satisfactory to the Chief Executive Horticultural and Health Officer.

11. Owners of fruit of any kind grown in any orchard, nursery, or place in which trees or plants are known to be infested with any insect or insects, or the germs thereof, namely, their eggs, larvæ, or pupæ, known to be injurious to fruit or fruit trees, and liable to spread contagion, and all persons in possession thereof, or offering for sale, gift, distribution, or transportation, are hereby required to procure a certificate of disinfection before removal for sale, gift, distribution, or transportation.

12. Any tree or plant cuttings, scions, plants, or trees of any kind, empty fruit boxes, fruit packages, or fruit baskets, or transferable material of any kind, offered for sale, gift, distribution, or transportation, in violation of the quarantine rules and regulations for the protection of fruit and fruit trees, approved by the Board of State Viticultural Commissioners, may be seized by the Chief Executive Horticultural and Health Officer, or by any of the local resident Inspectors appointed by him; said seizure to be the taking possession thereof, and holding for disinfection, or for an order of condemnation by a Court of competent jurisdiction.

13. Any person violating the above quarantine rules and regulations shall be deemed guilty of a misdemeanor, and upon conviction thereof shall be punishable by a fine of not less than twenty-five nor more than one hundred dollars.

MATTHEW COOKE,

Chief Executive Horticultural and Health Officer.
Sacramento, November 12, 1881.

(For proceedings in Court in relation to quarantine rules, see Chapter IV.)

At the Fruit Growers' Convention, held at San Jose on the fourteenth, fifteenth, and sixteenth of November, 1882, the following committee was appointed to prepare such bills as were considered constitutional, and required for the protection of the horticultural industries:

Wal. J. Tuska, San Francisco; L. M. Holt, San Bernardino County; F. C. De Long, Marin County; S. M. Leib, Santa Clara County; Dr. Chapin, Santa Clara County; Hon. J. H. M. Townsend, Santa Clara County; G. M. Gray, Butte County;

A. T. Hatch, Solano County; Hon. Wm. Johnston, Sacramento County; Matthew Cooke, Sacramento County.

The committee met at San Jose, December 27, 1882. Subsequent to this meeting two bills were prepared by one of the committee—Wal. J. Tuska, Esq., attorney-at-law, San Francisco—one bill creating a Board of Horticultural Commissioners, and the other to prevent the spread of insect pests, etc. These bills were presented to the Legislature as Senate Bills Nos. 2 and 3, by Senator Cox, of Sacramento County, and Assembly Bills Nos. 31 and 32, by Assemblyman Hollister, of San Luis Obispo County.

The bills were amended in the Senate. The work of amending to suit the views of Senators was done by Senator Whitney, of Alameda. The bills as amended passed the Senate. In the meantime the Assembly had passed the bill creating a Board of Horticulture, and appropriating for its expense seven thousand five hundred dollars ($7,500) per annum, the Senate bill only allowing five thousand dollars ($5,000) per annum. The Assembly then took action on Senate Bills Nos. 2 and 3, and both were passed. The bill creating a Board of Horticulture—Senate Bill No. 3—as passed by both houses, was approved by the Governor. Senate Bill No. 2, to prevent the spread of insect pests, etc., after passing both houses, was mislaid or stolen, so that it was not presented to the Governor for his approval. In connection with the passage of the bills referred to, I met with the greatest courtesy from the members of the Assembly and Senate from the time the bills were introduced until they were passed.

On the twenty-seventh of March, 1883, I called upon Governor Stoneman in relation to the appointment of a Board of Horticultural Commissioners. He expressed a desire to appoint only competent persons, irrespective of political consideration. If this promise was fulfilled *it is well.*

CHAPTER III.

Progress of the Warfare Against Insect Pests from January 1, 1879, to April 1, 1883.

Previous to January 1, 1879, the codlin moth, oyster-shell bark louse, San Jose scale insect, and black scale on deciduous fruit trees, and the soft orange scale, black scale, and red scale on citrus trees, were found to be spreading rapidly in some fruit growing districts. Excepting a few individuals, but little notice was taken of their presence. In 1879 some individual efforts were made to save the crops, which were successful. But, notwithstanding the great interests at stake, not one fruit grower out of every five hundred attempted to make any effort to prevent the spread of these enemies of their industry. Some fruit growers ridiculed the idea of any serious results occurring from the presence of such minute creatures; others seemed satisfied that the pests would disappear as they came, without any effort of the fruit growers.

On the 29th of October, 1879, I visited an orchard seriously infested by San Jose scale (*A. perniciosus*). I recommended the owner to spray his trees with an alkaline solution. On the morning of the 30th he met the gentleman who accompanied me to his orchard, and said the idea of washing trees was only a *hobby*. Had he done, as advised, in November, 1879, what he afterwards did in the Spring of 1882, he could have saved one half of his orchard, which had to be dug out.

In 1880 matters became so serious in some localities that efforts were made to destroy the pests. Such a difference of opinion existed as to the proper remedies to be employed, that a great deal of the work done did not produce very favorable results, thus leaving the matter in a more hopeless condition that in 1879.

The securing of legislation for the protection of the horticultural interests of the State, in the Winter of 1880 and 1881 (the laws passed taking effect on March 14, 1881), brought the subject of insect pest prominently before the fruit-growers. The results of the work done in 1881, 1882, and in the Spring of 1883, may be stated as follows:

1. It has been fully demonstrated that the insect pests can be exterminated in orchards at a profit to the owner of from one hundred to five hundred per cent on the amount expended for remedies thoroughly applied.

2. Remedies of undoubted utility have been discovered, that can be readily and thoroughly applied with the improved facilities that the necessity of the times brought forth.

3. We have become better acquainted with the natural history of the injurious and beneficial insects of California, and standard works on economic entomology can be readily procured by those wishing to get them.

4. In some of the districts that were seriously infested confidence has been restored, and land has increased in value from one hundred to two hundred per cent.

5. That the work became popular may be shown by the correspondence, etc., of the Chief Executive Horticultural Officer, of which the following is a sample :

From April 5, 1881, to April 5, 1883 :—

Number of letters, postal cards, and packages received, 5,581
Number of packages and letters containing specimens
   of insects, etc., - - - - - - - 491
Number of visitors at the office on business connected
   with insect pests, - - - - - - 611

The information derived from the experience of the two years' work will not only benefit the fruit-growers of California, but those of our sister States, and also of foreign countries. For instance, at a recent meeting of the Royal Agricultural Society of South Australia, Frazer S. Crawford, Esq., of Adelaide, read a paper entitled, "California Legislation Against Insect Pests," etc., which concluded as follows :

"In conclusion, let me again point out the importance of carefully watching this California experiment. If we find, in spite of this expense and trouble the Californians are willing to put themselves to, that these pests do not decrease, or that the results bear no comparison to the cost of the means employed, it would be folly to follow their example. If, on the other hand, their vineyards and orchards yearly become more prolific and free from insect and other plagues, while our orchards are being destroyed with fungoid pests, our vine-

yards deteriorated with odium, our orange, olive and other
trees ruined by scale insects, we may conclude that our American cousins are reaping the benefit of their wise legislation,
and that the sooner we follow their example the better,
remembering the truth of the old proverb : 'Providence helps
those who help themselves.'"

For the benefit of all interested in the warfare against insect
pests, positive assurance can be given of the benefits to be
derived from legislation against the spread of insect pests,
etc.; but success can only be achieved under certain well-
defined conditions, as follows :

First—The appointment of a Commission, each member of
which is required to have a thorough experience of the best
established methods used in preventing the spread of and in
exterminating the insect pests, and sustained in the performance of their duties by well-considered laws.

Second—Adequate compensation to secure competent officers.

Third—The appointment of officers to execute the laws of
the State and such regulations as are made by the Commission ; said officers to be in all respects competent to perform
the duties required, especially being thoroughly acquainted
with the science of entomology.

A Commission and officers appointed under such conditions
cannot fail to benefit the State tenfold the amount expended.

CHAPTER IV.

Disinfection of Return Packages versus Free Packages.

During the months of April, May, and June, 1882, there
arose considerable discussion amongst the fruit growers who
were selling their fruit in local markets, as to the utility of
disinfecting packages in which fruit was shipped to market
before the packages were reshipped to the orchards for further
use, as required by the quarantine rules, etc. Had the settle-
ment of the question been left entirely to the fruit growers, the
matter would have been amicably arranged, and successful

progress made in preventing the spread of the codlin moth
and peach moth. Unfortunately for those directly interested
(the fruit growers), an issue was made by the non-producers
(i. e. buyers and dealers of fruit), under the cloak of friendship
for the fruit-growers, but in reality for personal gain, and for
forcing the fruit-growers to use the so-called free package.
The time may be stated about the first of June, 1882. Lest
any doubt should exist that such was the intention of the
opposition to the disinfection of packages, the following letter
now in my possession, from a very extensive buyer of fruit in
San Francisco to a member of the State Board of Viticultural
Commission, will perhaps be sufficient proof:

SAN FRANCISCO, January 3, 1882.

DEAR SIR: The box factories tell me they are making a
larger quantity of peach baskets than ever before. Now, this
is very bad, because the Fruit Growers' Convention, through
the proper committee, particularly recommended the abolition
of baskets and the substitution of boxes; and furthermore, the
basket is as bad a package as can possibly be found for dis-
seminating pests. Why cannot the fruit grower help himself
in this matter? Nothing will do him good if he does not. You
can show this to Mr. Cooke, if you think best. Will you get
the sample of free packages into shape and put them on show
here, somewhere, and oblige,

Yours, Etc., ⸺ ⸺

[The above reveals the secret of the opposition to disinfection of packages.]

The State Board of Viticultural Commissioners had framed
the quarantine rules, so as to give an opportunity to those who
preferred the use of the return packages, to do so. The following circular was issued explaining the reason:

*To the fruit growers of California, and all whom it may con-
cern:*—

I take the liberty of calling your attention to the necessity
of disinfecting empty fruit packages before being returned from
market to the orchards, as required by the Quarantine Rules
and Regulations for the protection of Horticulture.

RULE 6. Fruit boxes, packages, or baskets, used for shipping fruit to any destination, are hereby required to be disin-

3

feeted, as hereinafter provided, previous to their being returned to any orchard, storeroom, salesroom, or place to be used for storage, shipping, or any other purpose.

RULE 9. Empty fruit boxes, packages, or baskets, may be disinfected by dipping in boiling water, and allowed to remain in said boiling water not less than two minutes; said boiling water used as such disinfectant to contain, in solution, not less than one pound (1 ℔) of commercial potash, or three-fourths ($\frac{3}{4}$) of one pound (1 ℔) of concentrated lye, to each and every twenty gallons of water, or in any other manner satisfactory to the Chief Executive Horticultural and Health Officer.

In order to secure a general compliance with the above rules, arrangements will be made at San Francisco, Sacramento, San Jose, Stockton, and other places, where fruit is shipped to a market or for storage; and local Inspectors appointed who will disinfect such packages, if required by the owners thereof; or the owners of empty packages, or their agents, may disinfect them, subject to inspection.

In answer to inquiries, the following questions may be asked:

Are those requirements oppressive on the fruit growers who wish to send fruit to market? I answer, they are not, for the following reasons:

1. There are several counties which may be termed fruit growing counties, such as Sonoma, Napa, Yuba, Tuolumne, Calaveras, and others, where the fruit growers have failed in having a County Board of Horticultural Commissioners appointed; therefore there is not any organization in any of the above and some other counties, with which I can consult in relation to matters in which the fruit growers are interested.

2. The Commissioners in several counties have been retarded in their work by the requirements of the law, that complaint must be made before they can make an inspection of an infested orchard. Although many of the fruit growers have made every effort in their power to cleanse their orchards, yet they hesitate to enter a complaint against their neighbor; therefore, in a large number of orchards, the pests have been allowed to spread as heretofore. And it is only by a general enforcement of the Quarantine Rules that any good results can be secured.

3. A large number of orchards are rented, leased, or owned by parties that, judging by past experience, cannot be depended upon to comply with the Quarantine Rules, except they are compelled to do so.

4. It would require at least three hundred and fifty (350) local Inspectors throughout the State, to have the work of disinfecting empty fruit boxes or packages, at the orchard, steam-

boat landings, and railroad stations, done as effectually as by
the disinfecting at markets, etc., as required.

5. The local Inspectors, if appointed, are allowed fees by
law, and would be a heavy expense on the fruit growers.

6. The rapid spread of codlin moth, to at least thirty counties in this State, since its first appearance in one orchard in
1874, can chiefly be attributed to the use of the return packages; therefore, the disinfection of all return packages is a
necessity.

7. There is a difference of opinion among fruit growers as
to the style of package that should be used in sending fruit to
market. Some growers recommend and have adopted the so-
called free package, and others use the so-called return package.

8. The disinfection of the packages gives an opportunity of
using the return packages at a nominal cost.

9. Stores, commission houses, and places where apples,
pears, and quinces were stored or sold last season, are generally
infested by the larvæ, pupæ, or imago (perfect insect) of the
codlin moth, and are liable to be taken from such places in
return packages to orchards at this season of the year.

10. The disinfection of return packages at the place where
the fruit is sent to market (or for storage) is not only the most
effective method for the prevention of the spread of insect
pests, but is by far the most economical for the fruit growers.

MATTHEW COOKE,
Chief Executive Horticultural Officer.

The opposition brought the question of constitutionality
of the law into the courts; pending a decision, the enforcement of the quarantine rules was abandoned, and the pests not
only allowed to spread as heretofore, but the free package system was not generally adopted. (In January, 1883, a decision
of the Supreme Court declared that the Legislature, in giving
the State Board of Viticultural Commissioners *the right to
declare what would constitute a misdemeanor, gave the Board
legislative power, and was therefore unconstitutional.*)

After mature investigation, I can find no reason to change
my opinion in regard to the prevention of the spread of fruit
pests; that the best, safest, and cheapest method is the general
and thorough disinfection of all packages used in shipping
fruit.

The non-producer and free package advocate asserts (hypo-
thetically, of course,) that the cost of disinfecting one load of
boxes and baskets sent from his store—containing three hun-

dred boxes and three hundred baskets, belonging to thirty owners—would be nearly thirty-five dollars, or as follows :

Disinfection of 300 boxes, at 1 cent, - - - - - - $3 00
Disinfection of 300 baskets, at ½ cent, - - - - - 1 50
Thirty Inspector's certificates, at $1.00, - - - - - 30 00

    Hypothetical total, - - - - - - - - - - $34 50

    The actual cost would have been :

Disinfection of 300 boxes, at 1 cent, - - - - - - $3 00
Disinfection of 300 baskets, at ½ cent, - - - - - 1 50
Certificate of Inspector (free), - - - - - - - -

    Total cost of hypothetical load of return packages, $4 50

    Again compare figures :

300 boxes, each containing 30 pounds of fruit, - - 9,000 lbs
300 baskets, each containing 25 pounds of fruit, - - 7,500 lbs

                                16,500 lbs

    Total cost of disinfection so that they could be used again for shipping purposes, $4.50.

    Free packages, to carry 16,500 pounds of fruit to market—550 packages, 30 pounds each, 16.500 ; cost of 550 packages, at 6¼ cents each, cheapest style, $35.75. Thus showing a clear gain to the fruit grower of $31.25, less return freight, in favor of the plan of disinfection.

CHAPTER V.

Quarantine.

    In a warfare for the extermination of the insect pests of the orchards, etc., quarantine laws are a necessity. The codlin moth, peach moth, and other species of pests are spread by the shipment of infected fruits. The scale insect and woolly aphis have been and are being continually spread on nursery stock.

From past experience I have concluded that it is an impossibility to secure any general law that will be acceptable to the fruit growers, nurserymen and speculators combined. It is, unfortunately, an established fact that a majority of fruit growers, or persons growing fruit, prefer taking the chances of allowing pests to spread rather than to disinfect their packages when returned from market. The proof of this is beyond question. At this date I do not know of twenty persons in this State who are taking this precaution. Nursery stock infested by the scale insect and woolly aphis is sent broadcast through the State, and what are the results?

A planted two thousand Bartlett pear trees in 1880; in the Fall of 1881 he discovered that they were infested by the San Jose scale; in the Spring of 1882 he cut them off at the ground as a sure prevention of their spread on the 10,000 trees adjoining, and grafted them. By thorough application of remedies he has the trees cleaned, but lost two years use of his ground.

B bought, in 1882, Bartlett pear trees for two acres; in the Spring of 1883 he discovered that they were infested by the San Jose scale. For the protection of the balance of his orchard property, the trees, after being planted one year, were dug out and burned.

C, in 1881, sent to a distant portion of the State for pear trees, to escape any chances of getting infested trees; unfortunately the nursery-man had purchased the trees in an infested district. In 1883, the trees planted two years ago are found to be seriously infested by the so-called San Jose scale (*A perniciosus*).

D, E and F have had the same experience in buying and planting young apple trees, the roots of which were infested by the woolly aphis.

The true remedy of these evils is, let every owner of fruit trees, etc., liable to be infested by insect pests, make himself or herself thoroughly acquainted with the appearance and natural history of injurious insects; then constitute a home quarantine board, of one or more, for the purpose of preventing the importation of insect pests on the premises.

CHAPTER VI.

Danger of Importing Injurious Insects from Foreign Countries.

Dealers in fruit at San Francisco often have consignments of lemons and oranges arrive by steamships and sailing vessels from Tahiti, from Australia, and also from European ports.

In some cargoes the fruit is seriously infested by scale insects. Australia sends us the red scale (*A. aurantii*), and Australia, Tahiti, and Europe send the leaf and fruit scale)*A. citricola*). The latter has not been found, so far as I am aware, in any orchard on this coast.

A shipment of apples received lately from New Zealand was infested by the greedy scale (*A. rapax*). These species of scale insects generally arrive in a healthy condition. The greedy scale and red scale are both well located here. Care should be taken that the *A. citricola* be not allowed to spread. Remember, "eternal vigilance is the price of fruit."

CHAPTER VII.

Danger of Spreading Insect Pests by the Transportation of Infested Fruit and Nursery Stock.

It would be superfluous to give any extended repetition of the dangers which threaten husbandmen by having infested fruits, nursery stock, seeds, etc., brought upon his premises, as the danger is fully described in other chapters of this work. The following instances are given for the purpose of further calling the attention of those living in districts not yet infested :

The Novato Ranch orchard is isolated from any other orchard from which it could become infested : the pests were brought in return packages, and to this date (July, 1883), the amount expended by Mr. De Long in getting rid of his visitors (codlin moth) must reach nearly ten thousand dollars.

The peach worm, which was discovered last year (1882), has

been found in isolated orchards this year, so that undoubtedly they were brought in return packages.

The discovery made in various sections of the country this Spring (1883), of the presence in young orchards of the San Jose scale (*A. perniciosus*) is sufficient proof that the nursery stock was infested when bought.

The presence of the codlin moth in an isolated apple orchard in San Diego County, can readily be accounted for by the owner carrying home his groceries in empty apple boxes that were shipped to San Diego with infested fruit.

The presence of the red scale (*A. aurantii*) in some orchards in Southern California is attributed to the importation of two or three young trees from Australia.

The presence of the cottony cushion scale (*I. purchasi*) can be charged to importation.

The codlin moth was brought in an importation of five barrels of apples from States east of the Rocky Mountains, about 1873. The grain weevil (*C. granaria*) and other injurious species of the weevil family (*curculionidæ*) can readily be spread in grain and plant seeds.

CHAPTER VIII.

Notes on Experimental Work.

It is not only a duty, but a necessity, that every cultivator of the soil, irrespective of the line of industry in which he is engaged, if his premises are infested or likely to be infested by insect pests, that he should use his utmost efforts in experimental work to find the best and cheapest methods for their extermination, or for preventing their spread at least. In the course of such experimental work, should he think he had made an important discovery, he should be guided by the following rules :

1. It is well worth the care of any one who wishes to be sure—as every one should—of conferring a benefit upon his fellows rather than risk doing them an injury, to delay the expression of results of experiments until the correctness of their results are tested by repeated trials.

2. Before advising your neighbor to adopt a certain course or remedy, be sure by repeated investigations that you are possessed of a fact, not a fancy. Experiments, to be convincing, require that they be subjected to well defined, clearly perceived conditions.

3. In reporting results of experiments, it should be remembered that "it is the weakest link that determines the strength of the chain."

One of the obstacles with which the enterprising fruit grower has to contend in attempting to prevent the spread of insect pests, is the reliance their neighbors place in remedies which they have not sufficiently tested before adopting them. For instance: A's neighbor, Mr. B, has decided to clean his orchard to prevent the ravages of the codlin moth, and as a business principle makes an estimate of the cost, which will probably be from fifteen to twenty dollars per acre, according to the size of the trees. About this time he notices an article in his newspaper: "No more use for pumps, sprays, nozzles, and solutions. By placing a branch of an eucalyptus tree in your apple and pear trees, the codlin moth will not attack the fruit." Mr. B adopts the cheap remedy without further investigation. Result: the codlin moth is not destroyed, and A's orchard is placed in danger. Mr. C, who has been advised to use wide-mouthed bottles, with sweetened water, etc., tries the experiment and pronounces it a success, having captured by actual count nearly five hundred moths in one night. The success is announced, and others are induced to give up the application of all other remedies, and they are successful in capturing a large number of moths. Query—Are they codlin moths? In one case, where four hundred and eighty-three moths were captured, in an orchard as badly infested by the codlin moth as any that can be found in this State, not a codlin moth was found in the whole number. In another case, where locomotive head-lights were used in an orchard at night and surrounded by devices for capturing moths, such as pans of sweetened water, rum and molasses, coal oil, etc., of the immense number of moths captured, less than one fourth of one per cent were codlin moths, or one in every four hundred. Mr. B called at my office a few days ago, and stated that he was capturing immense numbers of the codlin moth in dishes and bottles of sweetened water, in his orchard. On being shown a codlin moth, natural size, he discovered his mistake, and remarked that the moths which he was capturing, the body of each was over an inch in length and about as thick as a lead pencil. Serious consequences have happened by the application of highly recommended solutions for destroying scale insects, such as tree wash—a cheap production made of coal oil, which proved an excellent insecticide, but unfortunately those who recommended it did not wait to find out the effect on the trees. That it contained some pernicious qualities which destroyed the tree on which it was used was discovered in three or four months after application. Only use well known remedies if you wish to be successful.

CHAPTER IX.

Alkaline Washes as Insecticides and Fertilizers.

Since the passage of the laws in March, 1881, for the protection of the horticultural industries of this State, a large amount of money has been expended in purchasing solutions, and the necessary labor of applying the same for the purpose of exterminating insect pests. In many cases the result of the work did not meet the expectation of the fruit grower, and in consequence of being disappointed a premature verdict was given against the utility of the solution recommended. The question never occurred: "Did I get the best material, or was the proper application made?" The object of this chapter is to explain how such failures have occurred.

No. 1. A proposes to wash his orchard trees with concentrated lye so that it will destroy insect pests, and at the same time invigorate his trees. He calls on a wholesale merchant (who knows nothing of the material required) and states what he wishes to purchase. A potash is offered at a very low price, but warranted equal to any in the market, if bought in quantities of a ton, more or less. An order is given, and the so-called potash delivered at five and a quarter cents per pound, instead of nine or nine and a half cents per pound, the price of the proper article. The work of application is commenced, and it is only after the work has been completed that the fruit grower learns that the article he had bought for potash was a poor quality of caustic soda.

No. 2 wishes to secure one ton of whale oil soap. He goes to a soap factory and purchases the amount required, but after application he ascertains that whale oil soap proper is the residue precipitated in bleaching whale oil, in which the fatty matter of the oil is saponified by liquid potassa, and can only be purchased at the bleaching works; also that the whale oil soap he bought was made from a low grade of whale oil in the same manner as soft soap, and instead of the saponifying matter being all potash, the greater part was caustic soda. In such cases the best results cannot be obtained; perhaps one or two cents per pound are saved, but the same amount of labor was required that would have applied the proper wash.

CHAPTER X.

Thorough Application of Remedies.

"I consider a great deal of my success in exterminating scale insects was due to the thorough application of the remedies used."—Elwood Cooper.

I do not wish to be considered an alarmist, but I state, frankly, that from this time forth, any fruit grower in this State (especially in infested districts) who wishes the produce of his orchard to be choice and marketable, will be compelled to expend money and labor to protect his crop from the ravages of insect pests, and the following rules should be strictly adhered to:

1. Procure the best quality of such articles as are required for insecticides and fertilizers.

2. If you are not a judge of the article, secure the assistance of some person who is.

3. Be sure that the formula for making any solution is properly prepared, before application to the tree.

4. Use only the best mechanical appliances, such as pumps, nozzles, etc.

5. Make thorough application in every respect the basis upon which the work must be done.

The following cases are related from personal observation:

1. A was advised to scrape his trees, put on bands, sew them, etc., and was given full information how to proceed to prevent the spread of the codlin moth.

APPLICATION.—He contracted to get his trees scraped for three cents each—about forty per cent of what it should have cost; bands were placed on the trees, and during the four months succeeding the fifteenth of May, they were partly examined twice, instead of every seventh day.

RESULT.—The crop of apples and pears was destroyed, as heretofore. The remedies recommended were denounced by A as worthless. The failure and loss of time and money were caused by the negligent manner in which the work was done.

2. B has an orchard infested by the scale insect, and is recommended to wash his trees with a solution of one pound of concentrated lye to each gallon of water used. He substituted two and a half gallons of water to each pound of the lye.

RESULT.—The scale insect was not destroyed. B was indignant, and denounced the remedies recommended as an imposition, but failed to state that he used two and a half gallons of water instead of one gallon to each pound of lye used—labor and money lost by not making application as directed.

3. C was advised to use the concentrated lye solution, and not to use coal oil of any kind. However, by advice of his neighbor, he used a wash made of a low grade of coal oil.

RESULT.—It proved an excellent insecticide, but killed the trees.

4. D's orchard was infested by the San Jose scale (*A. perniciosus*). He procured material, the best to be found in the market, and commenced a thorough warfare against the pests. He succeeded admirably in his work, but business matters caused his absence for three or four days from the orchard. On his return he found some of his trees scorched by the application. On investigation he found that the liquid was taken

from the casks without stirring; therefore, the trees washed from the liquid taken from the bottom of the casks stood an application about three times as strong as necessary; however, they came out all right. The mistake was rectified, and the experience noted for future use.

RESULT.—D's thorough application was successful in every respect. It cost him $900, but his orchard is clear of San Jose scale.

E owns an orchard containing seven thousand trees, for which he paid $32,000 in 1879. In October, 1881, it was seriously infested by the San Jose scale. In November he bought seven tons of concentrated lye, and commenced work. About new-year his neighbor condemned his work as being so thorough it killed all his trees. He wrote me what he had done. I advised him to "*go on as he had been doing*," and in a few days after visited his orchard. The work was completed as commenced. I visited the orchard on the thirty-first day of March following (1881); every tree was in bloom, and but few of the scale insects in a healthy condition. I asked the owner at what value he estimated the result of his perseverance and thorough work. He answered: "Last October, I would have sold the orchard, bugs and all, for $15,000. At a cost of a little over $1,200, I have conquered the bugs. Now, sir, you cannot buy it for $40,000, and you can so inform your friends."

MORAL.—Follow E's example.

First-class material to make the solution and thorough application will destroy the scale insect without fail.

Do the work as recommended, thoroughly, and the bugs must go.

CHAPTER XI.

Pumps and Nozzles for Spraying Trees.

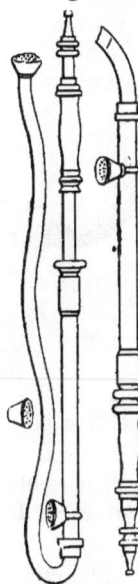

Experience has taught us that the most effective remedies for the extermination of insects injurious to fruit and fruit trees, especially those belonging to the *Coccidæ,* are those that can be applied in solution. For the application of such solutions the best method is by a pump and spraying nozzle. I have been frequently asked the question: What kind of pump, nozzle, etc., is the best? The accompanying illustrations represent the best apparatus we know of at present. Fig. 1 represents a fountain pump. The end of the hose is placed in a barrel or pail containing the fluid. Price, $7.50.

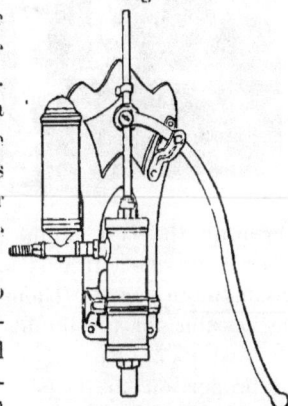

The fountain pump is designed for garden use, and is sold by Baker & Hamilton, Sacramento and San Francisco; and by H. P. Gregory & Co., San Francisco.

Fig. 2 represents a pump, known as the "Gould Pump," manufactured by H. P. Gregory & Co., Nos. 2 and 4 California Street, San Francisco. A large number of these pumps are now in use, and give excellent satisfaction, as the working parts are made to resist the chemical action of the disinfectants used.

Price of pump, - - - - - - - $17 00
Price, complete, with twenty-five feet of hose, suction,
    strainer, and spraying nozzle, - - - - 23 00
Or, with fifty feet of hose, double discharge, and two
    spraying nozzles, - - - - - - 28 00

Fig. 3 is an illustration of a new improved pump, manufactured by H. P. Gregory & Co., of San Francisco. This is

Fig. 3.

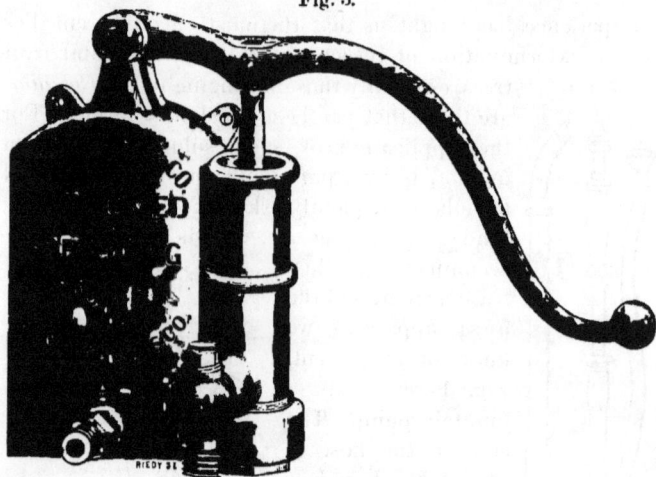

certainly the best pump for spraying trees that has been offered for sale. It is stronger made than most pumps, thus giving more power. There are only two joints where it would be possible for the liquid to leak out, and these, if properly screwed up, are perfectly tight. The air chamber is very large in proportion to the size of the pump, thereby giving an exceedingly steady spray. The cylinders are lined with brass, and the valves are made of the same metal as the body of the pump, including the handle. The outer parts are galvanized, thus preventing them from rusting, so that it will preserve the same neat appearance that it has when new.

This pump was designed by H. P. Gregory, of the above firm, and is certainly a great improvement.

FIG. 4.

FIG. 5.

FIG. 6.

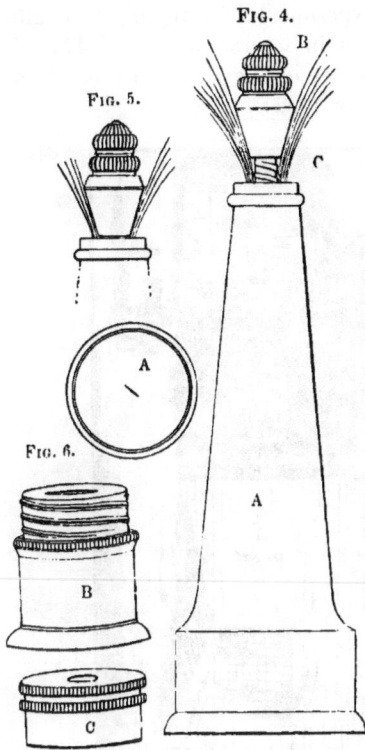

Of the spray nozzles in use, the most practical is a nozzle made by Wesley Fanning, No. 279 St. John Street, San Jose, and is known as the "Merrigot Nozzle." (Fig. 6.) A is a diaphragm, with a small opening in the center of any required size. This is placed on the piece B, on top of screw; the cap C is then screwed on, and the nozzle is complete.

With a force pump this throws a very fine spray over the tree.

A full set of diaphragms are sent with each nozzle.

Figs. 4 and 5 represents the Niagara Lawn Sprinkler.

CHAPTER XII.

DeLong's Moth-trap.

The apple house (Fig. 7) at the Novato Ranch, in Marin County, California, owned by Mr. DeLong, is a three story building about one hundred feet long by seventy feet wide; the lower story is built of stone twelve feet high, the second story of brick fourteen feet high, and the third or upper story is in the form of a hip-roof, and is about twenty feet from the floor to the ridgepole.

Mr. DeLong made this experiment of trapping the codlin moth, carried to his premises in return packages. The following if taken from the report of the Fruit Growers' Convention at San Jose, November. 1882:

Fig. 7.

Mr. Cooke—"I call your attention to one matter in reference to an orchard, of 250 acres, in Marin County, where there are 31,000 trees subject to the attacks of the codlin moth—the Novato Ranch. In June, 1881. I visited the ranch and, greatly to the dismay of Mr. DeLong, I found the pupa of a codlin moth. When I told him, he said: 'No sir, we have none here.' I replied: 'Mr. DeLong, there it is on this tree.' Now I would like to call for his experience since June, 1881."

Mr. DeLong—"All I can say is that I did not know that we

had one in the year 1879; we had none in the year 1880; we had none in the year 1881 until he discovered it, as he says, in the chrysalis form. That year we gathered our fruit and carried it into the apple-house, and I saw very little of it on the fruit. The apples did not seem to be much affected, but the suggestion was made to me by Mr. Cooke, the following Spring, that I had better scald all of my boxes. I asked him whether they had gotten out of these boxes into the cracks of the floor in the building, and how would it do if I should close this building up by putting mosquito netting over the windows, so that there would be no possible chance for the moths to get out. He agreed with me that it would be a good idea, and I did so. Having done that, I thought I would like to know the result of it. I nailed all the doors up so that it would be perfectly impossible for anything to get in and out without my knowing it, and I locked the door and took one of my men and put him in possession of the key. He commenced finding some of the moths about the middle of April and killed them, and up to the 27th of May, kept a running account; after that date he kept a daily account; he found that he had killed 15,627 up to the 27th of October. As the account is called for I will read it; it will give an idea. As the days grew longer and warmer the moths increased; as they grew colder the moths diminished. This is the account of the moths themselves, not the pupæ.

DESTRUCTION OF CODLIN MOTH IN FRUIT-HOUSE.

From beginning to May 27th, inclusive...	2671	June 20.........	221	July 15.........	62
		June 21.........	268	July 16.........	66
May 28.........	129	June 22.........	397	July 17.........	157
May 29.........	388	June 23.........	199	July 23.........	65
May 30.........	337	June 24.........	154	July 24.........	47
May 31.........	343	June 27.........	336	July 26.........	108
June 1.........	335	June 28.........	314	July 27.........	46
June 2.........	224	June 29.........	467	July 28.........	24
June 3.........	149	June 30.........	447	July 29.........	15
June 4.........	165	July 1.........	315	July 30.........	22
June 5.........	138	July 2.........	330	Aug. 1.........	36
June 6.........	218	July 3.........	415	Aug. 2.........	43
June 7.........	245	July 4.........	307	Aug. 4.........	14
June 8.........	146	July 5.........	238	Aug. 4.........	14
June 9.........	144	July 6.........	278	Aug. 4.........	16
June 12.........	582	July 7.........	75	Aug. 5.........	7
June 13.........	405	July 8.....No count.		Aug. 8.........	31
June 14.........	177	July 9.........	175	Aug. 13.........	14
June 15*.........	990	July 10.........	231	Aug. 15.........	14
June 16.........	484	July 11.........	143	Aug. 18.........	5
June 17.........	412	July 12.........	111	Aug. 20.........	3
June 18.........	116	July 13.........	112	Aug. 26.........	7
June 19.........	440	July 14.........	90		
*Cellar opened.				Total No.....15,627	

4

"I have kept ten or eleven men working continuously in the orchard all through the season. Nine of them going over the trees and pulling off all the fruit they could find that was infested with these worms, and others were picking them up and carrying them in. The means that I used to destroy them was to put them in a large boiler and boil them up, not trusting to the hogs or anything else to eat them; and I came to the conclusion that that was the surest way of exterminating them.

"I have carried that out until the apples got large, and they are now in the house. What the result will be, I don't know. I have worked the bands very effectually, killing some days thousands of larvæ. I never kept an accurate account."

Mr. Cooke—"How many did the bats eat?"

Mr. DeLong—"That is a question I do not know anything about. I don't know that I can give any further ideas. We were working under the mode Dr. Chapin spoke of, which I think is most effectual."

Mr. Cooke—"I merely wanted to call the attention of the convention to the facts; when Mr. DeLong covered all of the windows on the inside of the three story apple house (the upper story is formed with a high roof, and he had to keep the mosquito bar over both ends), he inclosed probably five hundred (500) bats. We know that a bat lives on insects, and it is certain that the bats lived on the codlin moth, and of these Mr. DeLong was unable to give any account. He had placed 34,000 boxes in this house; he neglected to say that he cleaned the boxes before they were put into the trap, and he got ten or twelve larvæ in some of these boxes. He got over 15,000 moths, the largest capture ever made in this line of warfare, for his Summer's work.

"I was present on May 29, 1882, at two o'clock P. M., and on the covering of the top window, marked A, on Fig. 7, one hundred and ten moths were captured. Extended notice is thus given in commendation of Mr. DeLong's practical work from the 15th of April to August 26th."

NOTE.—On Fig. 7 it can be noticed that there is a square cornice on gable of building; the bats could only get into this cornice from inside; the covering of mosquito bar kept them on the top floor, and as they appeared well fed during their confinement of four months and a half, it is presumed that they destroyed a large number of the codlin moths.

CHAPTER XIII.

Structure and Growth of Plants.

A great amount of damage has occurred in some fruit growing districts of this State by the applications of solutions, etc., as insecticides (i. e. a low grade of coal oil), which contained properties that were detrimental to plant-life. By examining Fig. 8, the outer bark is found to be represented by letter F, the green layer by letter G, the inner bark by letter H, and the cambium layer by the letter I. As the cambium layer is stated in the following description to be the seat of life of the plant, it may be readily understood in what manner a solution containing pernicious properties injures the tree when applied to the outer bark, viz. : by penetrating the outer bark, the green layer, and inner bark, thus reaching and destroying the cambium layer (or seat of life of the tree).

On the contrary, if proper solutions, such as are recommended in this work, are applied, they destroy insect life, and by penetrating the outer layers reach the cambium layer, and by means of the fertilization properties which they contain, invigorate the tree.

[The following extracts are taken from a paper by Colonel W. S. Clark, President of the Massachusetts Agricultural College.—Mass. Rep., 1873–4] :

"Every seed and every young plant consists wholly of cellular tissue, but with the development of leaves is combined the growth of fibro-vascular tissue."

"The first vessels to appear in the plantlet are arranged in a circle around a column of tissue, which remains loose and soft, and after the first season dries up and dies. This is called the pith, and seems essential to the life of every woody stem and branch during its infancy, although its special function is unknown. Between the vessels around the pith may be seen the rays of cellular tissue, which ultimately become hard and firm, and which unite in bonds, never broken except by some external force, the inside of the stem with the inside of the bark. These rays make up the woof and have much to do with the distinctive peculiarities of different sorts of timber."

"Immediately outside the vessels inclosing the pith grows a layer of woody fiber, upon which, in a more or less developed state, according to season, is a layer of organizable material, called cambium, which may be regarded as the seat of life of the plant."

" Investigation seems to demonstrate that the cambium layer is the seat of life, and that whenever the direct communication between the root and the foliage is cut off in this layer during one entire season of growth, the whole plant perishes. It has also been determined by experiment that if several rings of bark be removed from a growing shoot in such a manner that on one of the isolated sections of bark there be no leaf, while leaves remain on others above and below this, then the leafless section will fail to make any growth in any part. All other sections, if furnished with one or more healthy leaves, will increase in thickness by the formation of new leaves, of wood and bark. This seems to prove that the material for growth is elaborated by the leaves, and is transmitted only through the cambium, and has no power of penetrating the tissues of the wood."

" The peculiar vital and organic power of the cambium is remarkably illustrated in the structure and growth of grafted trees. Every person is aware that pear trees are grown upon quince roots, and that they often bear finer fruit than when cultivated as standards. This is doubtless owing to the fact that quince roots, being diminutive, furnish less water to the leaves, which thus elaborate a richer sap, and produce more perfectly developed wood and fruit."

" The apricot may be grafted on the plum, and the peach on the apricot, and the almond on the peach ; and thus we may produce a tree with plum roots and almond leaves. The wood, however, of the stem will consist of four distinct varieties, though formed from one continuous cambium layer. Below the almond wood and bark we shall have perfect peach wood and bark, then perfect apricot wood and bark, and at the bottom perfect plum wood and bark. In this curious instance we see the intimate correspondence between the bark and the leaf ; for if we should remove the almond branches we might cause the several sorts of wood to develop buds and leafy twigs, each of its own kind. Each section of the compound stem has its seat of life in the cambium, and the cambium of each reproduces cells of its own species out of a common nutrient fluid. Thus there is seen to be a flow of sap upward in the wood, and a flow of organizable material, essential to the life of the plant, proceeding from the leaf to the root, through the bark and the cambium layer. From this perfected sap the growth of the season is formed, and provision for the beginning of the next season's growth is also stored up, commonly in the root."

" Next to the cambium, and united to the wood by the rays from the pith, is the bark, consisting of three layers."

" The inner, or fibrous layer, is formed by bast cells and firm cellular tissue. Surrounding the inner bark is a layer of cellular tissue in which the rays from the pith terminate, and which is named the green layer, because it often exhibits this color in young shoots, and then performs the same function with the green tissues of the leaf. Outside of all this is the corky layer, consisting of dry, dead, cellular tissue, and developed annually from the green layer. This is not usually of much thickness, or consequence, but sometimes, as in the cork oak of Spain, it becomes an important article of commerce."

" The growth of our trees goes on in the cambium layer, from which is produced annually a layer of wood and a layer of bark, each formed of longitudinal fibro-vascular tissue and horizontal cellular tissue."

" As the trunk expands, the outer bark cracks and falls off, as in the shag-bark hickory, or distends and envelops it with a somewhat smooth covering, as in the beech and birch. In these latter cases the annual cortical layers are quite thin, and the outer layer very gradually wastes away, under the influence of winds and storms. In the cork oak the outer layer is specially thickened, and if removed every eighth year, may be obtained in stout, elastic sheets, which would crack and fall to the ground in the process of time if not harvested. The structure of the root is not unlike that of the stem, except that the pith is usually wanting, as well as the green layer of the bark, which could not be formed nor be of any use in the dark earth where the root makes its home."

Fig. 8 represents a section, both vertical and horizontal, of a branch of sugar maple, two years old, as it appears in December. The portion included in the lines marked A is of the first year's growth; those marked B indicate the wood of the second year; while those marked C include the three layers of the bark. D represents the pith of loose cellular tissue; E the pith rays of silver grain of hard cellular tissue connecting the pith with the green or middle layer of bark, which also consists wholly of cellular tissue. F marks the outer or corky layer of the bark, which is composed of dry, dead cells, which are formed of consecutive layers from the outer portion of the living green layer; G is the green layer of cellular tissue. H shows the liber or inner bark, made up of cellular tissue penetrated by

long bast cells, arranged parallel with the axis of growth. I
represents the place of the cambium or growing layer of organizable material which descends from the leaves between the
liber and the sap-wood during the period of growth. K is

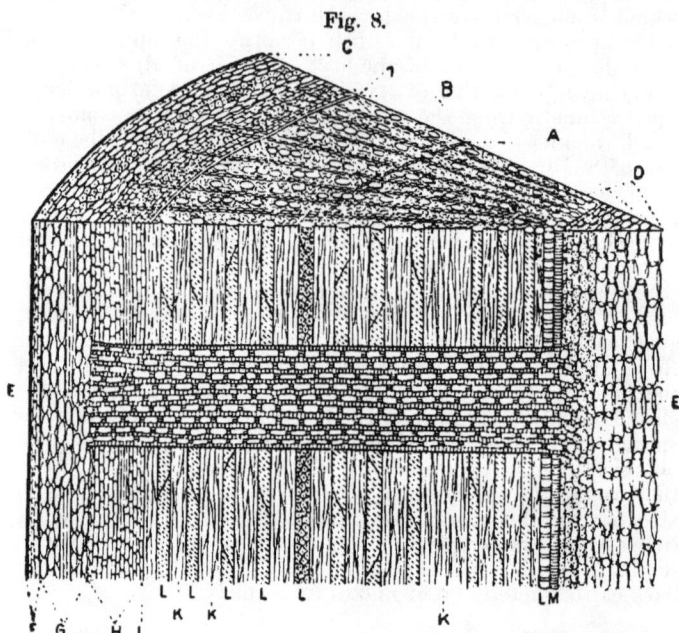

Fig. 8.

woody fibre, which gives strength to the stem, and through
which the crude sap rises. L indicates the vessels or ducts,
with various markings, such as dots, rings, and spirals, which
are formed most abundantly in the Spring, and usually contain no fluid. They convey gases and aqueous vapors, and it
may be that a large proportion of all the water ascending from
the roots to the leaves passes through them as a vapor. M is
the layer of spiral vessels or ducts which always inclose the
pith, and in the young shoot extend into the leaves and unite
them to the pith during its life, which ceases with the first
season.

"This part of the plant develops an annual layer of wood
and bark, with rays of cellular tissue like the stem. The num-

ber and extent of root branches in the soil depend much upon
its fertility and adaptation to the plant."

"As the vigor of vegetable growth depends chiefly on the
action of the roots, the importance of thorough tillage is
apparent."

"The striking peculiarity in the structure of the root is the
absorbent power of the young rootlets, which are either covered with a thick, spongy layer of cellular tissue, or furnished,
as is commonly the case, with exceeding minute but innumer-
able hairs, which penetrate the crevices of the earth in every
direction in search of food. The extreme tips of the rootlets,
about one sixth of an inch in length, are not clothed with
hairs, nor capable of absorption, but serve as entering wedges
for the advancing root, which lengthens only near the extremity."

"The bark of the larger roots becomes thick and impervi-
ous, like that of the trunk and its older branches, and the
inner portion of the wood, both above and below ground,
gradually solidifies, and becomes unfitted for the free trans-
mission of fluids. It is then called heartwood, in distinction
from the sapwood, through which fluids are transmitted freely.
The farther any layer of wood or bark is removed from the
living cambium the less vitality does it retain, and consequently the less useful is it in the economy of the plant."

"The leaf has been said, with some propriety, to be an
extension of the bark, and consists of a framework of fibro-
vascular tissue forming the stalk and veins, with a double
layer of loose cellular tissue covered with a distinct epidermis
or skin. The vessels in the leaf stalk and the veins, which are
its branches, are also in two layers, the upper connecting the
leaf with the vessels surrounding the pith, which are called
spiral on account of their peculiar markings, and the lower
which are united to the cambium layer through the tissue of
the inner bark."

"The distinctive features of the leaf is the presence of stomata or breathing pores, which are usually more numerous on
the under side. These stomata are furnished with openings,
so constructed as to close in very dry air, and open in that
which is moist, but they always remain shut, except under the
stimulus of light. As the chief function of the rootlets is to
absorb the liquid food of the plant from the earth, so it is the
special work of the stomata to transpire the surplus water of
the crude sap, which has been employed as a carrier of food
from one extremity of the countless series of cells which build
up the plant, to the other, in some cases a distance of five hundred feet, through imperforate membranes, and against the
force of gravitation."

CHAPTER XIV.

Entomology should be made a Permanent Study in the Public Schools.

That many orchards, vineyards, etc., in this State, are infested by insect pests is an assertion that cannot be successfully contradicted. How can this evil be remedied? So long as each fruit grower considers himself a competent authority as to the necessity of extirpating from his premises such insects as are known to be injurious, and eradicating such diseases as impair the value of the fruit crop, and decides to determine for himself the conditions upon which he will interfere with their spread for the protection of his own property or that of his neighbor's, there can be no uniformity of action, by which the evil can be thoroughly eradicated, until a general agreement of all parties interested in the fruit growing industry can be perfected.

The dissemination of such information as will give a thorough understanding of what is required, will require time. As the fruit growers of California are scattered over a large range of territory, and, as stated above, each individual having his own theory to cling by, any advice offered that is not in accordance with such a course as he has determined to adhere to, is looked upon as a malignant interference with his private affairs.

There are sources of power by which the obstacles mentioned will be overcome, to wit: THE PUBLIC AND PRIVATE SCHOOLS, combined with the intellect of Young America and the teaching in the family circle.

The query may be reasonably made by persons who have given but little attention to the teachings of entomology: Why is it necessary to introduce the study of economic entomology into the public schools at the present time? In answer to this inquiry, it is stated in the first part of this chapter that there is a conflict of opinion existing among fruit-growers, etc., whose industries are threatened by the invasion of insect pests. However, it is a matter of common agreement of those persons who have considered the subject, that it is only by a united warfare, or by united action, that the insect pests can be exterminated.

Therefore, under present circumstances, it is absolutely necessary that the enterprising husbandman should be fully acquainted with the teachings of the science of entomology, or in other words, with the natural history and habits of the insect pests which destroy his property.

Another inquiry may be made : Why has the study of this science been neglected? Until within the last few years the orchards in this State were free from codlin moth and scale insects ; the vineyards were free from phylloxera, vine moths and flea beetles ; the vegetable garden from cabbage worms and cabbage bugs ; the granaries and storehouses from the grain weevil, etc. ; therefore our educators introduced such studies as they thought best for the requirement of the times. Besides, there were obstacles in the way of introducing this science not easily overcome, viz : the text books relating to entomology were generally written in technical or scientific language, with which the masses of the pupils attending the public schools were unacquainted. In many cases mechanical accessions have to be employed, such as the pocket lens and microscope to detect the presence of these creatures, they being amongst the minutest works of creation, and seemingly endowed with an instinct to avoid the enquiring eye of man. Again, the student who attempted to make insect life a study was treated as a mere trifler and a dabbler in childish pursuits. It may therefore be readily imagined why the study of this science has been neglected in the past.

In order to promulgate information in relation to the natural history, etc., of injurious insects, it must not be treated as a matter of only local importance. The damage done to property by injurious insects is not confined to any one locality, but extends from the valleys to the hillsides and mountain tops, from the northern line of the State to the southern line, and from the Sierras westward to the sea ; so that the promulgation of information in relation to their natural history and habits must necessarily be general.

This requirement can only be attained by introducing the study of economic entomology into the public schools, and by discussing its teachings in the family circle.

Among the results that will follow, the husbandman will

thoroughly understand the natural history and habits of those parasites that prey upon his industry, and can therefore make a successful warfare for their extermination, having a complete knowledge of how and when to strike the blow for victory.

CHAPTER XV.
Economic Entomology.

Horticulturists and all other persons engaged in cultivating the soil should make the study of insects (especially those known to be injurious to fruit and fruit trees, grain, vegetables, etc.), a part of their every day work. It is as necessary for them to understand the natural history and habits of such pests of the orchards, etc., as it is to understand how to plant, prune, etc.

Some fruit-growers think that for such practical investigation of insect life, a scientific knowledge of the anatomy and physiology of these creatures is indispensable. Such, however, is not the case. It requires some apparatus (which can be procured at a nominal expense), a little patience, and an interest in the subject investigated, to learn practically what they have depended on others to furnish.

The following apparatus is necessary: (I am under obligations to Mr. C. Muller, optician, No. 135 Montgomery Street, San Francisco, for illustrations of this chapter.) Fig. 9 represents a one-inch focus watchmakers' glass; price, from 75 cents to $1.25 each. Fig. 10 represents a small microscope; price, from $10 to $20. Such instruments have sufficient power for any practical investigation required on the farm or its surroundings.

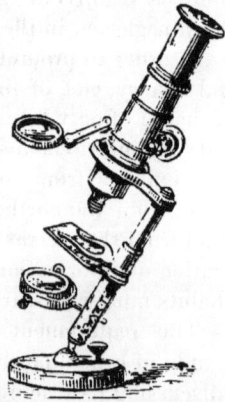

Fig. 9.

Fig. 10.

ings. One dozen glass slides, about $1.00; one dozen glass

covers, about 75 cents; Canada Balsam, 25 cents. A neat and useful outfit can be had for $15.00, at C. Muller's, 135 Montgomery Street, San Francisco. Should the fruit grower be unable to give half an hour of his time each day for such investigations, the ladies and children of the household should be trained to make observations that, when compared with those of others, such information will be obtained as will repay them for their time and labor. When the life history and habits of any of these insects is learned to such an extent as to be familiar with the metamorphoses (changes) as larva, pupa (chrysalis), imago (perfect insect), the fruit grower can then go to work intelligently to exterminate them. By following the above recommendations, the result gained will be replacing theoretical by practical information.

CHAPTER XVI.

Mildew or Scab on the Foliage and Fruit of Apple and Pear Trees.

(*Fusicladium dentriticum.*—F. K. L.)

For a number of years past the presence of what is commonly called mildew has been detected on the leaves of the apple and pear trees, but so far as the leaves are concerned, more abundant on the former. When the apple tree is attacked by this fungus early in the season, the young fruit is generally destroyed, and the leaves attacked hardly ever come to perfection, as they appear to dry up and crumble to pieces. On the pear leaf it is in the form of a brownish blotch.

When the attack is first noticed on the young pears, it is in the form of an irregular brownish spot on the skin; this dark spot or scab, as it is commonly called, does not penetrate into the fruit to a great extent, but destroys the skin and forms a hard surface, thus preventing the growth of the fruit on and immediately around the place attacked; consequently when the skin of the young fruit is attacked in one or more places, when it is full grown the surface is not uniform, and the market value is thus decreased. That the fungus spores which cause this fungus, or mildew, on the leaves and fruit of apple and pear trees in the Spring is on the trees from one season to another, I think is beyond question.

There is also a species of fungus that attack the apricot, giving it a speckled appearance and destroying its market value. It is also noticed on the peach, nectarine and prune this season, 1883.

To destroy mildew, fungus, etc., on fruit and foliage of apricot, peach, prune, nectarine and almond, use Remedies No. 5 or 7, one pound of the mixture to each one gallon of water.

On apple, pear and quince, use Remedy No. 6, one pound to each five quarts (1¼ gallons) of water.

NOTE.—The above remedies should be applied by spraying the fruit and foliage as soon as the fruit is well set from the blossom, or about the size of a small marble. The spraying should be repeated in two weeks.

CHAPTER XVII.

Birds—Beneficial and Injurious.

It would be a very difficult task for any person to ascertain what birds are injurious and which are beneficial. I do not think there is any bird that is *wholly* injurious, because those which are usually regarded as being injurious, such as the robin, blackbird, etc., are partially beneficial, since they sometimes feed upon injurious insects. And it would be about as difficult to name a bird that is *wholly* beneficial; the swallows are usually regarded as beneficial birds, and yet it is evident that they destroy more beneficial insects than injurious ones, since the former are mostly on the wing in the daytime, while the latter fly chiefly at night. The night-hawk and the whippoorwill are about the only birds that can be regarded as being *wholly* beneficial.

But, as any bird that feeds upon any part of our cultivated plants, shrubs or trees, or in any manner injures or destroys those animals which minister to our wants, is usually regarded as being injurious, I will endeavor to classify our more common birds into three classes: the beneficial, the injurious, and the doubtful. The first class will contain those birds which feed almost exclusively upon insects or small animals, and which are not known to injure fruits, grain, or anything of any value to us. The second class will contain those which feed principally upon fruits or grain, or which are known to cause extensive injury to some of the useful products of the soil, without making adequate return for their destructiveness by destroying noxious insects. In the third class will be placed those birds which sometimes depredate upon the useful products of the soil, or upon our domestic animals (including fowls and bees), and which also feed largely upon insects; so that it is doubtful as to whether we are to regard them as being beneficial or injurious.

BENEFICIAL BIRDS.

Bluebirds, pewees, flycatchers (except the beebird), swallows, martins, wrens, chickadee, vireos or greenlets, tanagers or redbirds, ground-robins, cuckoos, humming-birds, warblers, night-hawks, whippoorwills, meadow-larks, shrikes, butcherbirds, road-runners, vultures, turkey buzzards, gulls, plovers and snipes.

The meadow-larks and plovers, and perhaps a few other birds in this list, sometimes feed upon seeds, but only to a limited extent, their food consisting almost exclusively of insects; the shrikes and road-runners feed upon insects, snakes, small lizards, etc., and the former sometimes destroy small birds; gulls feed upon insects, frogs crayfish, etc.

INJURIOUS BIRDS.

House-finches or red-headed linnets, cedar-birds or waxwings, orioles, doves, wild geese and ducks. These are about the only birds that are considered as being very injurious, and even these partly atone for their injuries by feeding upon insects. The linnets sometimes occasion considerable damage by feeding upon the buds of fruit trees. (Early in Spring use Remedies Nos. 5, 6, or 7, one pound to each gallon of water used, and the birds will not eat the buds.) The cedar-birds and orioles feed upon fruit and berries, and the latter also feed upon green peas; the doves feed mostly upon grain, while the wild geese and ducks are sometimes very injurious to growing grain.

DOUBTFUL BIRDS.

Thrushes, robins, catbirds, blackbirds, beebirds, finches, sparrows, bluejays, magpies, crows, hawks, owls, quails, woodpeckers, and mocking-birds.

As stated above, these birds occasionally feed upon some of the useful products of the soil, or upon domestic animals, barnyard fowls, or bees, while they also feed upon injurious insects or other pests; and it is a matter of considerable doubt as to whether their good deeds do not counter-balance their evil ones. The thrushes, robins, catbirds, quails and woodpeckers sometimes feed upon fruits; the blackbirds, crows, bluejays and woodpeckers occasionally feed upon corn; the sparrows feed principally upon seeds, while the finches feed upon seeds and buds; the beebirds are sometimes quite destructive to bees, but also feed largely upon other insects.

INSECTS INFESTING THE APPLE TREE.

CHAPTER XVIII.

The Woolly Aphis.

(*Schizoneura lanigera*—Hausman).

SYNONYMS.—*Aphis lanigera*—Hausman. *Coccus mali*—Bingley. *Eriosoma mali* (Leach, M. S.)—Samoulle. *Myzoxylus mali*—Blot. *Schizoneura lanigera*—Hartig. *Pemphigus pyri* —Fitch. *Aphis (Schizoneura) lanigera*—Ratz. *Eriosoma lanigera*—Ruricola.

Order, HEMIPTERA;
Sub-order, HOMOPTERA; } Family, APHIDIDÆ.

[Living in hollows on the trunk or limbs of apple trees, a small plant-louse, which is more or less covered with a white, cottony matter.]

The presence of this insect can readily be detected from the appearance of the tree infested; the branches appear knotty, the wood dry, hard, and brittle, and the general appearance is that of over-age and decay. Its distribution may be said to be general in this State on apple trees, and it is very generally distributed on apple nursery stock. The opinions of writers differ as to whether the insect found on the roots, and those on the trunk and limbs, are the same species, some contending that they are similar, and that those on the trunk and branches go to the roots and hibernate during the winter, or deposit their eggs for the next season's brood, while others contend that they are distinct species. For the present purpose it is sufficient to know that one or more species infest our trees. The climate of California may be favorable to their hibernating throughout the winter season on the trunks and limbs, as I have found them in crevices of the bark throughout the winter season, and at the same time finding them on the roots. To destroy these insects effectually they must be attacked on the roots, and on the trunk and branches at the same time.

Fig. 11.

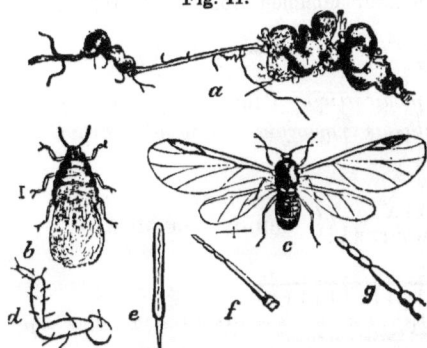

Fig. 11.—W o o l l y
Aphis; *a*, an infested
root; *b*, the larva—
color, brown; *c*, winged
adult—colors, b l a c k
and yellow; *d*, its leg;
e, its beak; *f*, its antennæ; *g*, a n t e n n æ
of the larva (*b*)—all
higly magnified.

The woolly aphis
(Fig. 11) is of a dark
russet-brown color, with the abdomen covered with a white
down, of a cottony appearance.   It is said by some writers that
it can only live on the apple tree.   I have found it on the pear
tree, and on pear nursery stock; also, one colony on a cherry
tree; and in each case they seemed to thrive well above ground.
This insect is only to be found on the roots, branches, limbs,
and trunks of the trees; it does not infest the leaves.

When the woolly aphis begins to spread, it appears in
blotches on the trees, of a white cottony appearance, which, if
rubbed with the finger, will produce a blood-colored fluid.

REMEDIES.—For destroying wolly aphis on roots of trees,
No. 39 or No. 40.   These remedies should be applied early in
the Fall season, so that the rains will carry the solution to the
roots.   No. 40 or 41 may be used, but No. 39 or 40 are preferable.

For destroying wolly aphis on the trunk of the tree, use
No. 43, when the tree is dormant.   (Spray).

For destroying wolly aphis on limbs and branches, when
the tree is in leaf, use No. 6, one pound of mixture to each five
quarts (1¼ gallons) of water used; or, No. 51.   (Spray.)

For destroying wolly aphis on roots of nursery stock
(young apple trees), use No. 42; or, No. 5 or 7, one pound of
mixture to each gallon of water used.

5

CHAPTER XIX.

Oyster Shell Bark-louse, or Common Apple Scale Insect.

(*Aspidiotus Conchiformis*—Gmelin.)

SYNONYMS—*Mytilaspis pomicorticis*—Riley. *Aspidiotus pomo-rum*—Bouche. *Mytilaspis pomorum*—Bouche. *Aspidiotus pyrus-malus*—Rob.

Order, HEMIPTERA ;   }   Family, COCCIDÆ.
Sub-order, HOMOPTERA : }

[The measurements of insects in this work are given in inches and lines. The above cut represents one inch divided into lines and fractions thereof.]

[A slender, slightly-curved scale insect, infesting deciduous fruit trees.]

This species of scale insect can be found in orchards in nearly all the central counties of California, and is very destructive to apple trees, and also infests the pear and other deciduous fruit trees.

Fig. 12.

Fig. 12.—A piece of bark infested by Oyster Shell Bark-lice—colors of scales, brown, yellow and gray.

This species can be readily distinguished from the other species of scale insects that infest the apple tree and described elsewhere in this work. The scale of the female is long and narrow, and more or less curved, and widened at the posterior end; it measures from one line to one and one quarter lines in length. Color—Dark yellowish-brown; exuviæ, amber-yellow.

The scale of the male is smaller than that of the female, and nearly straight; it is not so dark in color, and has a mottled appearance.

NATURAL HISTORY.

The eggs found under the female scale number from thirty-five to seventy-five; the young are hatched about the tenth of

May, and, so far as known at present, there is only one brood in each year. Larva length, one seventieth of an inch; color, pale yellow; form, ovoid; antenna, seven jointed, two anal setæ. In a few days after it is hatched the larva fixes itself on the wood, leaves or fruit, and perfects its change, or metamorphosis, as shown in Fig. 13.

Fig. 13.

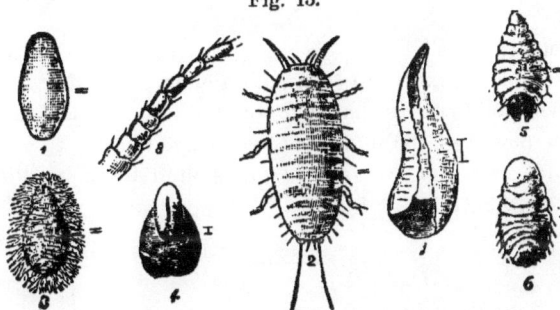

Fig. 13.—Oyster Shell Bark-louse, highly magnified. *1*, the egg—color, white or yellowish; *2*, the newly hatched larva—color, yellow; *3*, the larva after becoming fixed; *4*, the scale after the second plate is formed; *7*, fully formed scale, ventral view; *5*, ventral view of larva; *6*, adult female—color, pale yellow; *8*, antenna of the larva (2).

The male (perfect) insect is winged, as shown in Fig. 14.

Fig. 14.

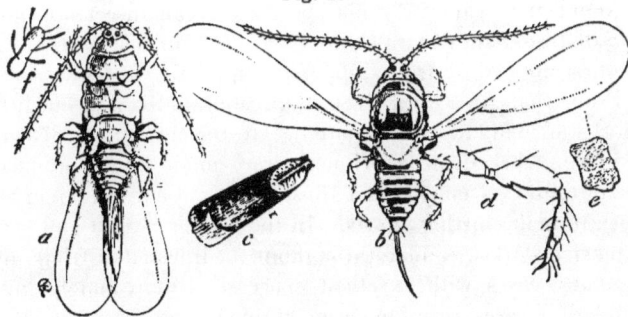

Fig. 14.—Oyster Shell Bark-louse (male), highly magnified; *a*, ventral view of winged male—color, gray; *f*, a joint of his

antenna more highly magnified : *c*, a male scale—colors, yellow
and brown ; *b*, a winged male, with wings expanded—color,
gray ; *d*, one of his legs, more highly magnified : *e*, upper surface of the wing, more highly magnified.

It has been found very difficult to destroy this species, as
the scale or shell is fastened very securely to the wood, etc. ;
but late experiments have proven that the pest can be eradicated.

REMEDIES.—When the tree is dormant, use No. 13—one
pound of the mixture to each gallon of water used. (Spray.)

When the larvæ are hatched and the tree is in leaf, use No.
6—one pound of the mixture to each five quarts (1¼ gallons)
of water used ; or, No. 5 or 7—one pound to each gallon of
water used. (Spray.)

CHAPTER XX.

The San Jose Scale.   (Cal.)

(*Aspidiotus perniciosus.*—Comstock.)

Order, HEMIPTERA ;   }
Sub-order, HOMOPTERA ;   } Family, COCCIDÆ.

[A small, nearly circular, and flattened scale insect infesting
deciduous fruit trees.]

About the year 1873, this species of scale-insect appeared
in San Jose, Santa Clara County; at least, in that year it was
the first noticed by fruit-shippers as infesting the fruit.

From that time until 1880 it spread rapidly, and but little
effort was made to exterminate it. In the Winter of 1879 and
1880, some practical experiments were made which produced
such results as encouraged those who had entered upon the
work to make further efforts. In the Winter of 1881 and 1882,
extensive work was done throughout the infested districts, and
in many cases with excellent success. Unfortunately, in a
number of cases, solutions were applied, such as a low grade
of coal oil that was in the market under the name of *tree wash*,
which sold at about fourteen cents per gallon : and another

under the name of *crude petroleum;* these were recommended
by those who had used them as excellent insecticides.
However, in a few months it was discovered that these solutions contained pernicious properties which were destructive
to plant-life, consequently a large number of trees were destroyed or killed. This dangerous insect has been introduced into many fruit growing districts on nursery stock, and
is found in over twenty counties at the present time (August,
1883). Great progress has been made in perfecting remedies
for the extermination of this pest, so that if the work is thoroughly done the orchardist need not fear this most dangerous
of the aspidiotus scale insects.

The *A. perniciosus* infests all the deciduous fruit trees, excepting, perhaps, the Black Tartarian Cherry; it has also been
found on the currant bush, and on tomatoes grown in the
vicinity of infested trees. It also infests the poplar, osage
orange, wild cherry, eucalyptus and other ornamental trees
and shrubs.

NATURAL HISTORY.

The females of the family coccidæ, to which the genus *aspidiotus* belongs, are described by Westwood as follows: "That
without referring to their singular habits we find some of them
on arriving at their last state are not only wingless, but also
footless and antennæless, and in which all appearance of
annulose structure is lost—the creature, in fact, becoming an
inert mass of animal matter—a slender *seta* arising from the
breast and thrust into the stem, or leaf, or fruit on which the
animal is fixed, being the only external appendage of the
body."

Prof. Comstock, in speaking of the *metamorphosis* of the sub-
family *diaspinæ*, to which the genus *aspidiotus* belongs, says
that "members undergo a remarkable change at the time of
the first molt, losing their legs and antennæ, and thus becoming apparently less highly organized than in the larval state."

DESCRIPTION.

The scale of the female is circular, and in color blackish
gray, excepting the exuviæ in the center, which is of a deep

straw color; sometimes it has a reddish hue: it measures from one line to one and a quarter lines in diameter. The scale of the male is oval in outline, and nearly black with the exuviæ between the center and the anterior margin of the scale, but is darker in color, and more obscure than that of the female.

The female insect (Fig. 15), is primrose yellow, and sometimes ochre yellow in color, and measures about half a line in diameter. Each female produces from thirty-five to fifty eggs.

[Fig. 15.—San Jose Scale, insect (adult female) enlarged.]

Fig. 15.

Fig. 16.

[Fig. 16.—Larva of San Jose Scale, enlarged, ventral view.]

The larvæ (Fig. 16), are yellowish, form oval, antennæ six jointed, two anal setæ, length about one seventy-fifth (1-75) of an inch. The larva creeps around for two or three days, then finding a suitable place it fastens itself to the wood (Fig. 17), leaf, or fruit (Fig. 18), and undergoes its change, or *metamorphosis*.

[Fig. 17.—Portion of a branch infested by San Jose Scales.]

Fig. 17.

Fig. 18.

[Fig. 18.—A pear infested by San Jose Scales.]

The male insect (Fig. 19), (perfect), is winged; wings nearly transparent—color, body light amber, with dark markings; antennæ ten-jointed (hairy): posterior stylet nearly as long as the body. Pupa of male insect (Fig. 20), fifteen days after the scale is formed; Fig. 21, thirty-five days after the scale is formed. The sting or bite of the female insect produces a dark red mark on the wood or fruit.

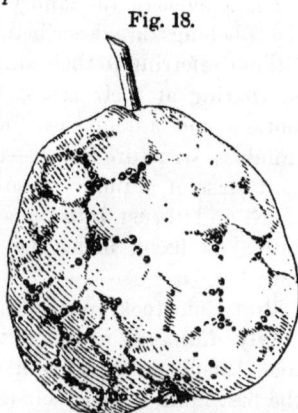

Fig. 19. Fig. 20. Fig. 21.

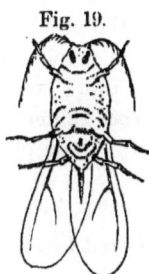

[Fig. 19.—San Jose Scale, insect (adult male) enlarged.]

Fig. 20.—Pupa of San Jose Scale, insect enlarged.

Fig. 21.—San Jose Scale, insect (male pupa) enlarged.

This species produces three broods each year; the first may be expected about the time that cherries begin to color, the second in July, and the third in October.

REMEDIES.—When the tree is dormant, use No. 11, No. 12. Or No. 13—seven pounds of the mixture to each eight gallons of water. (The latter is preferable.) (Spray.)

When the tree is in leaf use No. 6—one pound of mixture to each five quarts of water (or 1¼ gallons)—or No. 5, or No. 7.

CAUTION.—Beware of mineral oils.

CHAPTER XXI.

The Greedy Scale Insect. (Cal.)

(*Aspidiotus rapax*—Comstock.)

Order, HEMIPTERA ;   } Family, COCCIDÆ.
Sub-order, HOMOPTERA ; }

[Infesting apple, pear, quince, peach, plum, apricot, almond, and olive trees; also, acacia, willow, eucalyptus, locust, and other ornamental and forest trees.]

This species was given the specific name *rapax*, or Greedy Scale, by Prof. J. H. Comstock, in 1881, on account of the great number of plants upon which it subsists. I found it on the fruit trees at Santa Cruz, in May, 1881, and wrote of it under the common name, Santa Cruz Scale, but have since found that it is generally distributed throughout the fruit dis-

tricts of California. It infests the wood (limbs and branches),
foliage, and fruit of deciduous trees, including the peach, apricot, plum, almond, quince, and fig, and especially the apple,
pear, and olive. In some cases apples and pears were so
infested with this scale as to destroy their market value. The
acacia, willow, eucalyptus, locust, etc., are also its food plants.

NATURAL HISTORY.

Fig. 22 represents a portion of a branch infested by Greedy
Scales; at the left are two of the scales,
enlarged—colors, gray or yellowish, and
brown.

Fig. 22.

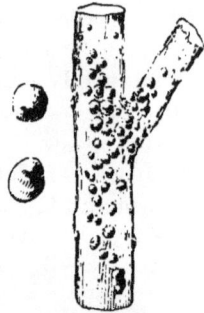

The scale of the female is nearly circular,
or slightly oval; yellowish in color when it
covers a living, mature insect, but is generally a light gray. The full-grown specimens measure nearly one line in diameter,
and its form is more convex than the other
species of *aspidiotus* described in this work;
the exuviæ is between the center and one
side, or edge, of the scale. Eggs, ovate—
color, yellow.

Fig. 23.

[Fig. 23.—Larva of Greedy Scale, enlarged;
ventral view.]

[Fig. 24.—Female Greedy Scale Insect, enlarged.]

Larva (Fig. 23)—color, yellow;
length, one-eightieth of an inch;
antennæ, six-jointed; two anal
setæ.

Fig. 24.

Female full grown (Fig. 24) is circular in form—
color, yellow, with clear or nearly transparent
blotches. She deposits from thirty-five to eighty eggs.

Male (perfect) insect, winged. (I have not been able to procure a perfect specimen.) There are probably two broods each
year, as I have found the eggs in May and August.

Use remedies as described for San Jose Scale (*A. perniciosus*),
Chap. XX.

CHAPTER XXII.

The Round-headed Apple-tree Borer.

(*Saperda candida*—Fabricius.)

SYNONYM.—*Saperda bivittata*—Say.

Order, COLEOPTERA; Family, CERAMBYCIDÆ.

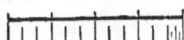

[The measurements of insects in this work are given in inches and lines. The above cut represents one inch divided into lines and fractions thereof.]

[Boring into the trunks of apple, pear, quince, and similar trees, a nearly cylindrical, yellowish-white, footless grub, which is finally changed into a beetle of a brownish color, having two white stripes on its back].

The greater number of injurious insects live exposed upon the plants which they attack; but there are several kinds which live concealed from view in the stems or branches of various kinds of plants, shrubs, and trees, and thus hidden from view, they carry on their silent work of destruction. Prominent among this latter class is the round-headed apple-tree borer, which is found over the greater part of the United States.

Fig. 25.

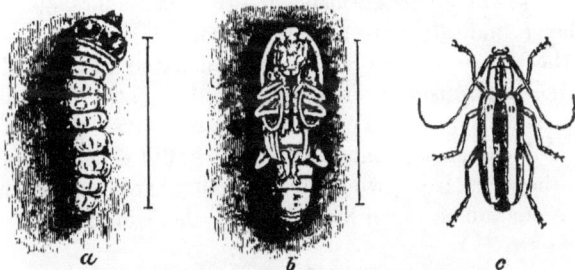

[Fig. 25.—Round-headed Apple-tree Borer; *a*, the larva—color, yellowish-white; *b*, the pupa—color, yellowish-white; *c*, the beetle—colors, brown and white.]

The beetles, or perfect insect (Fig. 25*c*) first make their appearance a few weeks after the apple tree has put forth its leaves. They soon pair, and in the course of a few days the females deposit their eggs. This operation consumes considerable time, so that about three months elapse before all of the beetles have finished depositing their eggs.

The latter are commonly deposited in the crevices of the bark, and usually near the surface of the ground, but sometimes they are placed in the axil of the lower branches, or the place where the branches start out from the trunk. In about a fortnight, from each of these eggs is hatched a minute, footless grub of a whitish color, with a yellowish head. These grubs eat their way obliquely downward through the bark, and for the first year of their lives they live upon the inner bark or sapwood, forming flat, shallow cavities. In their passage through the bark they push their excrements and refuse through the opening of their burrow, and being of a glutinous nature, it collects around the mouth of the burrow in a small mass, which, being usually of an orange color, is readily detected by the experienced eye. The following season the borer enlarges its burrow, pushing its castings out of the openings of its burrow in pellets, resembling in shape a grain of oats, but larger. These are commonly found in pairs, lying parallel, with their points toward the tree.

During this, the last Summer and Fall of their lives, they do their principal damage by widening their burrows on every side, destroying the alburnum deposited the year before, and often the layer under it. If there is only one in a tree at this age, and the tree is not more than one inch and a half in diameter, the borer usually kills it by girdling entirely around it, except about one fourth of an inch, on one side. One borer in a large tree does not materially injure it, but generally in such trees there are from two to five, and they girdle all around to within one fourth of an inch of each other's burrow, and thus kill the tree.

The borer or larva (Fig. 25a) during the last Fall of its life, eats voraciously until cold weather sets in, when it carefully houses itself away until the following Spring. As soon as the weather becomes mild, it begins to cut a cylindrical burrow from three to six lines long, usually up the trunk of the tree, but sometimes directly through it, ending it just under the outside bark, leaving the bark about the thickness of writing paper; it then draws back about an inch, places some coarse chips before and behind it, and soon assumes the pupa form. (Fig. 25*b*). After remaining in this state for from two to six weeks, it is changed to a beetle, which soon afterwards gnaws a hole through the bark that covers the end of its burrow, and thus effects its escape. The head of this grub is small, horny, and brown; the first ring or segment is much larger than the others; the next two are very short, as are also the eleventh and twelfth; the rings, from the fourth to the tenth, inclusive, are each furnished on the upper side with two fleshy warts, which are situated close together, and are destitute of the rasp-like teeth which are usually found on the grubs of the other kinds of borers; no appearance of legs can be seen, even with a magnifying glass of high power. When fully grown, it measures about one inch in length.

The beetle, or perfect insect, measures from six to nine lines, or one half to three fourths of an inch in length, and is of a cinnamon-brown color, marked with two white stripes, which extend from the head to the tips of the wing-cases; the face, antennæ, and legs are white, the antennæ being nearly as long as the body.

REMEDY.—Use No. 37.

CHAPTER XXIII.

The Flat-headed Apple-tree Borer. (Cal.)

(*Chrysobothris femorata*—Fabricius.)

Order, COLEOPTERA; Family, BUPRESTIDÆ.

[Boring into the trunks of apple, pear, peach, and similar trees; a pale, yellowish, footless grub, having the forepart of the body greatly widened and flattened; finally transforming into a greenish-black or bronze colored beetle, which is copper-colored on the under side.]

While the round-headed apple-tree borer, *S. bivitata* (Fig. 25), usually infests healthy, growing trees, the present species seems to prefer those which are more or less diseased, in such places where the bark is bruised, sunburned or dead, on which account it is not to be so much dreaded as the round-headed species.

The flat-headed apple-tree borer first makes its appearance in the month of April or May, soon after which it deposits its eggs; these are usually deposited in crevices or beneath the loose bark, several of them being not unfrequently found together. They are of an ovoidal shape, pale yellow, irregularly ribbed, wih one end flattened. The young larvæ hatched from these gnaw their way through the bark until they reach the green cambium layer, and gradually extend their broad and flattened channels beneath the bark. At length, when they have grown stronger and their jaws firmer, they bore into the more solid wood of the tree, working upward until about to undergo their transformations, when they cut a passage to the outside, leaving a thin covering at the surface through which the beetle (Fig. 26) afterwards forces its way. It is supposed to pass through its various changes with inthe course of one year. The grub or larva (Fig. 27) of the borer measures nearly nine lines or three fourths of an inch in length, when full grown, and is of a pale yellow color; it is entirely destitute of feet, and the second segment is very broad and flattened, by which character this grub may at once be distinguished from that of the round-headed borer (Fig. 25a). The beetle measures about six lines, or one half an inch in length, and is of a dark, dull, greenish color with a strong, coppery luster, which is deepest on the forehead and at the tip of the wing cases; the head is sunken up to the eyes in the thorax. On each wing-case are two irregular impressed spots, which are generally of a deeper green or coppery color than the surrounding surface, and sometimes appearing double. The under side and limbs are of a brilliant copper color. The portion of the abdomen covered by the wing cases is a light, blue-green.

Fig. 26.—Flat-headed Apple-tree Borer—color, dark gray or dull greenish.

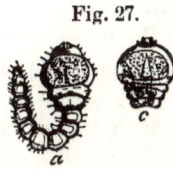

Fig. 26.   Fig. 27.   Fig. 28.

Fig. 27.—Larva of Flat-headed Apple-tree Borer; *a*, the larva; *c*, the underside of the head and forepart of the body—color, white.

Fig. 28.—Pupa of Flat-headed Apple-tree Borer—color, white.

REMEDY.—Use No. 37.

CHAPTER XXIV.

The Apple-twig Borer.

(*Bostrichus bicaudatus.*—Say.)

SYNONYM—*Amphicerus* (*Bostrichus*) *bicaudatus.*

Order, COLEOPTERA ; Family, PTINIDÆ.

[Living in grapevines, a small, whitish, nearly cylindrical, grub, which is thickest anteriorly; finally transforming into a brown beetle which burrows in the twigs of the apple, pear and peach trees.]

The natural history of this insect has never been clearly traced. Dr. Shimer states that he bred it from the larva found burrowing out the central pith of a grapevine, while other authors contend that it spends its larval stage in some forest tree, and that the beetle forms a burrow into the grapevines, apple-twigs (Fig. 29), etc., merely for the purpose of getting a sheltered place in which to pass the Winter.

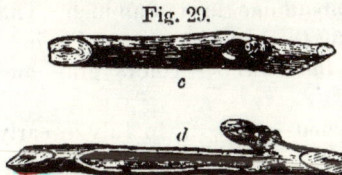

Fig. 29.

Fig. 29.—Twigs showing burrows of apple-twig borer; *c*, the entrance; *d*, the burrow cut open.

The beetles (Fig. 30) are found in their burrows from early Fall till late in the following Spring. They measure from four to five lines, or about three eighths of an inch in length; are of a dark brown color, the thorax nearly black, and the anterior half is covered with spine-like points; at the apex of each

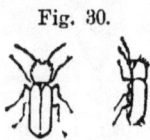

Fig. 30.   of the wing-covers of the male is a little horn from which the species derives the specific name: *bicaudatus*, or two-tailed.

Fig. 30.—Apple-twig Borer; back view and side view—color, brown.

REMEDIES.—If seriously infested, when the tree is dormant. use Nos. 11 and 12—one pound to each gallon of water used. (Spray.) In the Spring when the tree is in leaf, use No. 5 or 7—one pound to each gallon of water used. This will prevent the insect from boring into the tree. Also. prune as directed in No. 27.

* * *

CHAPTER XXV.

The Prickly Bark Beetle.

(*Leptostylus aculiferus.*—Say.)

Order, COLEOPTERA ; Family, CERAMBYCIDÆ.

Living under the bark of apple trees, small, whitish, footless grubs, similar in appearance to the young apple tree borers, occurring sometimes in multitudes, forming long, narrow, winding burrows upon the outer surface of the wood. these burrows becoming broader as the worm increases in size ; finally transforming into a rather short and thick brownish gray beetle (Fig. 31) with small prickle-like points upon its wing-covers, and back of their middle is a white curved, or V-shaped band,

Fig. 31.   with a black streak on its hind edge; length from three to four lines, or about one third of an inch. The perfect insect appearing the last of August.—Fitch.

Fig. 31.—Prickly Bark Beetle—colors gray and black.

REMEDY.—Use soap as directed in No. 37, in July or early in August.

CHAPTER XXVI.

The Gray Bark Eating Weevil. (Cal.)

(*Thricolepis simulator.*—Horn.)

Order, COLEOPTERA; Family, CURCULIONIDÆ.

[A small, light grayish snout beetle, feeding upon the bark of the small branches or twigs of apple trees early in the Spring.]

Early last Spring (1883), specimens were received from fruit growing districts located at least one hundred and fifty miles apart, of a small, light grayish colored weevil, reported as eating the bark and buds of apple trees.

This beetle (Fig. 32, Plate 1,) measures from one and one half to one and three fourth lines in length; the color is grayish brown, but is covered with fine white scales, giving it a light grayish color. The larvæ probably feed on the roots of plants, but as this species is only reported this Spring its natural history is not known to me.

REMEDIES.—Early in the Spring spray trees infested the previous year with No. 5 or 7; repeat in two weeks. This will prevent the beetle from eating the bark.

CHAPTER XXVII.

The Buffalo Tree-hopper.

(*Ceresa bubalus.*—Fabricius.)

Order, HEMIPTERA; Sub-order, HOMOPTERA; } Family, MEMBRACIDÆ.

[Living upon the twigs of the apple, peach, apricot, almond and plum trees; a green leaf-hopper, shaped something like a beechnut, with two short spines jutting out horizontally from each side of the anterior end, having some resemblance to the horns of a bull or buffalo.]

These insects obtain their nourishment by puncturing the twigs with their beaks and imbibing the sap.

Fig. 33.   Fig. 33.—Buffalo Tree-hopper—color, green ; *a*, side view: *b*, back view.

Fig. 34.

Fig. 34.—Eggs of Buffalo Tree-hopper; *a*, an egg, highly magnified ; *b*, eggs natural size in a twig.

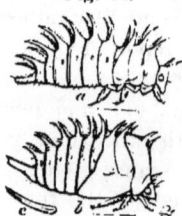

Fig. 35.   Fig. 35.—Larva and pupa of Buffalo Tree-hopper, enlarged ; *a*, the larva—color brownish ; *b*, the pupa—color greenish ; *c*, the ovipositor of the adult female, magnified.

From specimens of this insect (Fig. 33), and branches containing eggs (Fig. 34), received from several places in this State, and also from the State of Nevada, I am inclined to think that from eight to twelve eggs are laid in each puncture.

The young (Fig. 35) hatch in May.

REMEDIES.—No. 28 and No. 25.

CHAPTER XXVIII.

The Cicada, or Harvest Fly.

Order, HEMIPTERA ;   } Family, CICADIDÆ.
Sub-order, HOMOPTERA : }

[Making small slits in the under side of the lower branches of the apple, cherry, and similar trees, and depositing therein a row of pearl-white eggs ; a large, four-winged fly, the body marked with greenish.]

These insects are usually of a large size, their wings expanding from two to three inches. The males have a curious drum-like arrangement on each side of the body, behind the wings. It consists of convex instruments of fine parchment,

which are acted upon by small muscles; when these muscles contract and relax, which they do with great rapidity,
the drum-heads are alternately tightened and loosened, producing a rattling noise, like that caused by a succession of
quick pressures upon a convex piece of tin-plate. The body of
the female is provided with a piercer, with which she makes
numerous small slits in the under side of the branches (Fig.
36*d*,) of various shrubs and trees. The branches thus muti-
lated usually die back to the place where the slit nearest the
trunk occurs, and are frequently broken off by the wind.

Fig. 36.—Seventeen-year Locusts, pupa and eggs; *a*, the
pupa—color,
yellowish-
brown; *b*, the
cast pupa skin;
d, a punctured
twig, contain-
ing eggs; *c*, two
of the eggs re-
moved from
the twig—col-
or, yellowish;
c, the adult, or
perfect insect—
colors, black-
ish, and dull
orange.

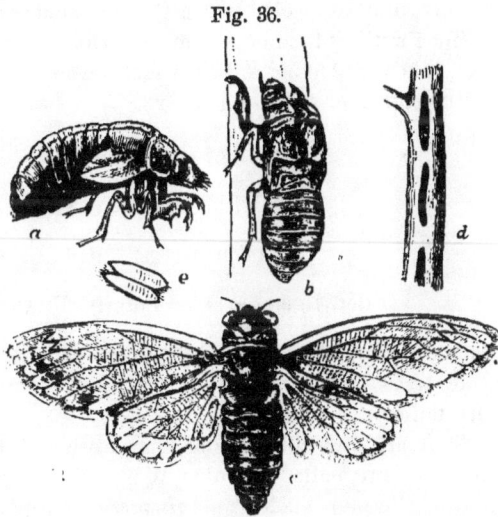

Fig. 36.

As soon as hatched, the young grub enters the earth, but
this is as far as its history is known with any degree of certainty, except that when about to be changed into a perfect
insect it comes out of the earth and ascends a plant (Fig. 36*b*),
to which it attaches itself firmly by means of its hooked claws.
In a short time the skin on its back splits open and the
included insect issues in its perfect or winged form.

Some kinds are known to live for thirteen and even seven-
teen years in the larva state. They do not pass through a
quiet pupa state as butterflies and many other insects are

6

known to do, but remain active from the time they issue from the eggs (Fig. 36c) until they die of old age, or some other cause.

A small species of *Cicada* is found in California, infesting apple trees. The larva is about one inch in length, when full grown; the pupa is from seven to ten lines in length; and only finding a skeleton of the perfect insect, a correct description can not be given. Its natural history and habits are evidently the same as the *C. septemdecim*, excepting the time it takes to mature the perfect insect, which at present can not be stated reliably. They were found at Dutch Flat, in Placer County, and are probably the *Cicada noræboracensis*.

Since writing the above, I am informed that a specimen has been taken at Anaheim, Los Angeles County.

REMEDIES.—No. 25 and No. 28.

CHAPTER XXIX.

Canker Worms.   (Cal.)

Order, LEPIDOPTERA ; Family. PHALÆNIDÆ.

[Feeding upon the leaves of the apple, cherry, and various other trees; yellowish or dark colored span-worm, provided with ten or twelve legs].

There are three specious of span-worms (geometers), which are commonly called " Canker Worms." (The Spring Canker Worm, Chapter XXXII. *Anisopteryx vernata*—Peck : *Palcacrita vernata*—Riley) ; the Fall Canker Worm, Chapter XXX. *Anisopteryx pometaria* — Harris : *Anisopteryx autumnata* — Packard) ; and the Yellow Canker Worm (Chapter XXXI. *Hibernia tiliaria*—Harris). The females of these species are wingless.

NATURAL HISTORY.

The eggs are deposited on the trees by the female moth. *A. pometaria* and *H. tiliaria* deposit their eggs in the latter part of December, and in January ; these are hatched about the time the apple tree has put forth its leaves.

The young caterpillars commence to feed on the new foliage, and in four or five weeks they attain their full growth; they then descend to the ground, which they enter to a depth of several inches; here each caterpillar forms a small cell, in which to pass the pupa stage. The *A. vernata* deposits her eggs early in the Spring; the caterpillars attain their growth in four or five weeks, and then enter the ground to pass the pupa state.

Use remedies as directed in Chapter XXXII—No. 10 or No. 89. (See note, Remedy No. 10).

CHAPTER XXX.

Fall Canker Worm.

(*Anisopteryx pometaria* — Harris; *Anisopteryx autumnata* — Packard.)

Fig. 37.—Fall Canker Worm; *e*, cluster of eggs; *a*, side view of one of the eggs, enlarged; *b*, view of upper end of same, enlarged; *f*, the worm—color, olive-green, with brown stripe and white lines; *c*, side view of a segment of its body, enlarged; *d*, back view of the same enlarged; *g*, female pupa—color, brown; *h*, tip of same enlarged.

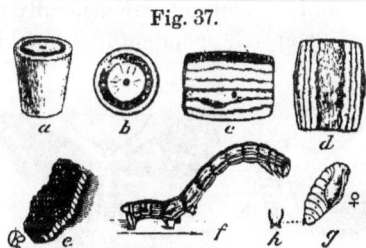

Fig. 37.

The caterpillar of this species (Fig. 37*f*) is about one inch in length, and is provided with twelve legs—-the pair on the eighth segment are very short; color, body pale greenish, marked on the back with a brown stripe (Fig. 37*d*), and with three white lines on each side of the body (Fig. 37*c*); below the spiracles is a brownish line, and below this a white line; the under, or ventral parts, are of a pale flesh color; head, brown; larva, when newly hatched, pale olive-green; head and cervical-shield pale.

Pupa.—On entering the ground the caterpillar spins a cocoon composed of silk, interwoven with particles of earth; in this cocoon it changes to a pupa (Fig. 37g and Fig. 38) in four or five weeks, and in this state it remains until the Fall.

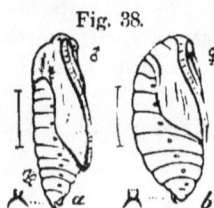

Fig. 38.

Fig. 38.—Pupæ of Fall Canker Worms, enlarged; *a*, the male; *b*, female—color, brown; at the left of each is shown the anal projection, as seen from below, enlarged.

IMAGO, OR PERFECT INSECT.

Toward the latter part of December, or early in January, the perfect insects (Fig. 39, *a* and *b*), emerge from the ground: the wingless females (Fig 39*b*) climb up the trees and lodge upon the branches, where they deposit their eggs in batches of from seventy-five to two hundred (Fig. 37*c*); they are placed side by side in regular rows, and can be readily recognized as belonging to this species, as they are flat upon the top, and marked with a brown ring or circle (Fig. 37, *a* and *b*). The eggs are generally deposited close to the bud, or in a crotch or indent of some kind.

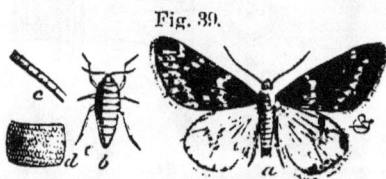

Fig. 39.

Fig. 39.—Fall Cankerworm Moths; *a*, the male moth—color, brownish-gray; *b*, the female moth—color, dark ash-gray; *c*, several joints of her antennæ, enlarged; *d*, a segment of her body, enlarged.

Female Moth (Fig. 39*b*)—color, dark ash-gray; body and legs smooth, and of a uniform color; antennæ over fifty jointed.

Male (Fig. 39*a*)—color, brownish gray—sometimes darker; the fore-wings are crossed by two whitish bands: the outer band is suddenly bent inward near the fore edge of the wing, forming a pale, quadrate spot: in some these bands are wanting, but in such cases the pale spot is nearly always present; expands one inch and three lines.

Remedies.—No. 32. No. 10. or No. 89. (See Remedy No. 10.)

CHAPTER XXXI.

The Yellow Canker Worm.   (Cal.)

(*Hibernia tiliaria*—Harris.)

The caterpillar (Fig. 40, Plate 1) of this species is about one inch and three lines in length, and is provided with ten legs. Color—body yellow, marked on the back with ten black lines, which sometimes impart a bluish tinge to the ground color; under or ventral parts, yellowish-white; head, yellowish-brown.

PUPA.—On entering the ground the caterpillar forms a cell, which it lines with a few silken threads, thus forming a cocoon. In from ten to fourteen days it changes to a pupa, in which state it remains until the following Fall.

IMAGO, OR PERFECT INSECT.

Female Moth (Fig. 41, Plate 1)—Color, whitish, dotted with black, and marked with two rows of black spots, and with a row of smaller black spots on each side of the body; the legs are ringed with black and white; the body is about half an inch (six lines) in length. The eggs are deposited in crevices and beneath the loose bark. Male moth (Fig. 42, Plate 1), fore-wings yellowish, dotted with brown, and crossed by two wavy brown lines, the line nearest the body being often indistinct. In the space between these lines there is usually a brown dot, placed nearest the front edge of the wing. The hind wings are pale-yellowish, with usually a brown dot near the center of each; expands about one inch and nine lines.

Use remedies as directed in Chapter XXXII—No. 10 or No. 89. (See note, Remedy No. 10).

CHAPTER XXXII.

Spring Canker Worm.

(Anisopteryx vernata—Peck.   *Palcacrita vernata*—Riley.)

The caterpillar of this species (Fig. 43*a*) is nearly one inch in length, and is provided with ten legs.   Color—body greenish, marked on each side with four whitish lines (Fig. 43*c*), the two lowest further apart than the others; head mottled and pale on the top, and marked on the top with two pale transverse lines.   The larva, when young, is dark green or brown, the head black.

Fig. 43.

Fig. 43.—Spring Canker Worm; *a* the worm—color, olive-green with paler stripes: *c*, a side view, and *d* a back view of one segment of its body, enlarged; *b*, an egg. enlarged; a small cluster of the natural size to the right.

Fig. 44.

Pupa.—On entering the ground the caterpillar forms a cell which it lines with a very few silken threads, thus forming a cocoon; in a few days it changes to the pupa state (Fig. 44) in which it remains until the following Spring.

Fig. 44.—Female pupa of Spring Canker Worm enlarged—color, brown.

IMAGO, OR PERFECT INSECT

Female moth (Fig. 45*b*)—color, pale gray, marked on the back with a black stripe, which is sometimes divided in the center by a whitish line or a row of whitish dots; legs ringed with black and gray; length of body, from four to six lines.

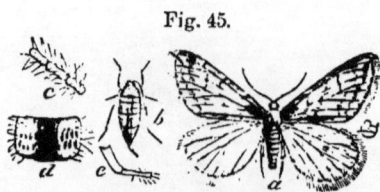

Fig. 45.

Fig. 45.—Spring Canker Worm Moth : *a*, the male moth– color, brownish-gray; *b*, female moth —c o l o r, pale-gray, a darker stripe on the back ; *c*, three joints of her antenna, enlarged ; *d*, one of her abdominal segments, enlarged; *e*, her ovipositor, enlarged.

The eggs (Fig. 43*b*) are deposited in irregular masses and secreted; they are elliptic-ovoid in form and can be readily distinguished from the eggs of the other species by the delicate shell being irregularly punctured.

Male (Fig. 45*a*)—color, brownish-gray ; the fore-wings are crossed by three jagged, dark colored lines, which are most distinct where they cross the larger veins, and at the front edge of the wing, where they divide the wing into four parts of nearly equal width; near the outer edge the wings are crossed by a pale, jagged band, which terminates at the apex, at which place there is a dark dash.   Expands one inch and three lines.

REMEDIES.—As directed in Nos. 22, 10 or 89. (See note, Remedy No. 10.)

CHAPTER XXXIII.

DeLong's Caterpillar. (Cal.)

(*Clisiocampa constricta.*—Stretch.)

Order, LEPIDOPTERA; Family, BOMBYCIDÆ.

[The measurements of insects in this work are given in inches and lines.   The above cut represents one inch divided into lines and fractions thereof.]

[Living upon apple and plum trees: a striped, slightly hairy, sixteen-legged caterpillar.]

The common name, DeLong's Caterpillar, is given this species, as previous to 1883 it was only found, in this State, at the Novato orchard, of which Mr. DeLong is the proprietor. In the early part of the month of May, 1881, the caterpillars of this moth infested the apple and plum trees in such numbers as to threaten the destruction of the entire crop of twenty thousand of the former and five hundred of the latter.   The caterpillar (similar to Fig. 55) is one inch and nine lines in length, nearly three lines in diameter, and is full grown by the twelfth day of May.   The body is sparingly clothed with soft and short hair, rather thicker and longer on the sides than elsewhere. The head is dark brown on each side, and dark brown above, leaving an inverted Y mark in the middle and front, and having much the appearance of a goblet, as one looks from above. The frontal mark is jet black, edged with a white stripe across and over the mouth parts and on each side of the inverted Y. The ground color of the upper part of the body is evidently light blue, with a dorsal row of oval orange spots, one on each segment; two sub-dorsal orange lines; also, two lateral orange lines. Between the sub-dorsal lines is a number of crinkled black and orange lines; between the sub-dorsal and lateral lines the space is blue, slightly variegated with fine orange and black lines intermingled. The lower part of the body and feet are dusky blue, with crinkled, orange and black irregular lines, and an amber-colored ring around base of pro-legs. The caterpillars do not make a tent or web, although they live in colonies on the tree.

Fig. 46.—Cocoon spun by cater-pillar of DeLong's Moth—color, yellowish.

Fig. 46.

The caterpillars spin their co-coons (Fig. 46) in folded leaves (Fig. 47) of the trees on which they feed, and on fences and build-ings in the vicinity of trees which they have stripped of foliage.

Fig. 47.

Fig. 47.—Fold-ed leaf contain-ing cocooon of DeLong's Moth.

PUPA.—Pupa elongate, poste-riorly attenuated. inclosed in a loose silken web, suffused with fine yellow pow-der. The moth appears in about sixteen days, or about the twenty-eighth of May; is reddish brown, with two transverse rust-brown nearly straight parallel lines on the fore-wings.

Fig. 48.—DeLong's Moth (male)—colors, light and dark brown.

Fig. 48.

Male (Fig. 48) antennæ, short, curved,
moderately bipectinated in both sexes,
the pectinations gradually decreasing in
length to the apex, and shortest in the
females; thorax, robust; pilose, (hairy);
abdomen, elongate—robust in female, and tufted in both;
femur (thigh) and tibia (shin) hairy. The male insect expands
one inch and three lines, the female one inch and six to seven
lines. The female is lighter in color than the male.

Fig. 49.

Fig. 49.—DeLong's Moth (female)—colors, light and dark
brown.

The female moth (Fig. 49) lays
her eggs—two to three hundred
in number—in rows around the
new growth of wood (Fig. 50),
and covers them with an apparent water-proof substance to
protect them through the Winter season.

Fig. 50.—Eggs of DeLong's
Moth—color, reddish brown.

Fig. 50.

About the time the leaves are
unfolding in the Spring, the
young hatch and feed on the foliage and young fruit. Mr.
DeLong had a block of two thousand apple trees completely
stripped of fruit and foliage by these caterpillars.

METHOD ADOPTED BY MR. DE LONG FOR DESTROYING THE
CATERPILLARS.

He placed a band of butter-cloth, about four inches wide,
covered with tallow, on the trees about two feet above the
ground. He discovered that the caterpillars could form
bridges over the tallow, especially at night. Over the tallow
he placed soft lard, which proved effective. Men with brooms
swept the caterpillars off the trunks, limbs, and branches of
the trees. The caterpillars attempted to ascend the trees
again, but would not cross the greased band. On every tree
they gathered in such immense numbers between the bands

and the ground, that they were easily destroyed by the use of clubs. By these means the crop of twenty thousand trees was secured from their ravages, although the orchard was not entirely cleaned.

GATHERING THE EGGS.

In the month of January, 1882, men were employed to pick the rings of eggs deposited on the branches, and succeeded in collecting eggs which would have produced sixty millions of caterpillars. The cost of collecting the eggs was one dollar and fifty cents per ounce. One ounce of the rings represented one hundred and fifty thousand eggs.

NATURAL REMEDIES.

In 1881, many cocoons were found infested by the larva of a Tachina fly (Fig. 51).

Fig. 51.

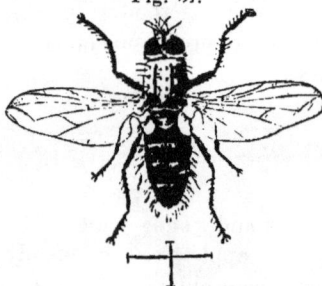

Fig. 51.—Tachina Fly (*Tachina doryphora*)—colors, gray and black.

In 1882, a small fly belonging to the *Braconinæ* (Fig. 51 *a*, Plate 1), a sub-family of the *Ichneumonidæ*, made its appearance and destroyed the caterpillars hatched from the eggs that escaped the egg gatherers. At the present writing (May, 1883), Mr. DeLong cannot find any of the caterpillars of this species in his orchard. The full description of the work done in this case is given to prove what can be done in destroying insect pests. Mr. DeLong's apple crop produces a revenue of thirty thousand dollars annually. Rather than to allow the caterpillars to destroy his crop, he exterminated them at a cost of about two thousand dollars.

Reports say that this species has been found this season in one of the Bay counties.

REMEDIES.—As described above. See Nos. 23 and 29.

CHAPTER XXXIV.

Orchard Tent Caterpillar. (Cal.)

(*Clisiocampa Americana.*—Harris.)

Order, LEPIDOPTERA ; Family, BOMBYCIDÆ.

[Living on apple, cherry and oak trees; a striped sixteen-legged caterpillar, sparingly clothed with hairs on the sides of the body.]

This insect infests apple trees, and is also found on cherry trees. Its presence can be easily detected by the web-like nests found on trees which they infest, and from which the insect derives its common name—Tent Caterpillar.

Fig. 52, *a* and *b*.

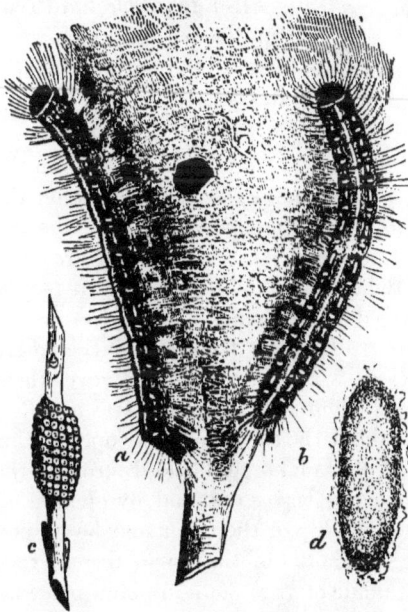

Fig. 52.—Orchard Tent Caterpillar; *a* and *b*, the caterpillars—colors, black, yellow, white and blue ; *c*. the eggs (poor figure ; Fig. 54 is more accurate); *d*, the cocoon--color, yellowish white.

The caterpillar (Fig. 52, *a* and *b*,) is about two inches in length, and nearly three lines in diameter; head black, frontal mark inverted Y, same as DeLong's caterpillar (*Clisiocampa constricta*); ground color of back apparently brownish-black.

A dorsal line of a yellowish-white color extends the whole length of the insect, on each side of which, on a yellowish or orange ground, are black crinkled lines, which on the sides

form a black lateral line, or when the caterpillar is stretched
appears as a large black spot on the side of each segment; in
the center of these spots is a small bluish mark; below this is
a yellow (orange) line, and lower are five crinkled lines, yellow
and black. Ventral parts a dark, dusky color; on one of the
posterior segments is a small blackish wart. The body is
clothed with soft short hairs, rather thicker on the sides than
on the back and ventral parts. When full grown it spins a
cocoon (Fig. 52d), in which it undergoes its transformations.

PUPA.—The pupa state is from fourteen to sixteen days.

MOTH.

Fig. 53.

Fig. 54.

Fig. 54.--Eggs of Orchard Tent
Caterpillar—color, brown.

Fig. 53.—Female Moth of Orchard Tent Caterpillar—(sometimes called the American Lackey Moth)—colors, yellowish or
reddish-brown and white.

The moth (Fig. 53) differs but little in appearance from the
C. constricta; the color is somewhat lighter; the lines on the
fore-wings are more oblique, and the apex shorter. In some
specimens the band between the lines of the fore-wing are dark,
or of the same color as the base and apex; in others it is very
light, or what may be termed a dirty white. The perfect insects
appear about the latter part of May. They deposit their eggs
(Fig. 54) on the branches on which they feed, and cover them
with a secretion to protect them in the Winter season. The
young caterpillars hatch about the time the leaves open. This
species can be exterminated by picking off and destroying the
bunches of eggs before the tree leaves out, and by picking off
and destroying tents when made; or the latter may be burned
with their occupants, at such hours of the day as the caterpillars are at rest. A torch made of rags bound upon a pole and
saturated with kerosene is a useful weapon for this kind of
warfare.

Remedies as above described. Also, see Nos. 23, 29 and 31.

CHAPTER XXXV.

The Forest Tent Caterpillar. (Cal.)

(*Clisiocampa sylvatica.*—Harris.)

Order, LEPIDOPTERA ; Family, BOMBYCIDÆ.

[Feeding upon the leaves of the apple, oak, cherry, and various other trees, a bluish, slightly hairy caterpillar, sprinkled all over with black dots, and with a row of oval or diamond-shaped white spots on the back.]

When young, these caterpillars live in communities under a web which is spun against the trunk of one of the larger branches of the tree, but as they grow older they disperse and live singly, unprotected by a web.

Fig. 55.

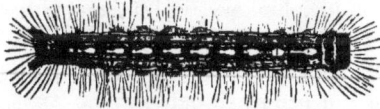

Fig. 55.—Forest Tent Caterpillar—colors, bluish-gray, white and yellow.

When fully grown (Fig. 55) they measure about one inch and six lines in length ; they then seek some sheltered place in which to spin their cocoons, which are filled with a yellow, mealy powder.

The moths appear early in June ; their forewings are of a grayish color, crossed by two parallel brown lines, the whole space between them sometimes being of the same dark brown color ; the male moth expands about an inch and three lines, while the female expands about an inch and nine lines.

Fig. 56.

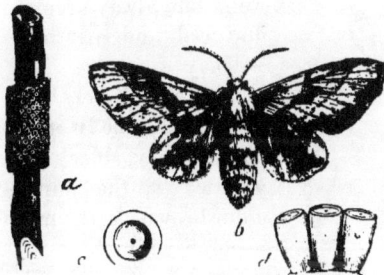

Fig. 56.—Moth and eggs of Forest Tent Caterpillar: *a*, the eggs—color, brown ; *c*, an egg highly magnified, top view ; *d*, three eggs highly magnified, side view ; *b*, the female moth—color, brownish-yellow.

The female (Fig. 56*b*,) deposits her eggs in rings

around the small twigs (Fig. 56a), and these do not hatch out until the following Spring.

The rings of eggs are similar in appearance to those of DeLong's moth, *C. constricta*, excepting that the eggs of the former are somewhat larger.

REMEDIES.—Destroy web or tent as described in Chapter XXXIV. (Orchard Tent Caterpillar.) Also, as in Nos. 23 and 29.

CHAPTER XXXVI.

The Fall Web Worm.   (Cal.)

(*Hyphantria textor.*—Harris.)

Order, LEPIDOPTERA;   Family, BOMBYCIDÆ.

[Living under a web on apple, hickory, walnut and other trees, and feeding upon the upper surface of the leaves; sixteen legged caterpillars of a yellow and black color, their bodies sparsely covered with whitish or brownish hairs.]

"This insect passes the Winter in the pupa state, and the moth emerges during the month of May.   The female deposits her eggs in a cluster on a leaf, generally near the outer end of a branch.   Each worm or caterpillar (Fig. 57a) begins spinning the moment it is hatched, and by their united efforts they soon cover the leaf with a web, under which they feed in companies, devouring only the pulpy portion of the leaf.   As they increase in size they extend their web, but alway remain and feed underneath it."—Riley.

Fig. 57.

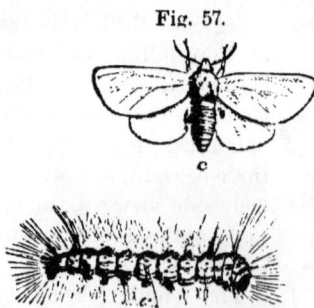

Fig. 57.—Fall Web Worm; *a*, the worm—colors, gray, black and yellow; *b*, the pupa—color brown; *c*, the moth—color, white.

"The web sometimes reaches a length of fully seven feet."—

La Baron. The young worms are of a pale yellow color, with black heads. When fully grown they are a trifle over an inch long; of a yellowish or bluish-gray color, the back usually black; the body is sparingly clothed with whitish, reddish or mouse-colored hairs, which grow in clusters from warts which are usually yellowish-brown, or the two rows on the back are frequently black, or reddish-brown marked with black. When fully grown these caterpillars descend to the ground, which they enter, and form small cells in which to pass the pupa state (Fig. 57b). They spend the Winter in this latter state, and the moths (Fig. 57c), which issue the following Spring, expand about an inch and three lines and are of a pure white color and without spots, except on the legs. "The proper time to destroy these caterpillars is while they are young; at such time the branch containing the nest can be removed and its contents easily destroyed."—Riley.

REMEDY.—Use No. 97.

CHAPTER XXXVII.

The Tussock Moth. (Cal.)

(*Orgyia leucostigma*—Abbot and Smith.)

Order, LEPIDOPTERA; Family, BOMBYCIDÆ.

[Feeding upon the leaves of the apple and various other trees; a black and yellow caterpillar having large bunches and plumes of hair on its body.]

The pretty caterpillar of this moth (Fig. 58) is found on the apple, pear, plum and horse-chestnut; also on the walnut and oak, and perennials in the flower garden, especially the rose.

Fig. 58.—Caterpillar of Tussock Moth — colors, yellow and black, or brown; hair white, the pencils blackish.

When full grown it measures from one

Fig. 58.

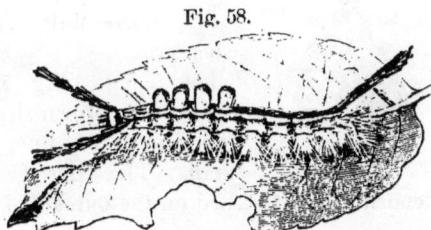

inch and three lines to one inch and six lines in length. Color, cream-yellow, a black dorsal stripe extending the whole length of posterior of the third segment; next to the dorsal stripe is a yellowish line, then a greenish-blue stripe on which is sometimes a black line; stigmatal line black, and below this is a yellow line. On dorsal section of segments 4, 5, 6 and 7, is a wide tussock of whitish hair; on each side of the segments next to the head, and on the dorsal part of the eleventh segment, is a pencil of long black hairs, which are knobbed at the apex; on top of segments nine and ten is a small red wart; ventral parts yellowish-white, tinged with blue. Head reddish-brown or dark red; cervical-shield red.

Pupa.—Pupa of male (Fig. 59d) elongate, posteriorly attenuated, inclosed in a coarse silky cocoon. Pupa of female (Fig. 59c), ovate in form.

Fig. 59.

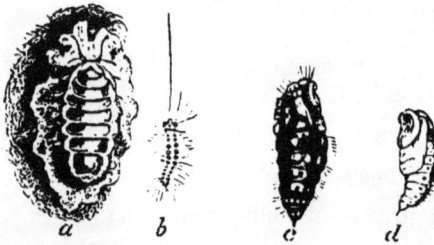

Fig. 59.—Tussock Moth, Caterpillar and Pupa; *a*, the female moth on her cocoon — color of former, whitish or gray; of the latter, gray or yellowish; *b*, a young caterpillar; *c*, the female pupa—color, brown or gray; *d*, the male pupa—color, brown.

Imago.—The male insect (Fig. 60) is ashen-gray; the forewings are crossed by wavy bands of a darker color; on each wing is a small white crescent near the inner angle; antennæ pectinated; expands one inch.

Fig. 60

Fig. 60.—Male Tussock Moth—color, ashen-gray.

The female (Fig. 59a and 61) is wingless—color, brownish; form oval, and is from four to five lines in length. Antennæ small. The eggs (Fig. 62) are generally deposited and fastened on the outside of the cocoon.

Fig. 61.—Female Tussock Moth—color, gray.

Fig. 62.—Eggs of Tussock Moth on the cocoon—color, eggs white, cocoon pale yellow.

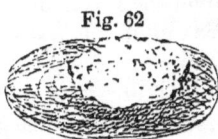

REMEDY—Use No. 30.

CHAPTER XXXVIII.

The Yellow-necked Caterpillar.

(*Datana ministra*—Drury).

SYNONYM.—*Pygœra ministra.*

Order. LEPIDOPTERA : Family. BOMBYCIDÆ.

[Feeding in communities upon the leaves of the apple, walnut, etc., a black or reddish brown caterpillar, which is usually marked with five yellow lines on each side of the body.]

Fig. 63.

Fig. 63.—Yellow-necked Caterpillar, Moth, and Eggs ; *a,* the caterpillar—colors, black and white ; *b,* the moth—colors, brown and yellowish ; *c,* the eggs—color, white ; *d,* an egg magnified.

These caterpillars usually live in communities, and when at rest have a habit of holding both extremeties of the body upwards. (See *a.* Fig. 63). They differ widely in their colorings, according to the kind of tree they infest. Those which feed

7

upon the leaves of the apple tree are usually of a black color, the top of the first segment yellow, and there are five yellowish lines on each side of the body, while those infesting the black walnut are destitute of the yellow stripes, and the top of the first segment is black; in both of these varieties the head is black.

When fully grown (Fig. 63a) they measure about two inches in length; they then descend to the ground, which they enter a short distance, where each one forms a cell in which to pass the pupa state, which continues throughout the Winter. Only one brood is usually produced each year.

The wings of the moth (Fig. 63b) expand nearly two inches, and are of a reddish brown color, crossed by four transverse lines of a darker brown; the hind wings are of a lighter color, and are unmarked. The moths from the caterpillars which infest the black walnut, are more of a smoky brown color. Owing to the ·fact that these caterpillars congregate in large companies upon a single branch, the latter can be easily removed from the tree, and the caterpillars can then easily be destroyed.

REMEDY.—Use No. 97.

CHAPTER XXXIX.

The Red-humped Caterpillar.   (Cal.)

(*Notodonta concinna*—Smith.)

Order, LEPIDOPTERA; Family, BOMBYCIDÆ.

[Feeding in communities upon the leaves of the cherry, apple, plum, and pear: a striped caterpillar, having two rows of black spines along the back].

These caterpillars live in large companies, and when at rest they elevate the hind part of the body. They are of a reddish color, and are striped lengthwise with yellow and white lines; on the fourth segment is a coral-red hump, on which are four black spines; scattered over the body are numerous black spines or points, those on the back the largest; head, coral-red. When fully grown (Fig. 64, Plate 1) they measure about one inch and three lines in length; they then leave the trees and conceal themselves beneath the fallen leaves, etc., where each one spins a whitish, parchment-like cocoon (Fig. 65, Plate 1); they remain in this cocoon a long time before changing to pupa, the moths not issuing until the following Summer. Specimens in breeding cages changed to pupa in February, and emerged from pupa March 20, 1883. As these specimens were collected the previous September, it is possible the Notodonta may be double-brooded.

The fore-wings of the moth (Fig. 66, Plate 1) expand from one inch to one inch and three lines; are of a brownish color along the hind margin, with the rest of the wing grayish, and marked with dark brown and whitish; the hind wings of the male are brownish, or dirty white, with a brown spot at the hind angle; those of the female are dusky brown. This insect is known to occur in large numbers in some portions of this State.

REMEDIES.—No. 97 and No. 20.

CHAPTER XL.

The Greater Leaf-roller. (Cal.)

(Loxotænia rosaceana.—Harris.)

Order, LEPIDOPTERA; Family, TORTRICIDÆ.

[Living in a rolled leaf on the apple, cherry, rose, etc.; a greenish worm with a black or brownish head.]

The larva brings the two opposite edges of a leaf nearly together, and holds them in this position by means of a great many silken threads; in this case, or nest, it remains during the greater part of the time, coming forth only to feed. It feeds upon the leaf which forms its nest, and after eating itself out of a habitation it repairs to another leaf and constructs a similar shelter. In this way it proceeds until reaching its full growth, when it lines the interior of its case with a fine layer of silk, and soon afterwards assumes the pupa form. The full grown larva measures nearly an inch in length, is of a green color, sometimes tinged with yellow; the head is yellowish-brown, with the regions of the jaws black, or entirely black in the young. Sometimes the upper part of the face is tinged with brown; on top of the first segment is a black spot tinged with green next the head, or in its place is a simi-circular black line.

Fig. 67.—Greater Leaf-roller — colors, light and dark brown.

Fig. 67.

The fore-wings of the moth (Fig. 67) expand about one inch, are much arched at the middle of the front edge, and curve in an opposite direction near the tip; they are of a light brown color, crossed by dark brown lines and bands. The hind wings are yellowish, with the part next the body blackish. Dr. Emmons (N. Y. Rep., 1854), states that this insect passes the Winter in the egg state, and is to be found in small clusters on the bark of trees infested the previous year.

In California there are several species found, which, if they do not belong to the genus *loxotænia*, are closely allied to it. The young caterpillars appear early in the season. and .make their nest under the blossom leaf. or petal, of the apricot, when the fruit is not larger than a garden pea, and feed upon the skin or epidermis of the fruit on which they lodge. As the fruit becomes larger and the insect gains in strength, it bores into the fruit, destroying the pulp or mesocarp, and in many instances eat part of the pit or stone.

REMEDIES.—When the tree is dormant wash or spray thoroughly with No. 53, to destroy any eggs deposited on the tree. All infested leaves and fruit should be picked off and destroyed. See No. 24.

CHAPTER XLI.

The Apple Leaf Crumpler.

(*Phycita nebulo.*—Walsh.)

SYNONYM.—*Acrobasis nebulo.*

Order, LEPIDOPTERA ; Family, PYRALIDÆ.

Living in a curved, black, silken tube (Fig. 68a), on apple and plum trees, a reddish-brown worm, having a roughened head. (Fig. 68c.)

Fig. 68.—A p p l e L e a f Crumpler ; c, the head and fore part of the caterpillar's b o d y e n l a r g e d—color, brown ; d, the moth, enlarged—colors, gray and brown ; a, a case in which the caterpillar lives, with the fore part of the latter's body protruding from the opening in the larger end ; b, several c a s e s fastened together—color, black.

Fig. 68.

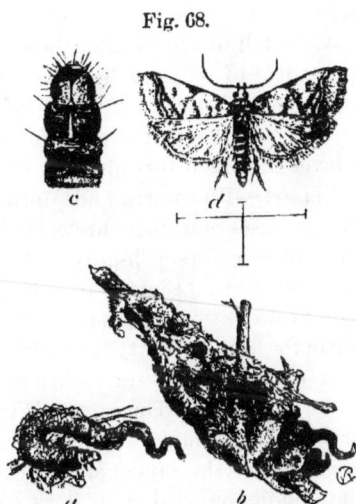

When fully-grown this worm measures nearly six lines in length ; it then closes the opening of the silken tube in which it lives, and soon afterward assumes the pupa form. The moth (Fig. 68d,) issues during the Summer season, and the worms or larvæ which are produced from the eggs she deposits pass the Winter inside of their silken tubes (Fig. 68b), there being but one brood produced in one year. It has the habit of fastening dead leaves to the outside of its case, which makes its presence very conspicuous during the Winter season. In order to lessen the ravages of this insect, it is only necessary to collect the silken tubes containing the larvæ and burn them. This can best be accomplished in the Winter season after the leaves have fallen from the trees, at which time the cases of this insect may be readily discovered.

CHAPTER XLII.

The Bud Worm.

(*Penthina oculana.*—Harris.)

Synonym.—*Spilonota oculana.*

Order, Lepidoptera; Family, Tortricidæ.

[Fastening together and devouring the leaves of the opening buds of apple trees; a small, brownish caterpillar.]

Although of small size, these worms sometimes occasion a great deal of damage by devouring the buds. They usually attain their full growth by the middle of Summer, when they prepare to assume the pupa form by lining their retreat with a layer of silken threads.

Fig. 69.—Bud-worm and Moth: lower figure, the worm—color, pale brownish; upper figure, the moth—colors, ash-gray and whitish.

Fig. 69.

These worms (Fig. 69) or caterpillars are of a pale or dull brownish color, with the head and top of the first segment shining brown : and there is a dark brown spot on the top of the eighth segment, which appears to be under the skin. The moths (Fig. 69), which appear in June or July, have the head and thorax dark ash color : the fore-wings are of the same color at each end, and grayish-white in the middle, mottled with dark gray : there are two little eye-like spots on each one, near the tip, consisting of four little black marks placed close together in a row : the second eye-spot is near the inner hind angle, and consists of three black dots arranged in the form of a triangle, sometimes with a black dot in the middle ; the hind-wings are dusky brown. The fore-wings expand from six to seven lines.

Remedies.—Use No. 65, and cut out infested branches. See No. 27.

CHAPTER XLIII.

The Many-dotted Caterpillar.

(*Brachytænia malana.*—Fitch.)

Synonym.—*Xolophana malana.*

Order, Lepidoptera ; Family, Noctuidæ.

[Eating the leaves of the apple. cherry, and peach tree : a naked, sixteen-legged caterpillar, of a green color, dotted with white, and marked with fine whitish lines.]

"These caterpillars live exposed upon the leaves of several different kinds of trees. When fully grown they measure an inch in length; they then roll up a leaf, and inside of this roll they spin a thin cocoon. Two broods are usually produced in a year, the last brood passing the Winter in the pupa state. The fore-wings of the moth expand about one inch, and are of an ashen-gray color, crossed by three zigzag black lines, which are connected in various places by black dashes.

"For the destruction of these worms it has been recommended to place blankets beneath the trees and jar the worms off, when they may then be easily destroyed. As the last brood remains in the pupa state within the leaves, by raking these into winrows in the Winter and then burning them, the greater number of these pests will be destroyed."—[Condensed from Fitch.

CHAPTER XLIV.

The Turnus Butterfly. (Cal.)

(*Papilio turnus.*—Linnæus.)

Order, Lepidoptera : Family, Papilionidæ.

[Feeding upon the leaves of the apple and cherry tree; a bluish-green, sixteen-legged worm, having an eye-like spot on each side of the third segment.]

This caterpillar spins a fine web upon the upper surface of a leaf, drawing the edges of the leaf slightly upwards ; when not feeding, it rests upon the upper side of this web. The young caterpillar is of a black or brownish color, marked with flesh color or white.

Fig. 70.—Caterpillar of Turnus Butterfly— colors, green, yellow, and black.

Fig. 70.

When fully grown (Fig. 70), it measures from one inch and six lines to two inches in length, and is of a bluish-green color; on each side of the third segment is a black spot, centered with blue and surrounded by a yellow ring, and this by a black one; on top of the fourth segment is a transverse yellowish ridge, in front of which is a row of four blue dots; on top of the last segment is a transverse yellowish ridge; head, pinkish-brown. When about to pupate, this caterpillar suspends itself by the hind-feet and a transverse band of silken threads which is passed around the forepart of the body. It assumes the chrysalis form in the Autumn, and the butterfly does not issue until the following Spring.

Fig. 71.

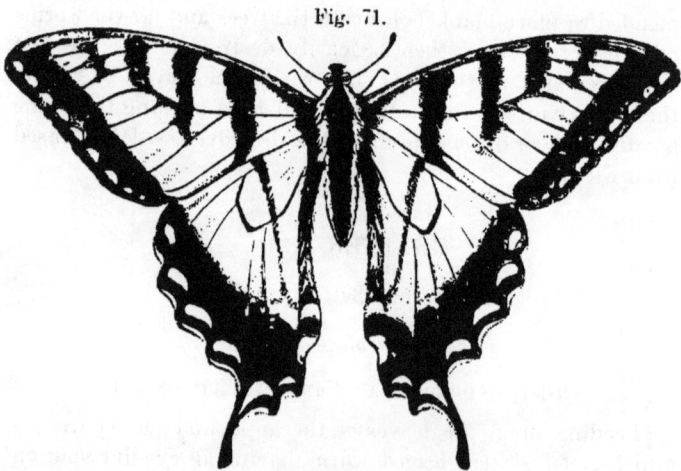

Fig. 71.—Turnus Butterfly—colors, yellow, black and orange.

This is one of the largest butterflies (Fig. 71) found in this State. The wings expand from four inches and six lines to five inches, and are of a pale yellow color, with a broad, black outer margin marked with yellow spots; the fore-wings are marked with four black bands, the one nearest the body extending across the hind-wings; the latter are tailed, and have an orange-colored spot near the hind angle.

REMEDY.—Use No. 14.

CHAPTER XLV.

The Apple Bucculatrix. (Cal.)

(*Bucculatrix pomifoliella*—Clemens).

Order, LEPIDOPTERA; Family, TINEIDÆ.

[Feeding upon the leaves of the apple tree, a small, dark, yellowish-green, sixteen-legged worm, nearly six lines long; spinning against the bark an elongate, dirty-white, ribbed cocoon].

This larva assumes the pupa form within its cocoon (Fig. 72 *a* and *b*], and before the moth issues, the pupa works itself part way out of the cocoon. Several broods are probably produced in one year; the last brood passes the Winter in their cocoons.

Fig. 72.—Apple Bucculatrix; *a*, the cocoons on a twig—color, dirty white; *b*, one of the cocoons, enlarged; *c*, the moth enlarged—color, gray and brown.

Fig. 72.

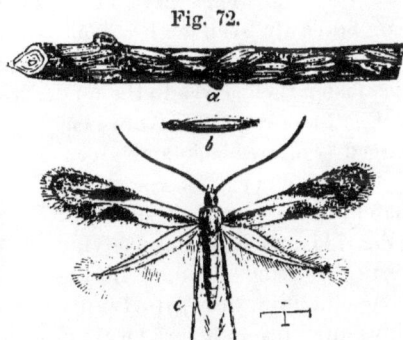

The perfect moth (Fig. 72*c*) is a dirty white or gray color, marked with brown, as in the figure.

REMEDY.—When the tree is dormant, spray or wash thoroughly the parts infested with No. 13--one pound to each gallon of water used.

CHAPTER XLVI.

The Apple-tree Aphis.   (Cal.)

(*Aphis mali.*—Fabricius.)

Order, HEMIPTERA; } Family, APHIDIDÆ.
Sub-order, HOMOPTERA; }

[Living on the leaves or twigs of apple trees: small green, or green and black, plant lice.]

This plant louse, as its name indicates, is frequently found on the apple tree. It may be easily detected by the black appearance of the ends of the succulent twigs and leaves, caused by a honey-dew emitted by this insect and others closely related to it. (See *Aphis malifolia;* Chap. 47.)

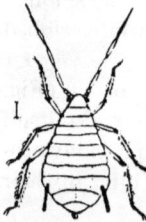

Fig. 73.

Fig. 73.—Apple-tree Aphis (young), enlarged—color, green.

The wingless individuals (Fig. 73) are small, green lice, and can be found in great numbers on the under side of the leaves and tender twigs. They measure about one line in length, and are often accompanied by winged individuals. As the Winter season approaches, the eggs are laid on the branches and twigs, and can be easily seen with a lens. About the time the leaves begin to open in the Spring, these eggs hatch, and the young lice fasten themselves to the tender leaves and extract the sap.

Fig. 74.—Apple-tree Aphis, enlarged—colors, black and green.

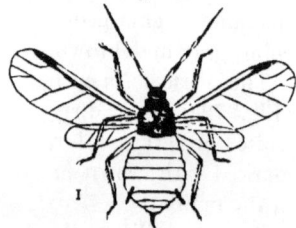

Fig. 74.

In about ten days after hatching the lice reach maturity (Fig. 74) and commence giving birth to living young. In the course of from fifteen to twenty days after reaching maturity, they die. The young, after reaching maturity, become parents, and it is generally conceded by naturalists that in the Summer season they reach maturity in five or six days. Therefore, it can be easily seen that their increase surpasses computation. This species is spreading rapidly in the vicinity of Sacramento. Trees infested emit a very disagreeable smell. Mr. Haywards, writing of this pest from British Columbia, says: "After night, in passing along the road I can tell an infested tree." In cases where trees are badly infested they produce a smell similar to that of decayed fish. I have noticed in handling infested branches that this loathsome smell remains on the hands.

Leaves infested by these plant lice curl or curve backwards until a roll is formed, thus furnishing the insects with shelter from the rays of the sun, or from rains, or dews, and makes it difficult to destroy the insects by spraying the trees.

During the Summer they are found grouped together on the leaves in all stages of their growth, and of various colors. The mature insects (Fig. 74.) are of a yellowish-green, and when half grown pale-yellowish, and when born the color is nearly white. In Autumn the color changes in many cases, either from the change in temperature or change in nourishment.

REMEDIES.—When the tree is dormant, spray thoroughly with No. 13; one pound of the mixture to each gallon of water used. When the leaves begin to expand, if the lice are present, spray thoroughly with No. 64 or No. 65, and repeat if necessary.

CHAPTER XLVII.

The Apple Leaf Aphis. (Cal.)

(*Aphis malifolia*—Fitch.)

Order, HEMIPTERA ;
Sub-order, HEMOPTERA ; } Family, APHIDIDÆ.

[Living upon the leaves of the apple tree, a small, blackish plant-louse.]

This species infests the leaves of the apple tree, and in their habits they are similar to and are often mistaken for the apple tree aphis (*Aphis mali*), though they are larger, and generally of a darker color. The winged insect of this species differ from the A. mali in being larger, and the thorax and abdomen are black; there is also a slight difference in the venation of the wings.

These insects emit a honey dew, which gives the leaves and branches an appearance similar to that caused by the apple-tree aphis (*A. mali*).

REMEDIES.—To be used as described in Chapter XLVI for the apple-tree aphis.

CHAPTER XLVIII.

The Ten-lined Leaf Eater.   (Cal.)

(*Polyphylla decemlineata—Say.*)

Order, COLEOPTERA : Family, SCARABÆIDÆ.

[The measurements of insects in this work are given in inches and lines. The above cut represents one inch divided into lines and fractions thereof.]

[Feeding upon the leaves of the apple and other fruit trees, a large, grayish-brown beetle, marked with white lines; or, feeding upon the roots of grass, a large, white, six-legged larva or worm].

The perfect beetle (Plate 1, Fig. 74, male; Fig. 75. female) measures from an inch to an inch and three lines, and is of a reddish brown color, covered with short yellowish hairs, which give it a grayish appearance; the thorax is marked with three white stripes, and on each wing-cover are three white stripes, and two or three less distinct whitish lines.

The habits and transformations of this species are similar to those of the common May beetle.

REMEDY.—Use No. 38.

CHAPTER LIX.

The Goldsmith Beetle. (Cal.)

(*Cotalpa lanigera*—Linnæus.)

Order, COLEOPTERA ; Family, SCARABÆIDÆ.

[Feeding upon the leaves of the apple, pear, and various other trees, a broad beetle of a rich yellow color, the top of the head and thorax having the appearance of burnished gold.]

Fig. 77.—Goldsmith Beetle—color, rich metallic yellow.

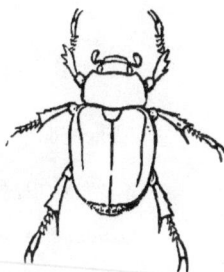
Fig. 77.

This beetle (Fig. 77) measures about an inch in length. The female deposits her eggs in the ground, and these hatch into white, six-legged grubs which closely resemble those common-ly known as *white grubs;* they feed upon the roots of various plants, and in this way are sometimes very injurious to strawberry patches. They spend sev-eral years in this their larval stage, and finally assume the pupa form in the Fall, and are changed to beetles in the following Spring.

REMEDY.—Use No. 38.

CHAPTER L.

The Robust Leaf Beetle. (Cal.)

(*Serica valida*—Harold.)

SYNONYM.—*S. robusta*—Leconte.

Order, COLEOPTERA ; Family, SCARABÆIDÆ.

[A reddish-chestnut colored beetle, feeding upon the leaves of the apple, apricot, plum, and prune trees].

This beetle (Fig. 78, Plate 1) has been reported damaging the foliage of apple, apricot, plum, and prune trees.

They feed on the foliage at night and hide themselves in the ground and dark places in the day time; recent reports state that in some sections they have damaged the foliage of young trees.

Description.—Length, four and one quarter lines; form, elongate ovate, narrowing toward the head; color, reddish-chestnut, but grows darker with age: antennæ, lamellate.

The larva of this species I have not found, but it probably lives in the ground, feeding upon the roots of grasses, etc.. near the roots of the trees on which the perfect insect feeds.

REMEDY.—Use No. 38.

CHAPTER LI.

The Codlin Moth, or Apple Worm. (Cal.)

(*Carpocapsa pomonella.*—Linn.)

Order, LEPIDOPTERA; Family, TORTRICIDÆ.

[Living in apples, etc.. a whitish, sixteen-legged worm.]

It is generally conceded that this insect was imported into this State in shipments of apples received from states east of the Rocky Mountains, and placed on exhibition at the State Fair in or about the year 1873. Its first appearance in an orchard in the vicinity of Sacramento was in the Spring of 1874. Since that date it has spread rapidly, and can be found at the present time infesting orchards in thirty-four counties.

The moth belongs to the family Tortricidæl, and is known to naturalists as *Carpocapsa pomonella* (the codlin or apple moth). It passes the Winter in the larva state, and in some instances in the chrysalis form. The larva can be found hibernating under the loose bark, in crotches or indents, or in cracks in the bark of the trees infested the previous year, or in the crevices of wood, or woodwork of rooms, or places where fruit infested by the larva was stored or packed, and in empty packages in which fruit was shipped or gathered. It is often found hibernating on bark of trees, from one to six inches below the surface of the ground, especially if the tree has smooth bark. In one case, where four hundred apple trees were dug up, the larvæ were found in large numbers in the roots of such trees as were decayed at or above the surface of the ground.

Fig. 79.—*a*, nest of larva as it appears on inside of bark when taken off tree—color, drab; *b*, pupa or chrysalis—color, dark amber; *c*, appearance of larva when cover is removed from Winter nests —color, body yellowishwhite, head dark brown; *d*, appearance of bottom of Winter nest on bark

Fig. 79.

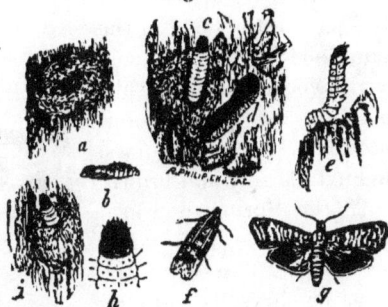

when larva is removed in the following Spring; *e*, a position the larva takes when looking for a tree or place to make its nest when ready to assume the pupa or chrysalis form. [Note: when the larvæ are full grown and ready to assume the pupa or chrysalis form, the color is light pink]. *f*, the moth, at rest carries its wings like a steep roof; *g*, moth with wings spread, length of body five lines, spread of wings nearly nine lines—color, body and legs rich bronzed light drab, fore-wings mottled with gray and drab, with dark copper bar across hind margin on which is a golden ocellated patch near inner angle, hind-wings plain drab a little darker than body (the moth after depositing eggs has assumed a light drab color on fore-wings, and copper bar changes to a very light color, scarcely perceptible, caused probably from flying among the branches and leaves); *h*, head of larva as seen through a glass magnifying nine times; *i*, in this figure it was intended to represent the pupa or chrysalis case protruding through nest prior to moth leaving it, but represented as larva to show better. [Note: the figure would be correct if the chrysalis (*b*) was represented instead of larva (*c*); the figures *a*, *b*. *c*, *d*. *e*, and *f*, are natural size; *g* is a little larger than natural size: *h*, as described]. See, also, Fig. 80.

Fig. 80.—Codlin Moth;
a, an infested apple; b, the
place where the larva entered the same: c, the larva—
color, whitish; h, head and
fore part of the body of the
same—back view, enlarged;
i, the cocoon—color. whitish; d, the pupa—color.
brown: g, the moth—colors.
light, dark gray and brown.

Fig. 80.

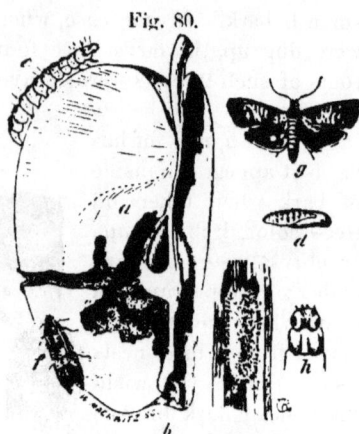

If the Spring is warm
and favorable, the larvæ
are ready to assume the
pupa or chrysalis form by
the fifteenth of April. The duration of the pupa or chrysalis
state depends on external circumstances: if warm Spring
weather, the perfect insect may appear in from fifteen to
twenty days, and may be prolonged to twenty or thirty days.
The Spring of 1881 has proven an exception. I found Mada-
lene pears on the sixteenth of May in which the larvæ had
matured and left: also, on May seventh found a pear with
larva about eight days old. (This is about eighteen days
earlier than usual.) On the seventh of April, 1883, I found an
empty pupa case from which the moth had escaped.

FIRST APPEARANCE OF THE MOTH.

The moth generally appears from April 25th to the fifteenth
of May—a few in favorable locations by April fifteenth. The
time at which the eggs arrive at maturity apparently coincides
with the ends or terminations of the pupa or chrysalis state,
so that the sexes are ready to unite soon after transformation.
The moths produced by the hibernating larvæ deposit their
eggs in the blossom end (or calyx) of the fruit, generally;
possibly because they cannot puncture the epidermis (or skin)
of the young fruit. Later broods deposit their eggs on any
part of the fruit. The eggs are attached to the fruit by a pasty
substance. It is rare to find more than one egg on any apple,
pear, or quince, or more than one larva. The larva is hatched

in from seven to ten days, and begins to eat eagerly and burrow toward the core.

Fig. 81.—A, blos-
som end or calyx of
apple, and where lar-
va is supposed to en-
ter the fruit; B repre-
sents an empty space
where carpellary ova-
rium or shell contain-
ing the seeds was lo-
cated before the en-
trance of the larva; C
represents the bur-
row made by the
larva through the pe-
ricarp by which it

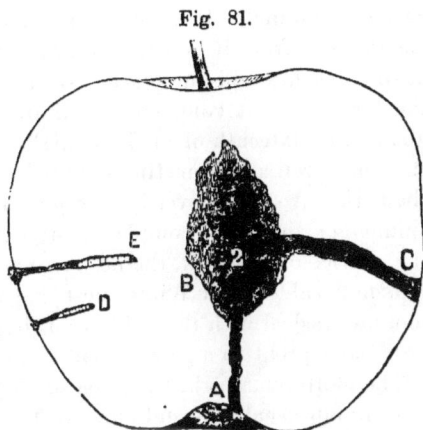

Fig. 81.

escapes from the fruit when it is ready to assume the pupa or
chrysalis form; D, appearance of larva in burrow when six
days old; E, appearance of larva in burrows when ten days
old.

The larva when hatched can scarcely be seen with the
unaided eye; at six days it measures nearly one quarter of an
inch in length, is about as thick as a fine silk thread, and
shows first signs of excrement at burrows (D Fig. 81); at ten
days three eighths of an inch, and about as thick as a number
twenty wire (E Fig. 81). It has burrowed by this time about
three fourths of the distance to the core (B Fig. 81). At
twenty days nearly full grown (*c*, Fig. 79), and often as large
(*e*, Fig. 79).

When the larva is ready to assume the pupa or chrysalis
form, it leaves the fruit by gnawing a hole through the pericarp (C Fig. 81). Nature has supplied it with a spinneret,
the opening apparently in the lower lip, from which issues a
viscid fluid in a fine stream and hardens into silk on contact
with the air. By this means it lowers itself to the ground or
intervening branches. If it reaches the ground, it immediately crawls toward the tree, and on its journey can often be

8

seen. as at *c* Fig. 79. On reaching the tree it searches for a nesting place under the loose bark, in the crotches, or in any cavity it can find. If it comes in contact with a branch when leaving the fruit, it generally crawls toward the crotches, or until it reaches a hiding place. If the place selected is under the loose bark, it commences building an oval-shaped wall about one sixteenth of an inch high, composed of silk from the spinneret, and sometimes mixed with pieces gnawed off the bark. A silken cover is then put on the nest by using the spinneret; the whole completed in twenty-four hours. If in the crevice of the bark, the nest is made in different shapes. It is noticeable in the Winter nest that the top, sides, and bottom are washed with the fluid from the spinneret, making the nest water proof to a great extent.

The moth remains in the pupa or chrysalis form about nine days in our usual May and June weather; a little longer if the weather is cool. At the proper time the pupa case is burst open, and the perfect moth appears. (Fig. 70, *f* and *g*.)

PROBABLE RATE OF INCREASE OF THESE MOTHS.

Each female lays from two hundred to two hundred and fifty eggs. Taking two hundred as the lowest number, twelve female moths in one orchard would produce two thousand four hundred caterpillars. If one half of these were females, they would produce two hundred and forty thousand. In proportion the third generation would reach twenty-four millions, supposing that no untimely deaths took place. Most of the books on this subject speak of the codlin moth as going through but one generation in a year. This may be true in colder climates and shorter seasons than ours, but in parts of our favored State there is no question that two or three generations or broods are common. From personal observations I know that the rule for the Sacramento Valley is three broods each year. In 1881, on account of the early appearance of the first moths, as noted above, we had four broods. These facts explain the exceptional importance of this insect in California.

I have in my possession a vial containing eighty-five eggs, deposited by one codlin moth. I have read statements by

fruit-growers that they have seen the codlin moth flying in
large numbers. In my investigations I have not seen more
than two at any one time. The moth will live in a glass vial
seven days. The female moths deposit their eggs within
forty-eight hours—these are deposited at night. The egg
cannot be seen plainly by the unaided eye.

The best time to see the moths at work is at the dawn of
day in the months of June and July. Part of the early fruit
falls prematurely when attacked by the larvæ; but little of
the late fruit falls until the larvæ escape.

The theory that the moth always deposits the egg on the
fruit blossom, and that it remains there until the fruit has
grown to natural size, is a mistake.

HOW TO PROCURE SPECIMENS OF THE MOTH.

When you find the larva, if it is on the loose bark, remove
the piece of bark or wood to which it is attached, place it
in a small vial, and if in the Summer time, inside of ten
days you will have a genuine specimen of the moth (*f* and *g*,
Fig. 79). Fruit-growers can get important information concerning the natural history and habits of the insect pests by
experiments of this kind. Every fruit-grower should unite in
the crusade against this pest. "In union there is strength."

Since writing the above, or on August 8, 1883, I bought a
box of bellflower apples which was offered for sale in market.
With few exceptions each apple contained two larvæ of the
codlin moth, and in nearly every apple infested by the later
broods the egg had been deposited near the stem. This is the
first instance that furnished me actual proof that the later
broods would deposit their eggs in fruit that had been infested by
the earlier broods of the season. In the absence of the seeds of
the apple, eaten by the larva of the earlier broods, the larva of
the later brood seemed to burrow (or honeycomb) the pericarp
in all directions. The larvæ in the apples containing more
than one specimen were apparently of different ages.

From investigation it is probable that there are more than
one species of codlin moth infesting fruit in this State; but I
am not prepared to report at present writing.

That the codlin moth infests the peach and apricot, I have abundant proof by rearing moths from both kinds of fruit.

REMEDIES.—For trees. No. 69; return packages (see Chap. IV), No. 70; packages stored from previous year, No. 71; store-rooms, sale-rooms, etc., No. 72; debris accumulating from preparing fruit for drying, No. 15.

CHAPTER LII.

The Apple Maggot.

(*Trypeta pomonella.*—Walsh.)

Order, DIPTERA; Family, TRYPETIDÆ.

[Eating the pulp of apples, causing them to decay, a white, footless maggot which, when fully grown, enters the earth and is finally transformed into a black and white two-winged fly.]

This maggot (Fig. 82, Plate 1,) may be easily distinguished from the larva of the codlin moth by being entirely destitute of legs. It also differs from the latter in its mode of operating, for while the larva of the codlin moth works for the most part in the core of the apple and vicinity, this maggot runs its burrows in all directions through the pulp; it also differs from the larva of the apple curculio (Fig. 85b), which, like itself, is also destitute of legs, by apparently having the hind end of the body obliquely cut off, the curculio larva having this part rounded.

The apple maggot usually appears rather late in the Summer, and after reaching its full size—about three lines in length—it deserts the fruit and enters the earth, where it forms a small cell in which to undergo its transformations. It assumes the pupa form (Fig. 83, Plate 1,) in the Autumn, and is not changed to a fly until the following Summer.

The body of the fly (Fig. 84, Plate 1) measures three and a half or four lines (or from one fifth to one fourth of an inch) in length, and is of a black color, the thorax marked with four whitish lines, and with a white dot next to the abdomen; the latter is marked with three or four whitish transverse lines. The wings, which are only two in number, are transparent, and marked with four black cross bands, which are more or less united with each other. The only remedy seems to be to gather the infested apples before the maggots have deserted them, and make such use of them as will destroy the maggots. This insect, so far as is reliably known, has not yet been found in California, but from descriptions given of the decay of late varieties of apples in 1882, it is thought necessary to give the above description. (See U. S. Agr. Rep. for 1881.)

REMEDIES.—Should this insect appear, it can be kept off the fruit by spraying, in July and August, with Nos. 5 or 7, but probably No. 4 would be better.

CHAPTER LIII.

The Apple Curculio. (Cal.)

(*Anthonomus quadrigibbus*—Say.)

Order, COLEOPTERA; Family, CURCULIONIDÆ.

[The measurements of insects in this work are given in inches and lines. The above cut represents one inch divided into lines and fractions thereof.]

[Living in apples, pears, and quinces; a curved, footless grub of a white color, marked with bluish-black; assuming the pupa form within the fruit, and finally producing a rusty-brown snout-beetle, having three pale lines on the thorax, and four humps on the wing-cases.]

The female curculio punctures the apple with her long snout, and after widening the puncture at the bottom, she deposits therein a single egg, from which is hatched a footless grub (Fig. 85b) which burrows still deeper into the fruit, and feeds upon the latter in the vicinity of the core. After attaining its full growth, it forms a small cell, with a burrow leading from it to the outside of the apple, and in this cell it soon casts off its skin and enters upon the pupa stage (Fig. 85a), from which the perfect beetle is evolved in the course of a few weeks.

Fig. 85.

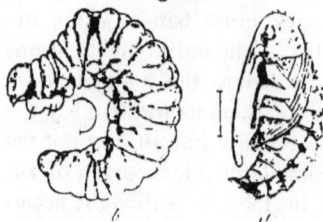

Fig. 85.—Larva and Pupa of Apple Curculio. enlarged; *a*, the pupa; *b*, the larva— color of each. white.

The full grown larva measures a little under five lines, or one third of an inch in length, and usually lies in a curved position: it is of a white color, sparsely covered with wrinkles, the spaces between which are bluish-black. of which color is the line on the back.

The perfect beetle (Fig. 86) measures from one and one half to two lines in length, and is of a dull reddish color, marked on the thorax with three indistinct whitish lines: on the hind part of the wing-cases are four prominent humps, and the snout is nearly as long as the body.

Fig. 86.—Apple Curculio; *a*. natural size; *b*, enlarged. side view: *c*. enlarged. back view— colors. brown and gray.

Fig. 86.

The perfect insect issues from the pupa state early in September, and passes the Winter in a semi-torpid state. It infests apples, pears, quinces, thorn-apples or haws, and crab-apples.

Previous to last Fall (1882), when apples were found with a single empty burrow, ants and some other insects were credited with capturing the larva of the codlin moth before maturity, but the presence of this insect being detected, explains the true facts in many of the cases mentioned.

REMEDIES.—See note at end of No. 69. In the Spring use No. 5, or No. 6, or No. 7.

CHAPTER LIV.

The Earwig.   (Cal.)

(Forficula auricularia—Linnæus.)

Order, ORTHOPTERA; Family, FORFICULARIDÆ.

[Feeding upon the flowers of various plants, and also upon fruits. A brownish or black six-legged insect, having a forceps-like appendage at the hind end of the body.]

The female Earwig deposits her eggs beneath stones, etc., and—what is very unusual among insects—she broods over them like a hen until they are hatched out, and afterwards manifests the most lively interest for the safety of her young. The latter (Fig. 87, *left*) closely resembles the adults, but are entirely destitute of wings. (Pupa, Fig. 87, *middle*).

Fig. 87.

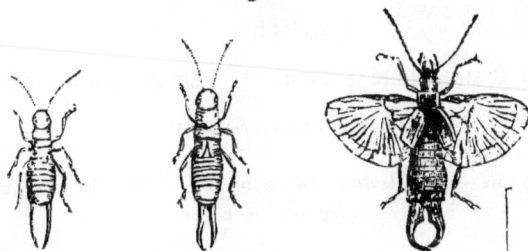

Fig. 87.—Earwigs, enlarged—color, brown; at the left, the larva; in the middle, the pupa: at the right, the perfect insect, with its wings expanded.

In the adults (Fig. 87, *right*) the wing-cases are very short, and the wings, when not in use, are folded in a very complex manner, and concealed beneath them. Although these insects have been reported as crawling into the ears of certain persons, yet no authentic instance of this kind is on record. These insects are sometimes quite destructive to various kinds of fruit, especially such as have been injured by some other insects.

INSECTS INFESTING THE PEAR TREE.

CHAPTER LV.

The Oyster-formed Scale of the Pear and Apple. (Cal.)

(*Diaspis ostreaeformis—*Curtis.)

SYNONYMS.—*Aspidiotus ostreaeformis—*Ruricola. *Aspidiotus circularis—*Fitch.

Order, HEMIPTERA ;   }
Sub-order, HOMOPTERA ; }   Family. COCCIDÆ.

[A small, circular scale insect, infesting pear trees].

In the United States Agricultural Report. 1880. Professor J. H. Comstock writes of this species as follows :

"This is a common species on the pear and apple in England. Although I do not know of its occurrence in the United States, it will be strange if it is not found here."

I received the following, dated—

"ITHACA, N. Y.. March 3, 1882.
"Your letter and specimens duly received. Without doubt
you are right. The scale is *Diaspis ostreæformis*. This is very
important, as it is the first instance of which I know of this
species in this country. Strong measures should be taken to
crush it out before it gains a foothold," etc.

"J. H. COMSTOCK."

This species is found on pear and apple trees in the vicinity
of Sacramento; it is not known what length of time the trees
have been infested, but serious damage has been done. In
many cases the bark is destroyed.

NATURAL HISTORY.

The scale of the female insect is nearly circular, or a broad
oval, and measures about three fourths of a line in diameter—
color, ashy-gray; exuviæ in the center, or nearly so, yellowish-brown; the inside of the scale and the venter is snowy
white.

The scale of the male is elongated. The eggs are pinkish-
red and ovate. The female insect is of a reddish-purple color,
and in form is somewhat elongated. There are at least two
broods each year, as I have found the eggs early in May, and
also in the latter part of July. The male insect is winged,
and is described by Curtis as being of a bright ochreous
color, with a black band on the thorax.

REMEDIES.—When the tree is dormant, a thorough spraying
with No. 13—one pound of mixture to each gallon of water used.
Summer wash, Nos. 5 or 7. Where trees are seriously infested,
when dormant No. 44 may be used as a spray, but followed
by No. 13 in about twenty-four hours.

CHAPTER LVI.

The Pear Tree Scale.   (Cal.)

(Lecanium pyri.—Schrank.)

SYNONYM—*Coccus pyri.*—Schrank.

Order, HEMIPTERA ;   }
Sub-order, HOMOPTERA ; }   Family. COCCIDÆ.

[A light brown scale insect or bark louse, about half the size of a pea, infesting the pear trees.]

This species of scale insect can be found in several of the fruit-growing districts of this State, and is injurious to the trees infested by it. It is only found on the pear, so far as known at present.

NATURAL HISTORY.

The insect is hemispherical, and is about the size of half a pea. It is of a bright brown color, and oval in form; the upper surface slightly indented. Its longest diameter is from one and one half to two lines; the width from one line to one and one half lines. Specimens sent me were received early in June, at which date the young were hatching out. The eggs are oval, and of a dirty white color; number produced by each female, from fifty to one hundred; length, one eighty third of an inch; color, brownish-white; antennæ and anal setæ present. In passing through the stages from the larva state to the mature insect, at first the color is greenish-yellow, but changes at the approach of maturity to a chestnut-brown, then to a bright brown.

After the eggs are deposited the insect dies; then the scale or outer covering is blackish. It is probable that there is only one brood in each year. (See Black Scale.)

Fig. 87½.

Fig. 87½. Portion of a branch infested by Pear-tree Scales; at the right are two of the scales—color, brown.

REMEDIES.—When the tree is dormant, spray thoroughly with Nos. 11 or 12. Or in Summer, No. 4, or No. 5, or No. 7. For Summer wash the latter are preferable, as the sulphur is an enemy to fungi.

CHAPTER LVII.

The Pear Tree Borer. (Cal.)

(*Ægeria pyri.*—Harris.)

Order, LEPIDOPTERA; Family, ÆGERIDÆ.

[Boring into the trunks of the pear tree, a pale-yellow, sixteen legged larva.]

This borer has about the same habits as the peach tree borer, but, as far as known, never infests any other kind of tree than the pear.

The perfect insect (Fig. 88—moth) usually issues from the pupa state in July. The wings expand about eight lines; they are transparent, but bordered and veined with purplish black, and across the tips of the front wings is a broad, dark band, showing a coppery reflection; the upper side of the body is purplish-black, with the edges of the collar and of the shoulder tufts, three bands across the abdomen, and the tuft at the posterior end of a golden-yellow; the under side of the body is mostly of this color.

Fig. 88.

REMEDY.—Use No. 37.

CHAPTER LVIII.

The Pear-tree Scolytus.

(*Scolytus pyri*—Peck.)

Order, COLEOPTERA; Family, SCOLYTIDÆ.

[Boring into the branches of apple, pear, apricot, and plum trees; a small, footless grub, finally transforming within its burrow into a dark-brown beetle.]

The egg from which this grub hatches is deposited in the latter part of the Summer, and is usually placed at the base of a bud; as soon as hatched, the grub gnaws its way into the branch and works around the central part, usually following

the course of the central part of the branch.   By this means
the vessels which convey the ascending sap is cut off, and that
part of the branch above the place where the insect is located,
withers and soon becomes dead wood.

The larva assumes the pupa form in its burrow, and the
perfect insect appears early in Summer.

The perfect beetle (Fig. 89) measures about one and one
quarter lines, or a tenth of an inch, in length, and is of a deep
uniform brown color.

Fig. 89. — Pear-tree Scolytus, natural size and en-   Fig. 89.
larged—color, brown or black.

For the destruction of this insect, it has been recommended to cut off and burn the infested limbs.
This should be done earlier than the month of April,
otherwise the beetles will have completed their transformations and made their escape.

I have not found this insect in this State, but from specimens of branches sent me for examination, there can be no
doubt of the presence of this beetle, or a closely allied species.

REMEDY.—Prune, as above described, and use No. 37 on the
branches as soon as the beetle appears, which is in the latter
part of July.   The branches, etc., may be sprayed with No.
4, or Nos. 5 or 7.

CHAPTER LIX.

The Branch and Twig Burrower.   (Cal.)

(Polycaon confertus.—Leconte.)

Order, COLEOPTERA : Family, PTINIDÆ.

[An elongate pitch-colored beetle, about half an inch in
length, burrowing into the branches and twigs of the apple,
pear, cherry, almond, apricot, peach and olive trees, and also
into grape canes.]

In June, 1881, Mrs. E. R. Thurber, of Vacaville, sent me
some olive branches (Fig. 92, Plate 1) which were infested by
this beetle.   In the Fall of 1881, grape cuttings (Fig. 93, Plate
1) were sent me from Sonoma County, with a beetle, which

proved to be this species, burrowing into the pith. In 1882, I received apple, pear, cherry, almond, apricot, peach, and olive branches, all of which were infested by this beetle. I also visited one pear orchard, and one orchard in which olives are grown, and found some trees damaged to a serious extent. The nature of the damage done by this beetle, is the burrowing into the branch and eating the center or pith. (Fig. 94, Plate 1.) The place selected to commence operations is generally in the axil of a bud, or small branch. The burrows made are invariably downwards, and measure from six lines to one inch in depth, and from two to three lines in diameter. A pear branch, thirty inches long, contained eleven of these burrows.

The damage done the trees is caused by their burrows being filled with water by the Winter rains, causing the branches to decay, and also by the branches burrowed breaking off.

The natural history of this beetle (Fig. 95, Plate 1) has not been fully studied, but it is supposed the eggs are deposited and the larvæ live in forest trees.

The perfect insect, both male and female, burrow into the branches of the fruit trees, but the eggs or larvæ have not been found in any of the varieties which they infest. This Spring, 1883, serious damage has been done by this species to trees planted last year.

REMEDIES.—Use No. 27, and early in the Spring spray with Nos. 4, 5, or 7.

CHAPTER LX.

The Pear-tree Psylla. (Cal.)

(*Psylla pyri.*—Linnæus.)

Order, HEMIPTERA; } Family, APHIDIDÆ.
Sub-order, HOMOPTERA; }

[Living in communities upon and puncturing the twigs of the pear tree; a small, yellowish or greenish louse.]

These insects possess the power of leaping, and hence in some localities are known by the name of flea-lice. They obtain their nourishment by puncturing the twigs with their beaks and imbibing the sap. The larvæ, or young, are of a dull-orange color, and are obtuse behind. (Pupa, Fig. 90). The perfect or winged insects (Fig. 91) are a little over a line long to the tip of the closed wings; the eyes are large and prominent; the head and thorax are of a brownish-orange color, and the abdomen is greenish; the wings are transparent. I have found this species in one orchard only, and not sufficiently numerous to do much damage to the trees infested.

Fig. 90.   Fig. 91.

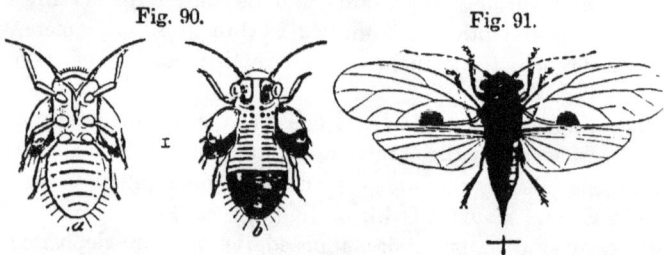

Fig. 90.—Pupa of Pear-tree Psylla, highly magnified—colors, orange-red and black : *a*, ventral view; *b*, back view.

Fig. 91.—Pear-tree Psylla, enlarged—colors, orange-red and black.

REMEDIES.—Trees infested the previous year should, when dormant, be thoroughly sprayed with No. 13—five pounds of mixture to six gallons of water. In April, spray with Nos. 5 or 7. Repeat the spraying in two weeks, if necessary.

CHAPTER LXI.

The Pear Slug.

(*Selandria cerasi*—Peck.)

Order, HYMENOPTERA ; Family, TENTHREDINIDÆ.

[A small twenty-footed caterpillar, covered with a sticky olive-colored slime, infesting the foliage of the pear and cherry trees.]

The specific name, *cerasi*, given to this saw-fly, places this insect as a pest of the cherry tree, but in this state it does most injury to the pear; therefore, it is placed in the list of pear insects.

The pear slug is found in many orchards in Central California; it feeds upon the foliage of the pear, the cherry, and the plum tree, but only eats the epidermis off of the upper side of the leaves, leaving the framework and under surface untouched.

"The trees attacked by them are forced to throw out new leaves during the heat of the Summer at the end of the twigs and branches that still remain alive; and this unseasonable foliage, which should not have appeared until the next Spring, exhausts the vigor of the trees and cuts off the prospect of fruit."—Harris.

The egg is deposited in a cut made in the leaf, by the saw-like apparatus or ovipositor of the female. Nineteen eggs have been found deposited in one leaf.

The larva (Fig. 96) is hatched from the egg in two days, and feeds upon the leaves, as described above; it attains its full growth in from twenty to twenty-five days. During the time it is feeding it exudes an olive-colored slimy substance, which covers the body and gives it the appearance of a tadpole. When it ceases eating it casts its skin and slimy coat, and appears with a clean, yellowish skin; the divisions of the segments of the body are plainly seen; it then descends to the earth and crawls beneath the surface, from one to four inches, and forms a cocoon, where it undergoes its *metamorphoses* or changes; in about fifteen days the perfect insect (Fig. 97) appears.

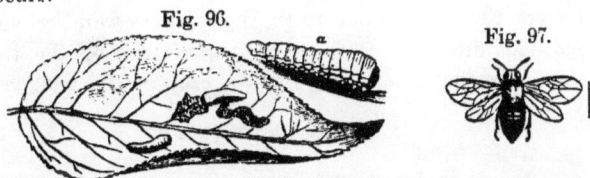

Fig. 96. **Fig. 97.**

Fig. 96.—Pear Slug; an infested leaf on which are two slugs; above it is one of the slugs, enlarged—color, olive-brown.

Fig. 97.—Pear-Slug Saw-fly, enlarged—color, black.

The first brood appears late in April, or early in May; the

second, early in July. When the larvæ of the second brood are full grown they enter the earth, and remain unchanged until the following Spring. Harris says, referring to the first brood :

"It seems that all of them, however, do not finish their transformations at this time ; some are found to remain in the ground unchanged till the following year, so that if all the slugs of the last hatch in any one year should happen to be destroyed, enough from a former brood would still remain in the earth to continue the species."

There are apparently but two broods in each year.

Larva length. five and one half lines. Perfect insect, body shining black, nearly three and one half lines long; expanse of wings, six lines ; wings transparent ; lower part of fore anterior legs, dirty-white.

REMEDIES.—Use No. 61 and No. 63.

CHAPTER LXII.

The Pear-leaf Caterpillar. (Cal.)

(*Nematus Sp?*)

Order, HYMENOPTERA ; Family, TENTHREDINIDÆ.

[A small twenty-footed caterpillar, feeding upon the foliage of the pear tree.]

Considerable damage has been done to pear trees in the Sacramento Valley by a small green caterpillar (of a saw-fly) eating the leaves. In some orchards the varieties of pear trees which were the first to put forth their leaves were seriously infested, and not only injuring the crop of fruit on the trees by destroying the foliage, but also the crop of the next year.

NATURAL HISTORY.

About the time the leaves begin to appear, the eggs are deposited in a small slit or opening made by the ovipositor of the female fly in the surface of the leaf. These are hatched in about ten days.

The young caterpillar (Fig. 98) commences to feed by eating a short track, apparently under the surface of the leaf; it eventually makes an opening in which it feeds until one fourth grown, or six days old; it then leaves this opening and commences feeding on the edge of the leaf, and as there are often from one to eight on a leaf, they move from one leaf to another until full grown, which is in about twenty-two days from the time they were hatched. When the larva, or caterpillar, ceases to eat, it descends to the earth and crawls below the surface and makes a tough, dark-brown oval cocoon. In this cocoon it hibernates, in the larval state, until the next Spring. Caterpillar (larva); length, six lines; color, green; head, yellowish-green; eyes, black; twenty legs.

Fig. 98.

Fig. 98.—Pear leaf, caterpillar and work.

9

Fig. 99.—Pear-leaf Saw-fly—colors, black and yellow.

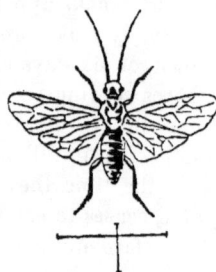

Fig. 99.

Perfect insect (Fig. 99)—length of body, three and three fourths lines; expanse of wings, about seven lines; color, head and thorax black, abdomen yellowish, with a black transverse band on the dorsal half of each segment; wings transparent, with a brownish hue.

There is, apparently, only one brood each year.

REMEDIES.—As soon as the larvæ appear, spray as directed in No. 61, or 65; also, see No. 62.

CHAPTER LXIII.

The Thrips.

Order, HEMIPTERA;
Sub-order, HETEROPTERA; Family, THRIPIDÆ.

[Feeding upon the upper surface of the leaves of the pear, peach, etc., minute black or yellow six-legged insects.]

Last year, 1882, the owners of a great number of orchards complained that the leaves of the pear, peach, and plum trees were attacked by some insect or disease which caused them to wither and fall off. On examination they were found to be infested by a species of *Thrips* not heretofore noticed on fruit trees in this State.

The leaves infested by these insects appeared marked all over their surface with minute black dots, either caused by the bite of these minute insects or were their excrements. Many of the branches, especially on the lower part of the tree, were denuded of their foliage.

Fig. 100.

Fig. 100.—Larva of Thrips—colors, bright and dark yellow.

The larva (Fig. 100) of this species of Thrips is one twenty-sixth of an inch in length; color, primrose-yellow, with narrow

transverse orange-colored markings.   (Fig. 100, highly magnified.)

The pupa (Fig. 101) is one twenty-fifth of an inch in
length; about the same color as the larva, only the transverse
markings are not so clear.   The legs and antennæ are
obscured by a film, the wings are in a sheath (see Fig. 101,
magnified), and the insect moves more sluggishly than when
in the larva state.

Fig. 101.—Pupa of Thrips—col-
or, yellow.

Fig. 101.

Fig. 102.

Fig. 102.—Thrips—color, black.

The perfect insect (Fig. 102) is
one twenty-second of an inch in
length; body black; wings black,
bordered with a silvery fringe;
there is a yellowish transverse bar
across the base of the wings.

REMEDIES.—As soon as the
thrips are noticed on the foliage,
spray thoroughly with No. 4, or No. 5, or No. 7; repeat spraying if necessary.

INSECTS INFESTING THE QUINCE.

CHAPTER LXIV.

The Quince Scale.

(*Aspidiotus cydoniæ.*—Comstock.)

Order, HEMIPTERA ;
Sub-order, HOMOPTERA ; } Family, COCCIDÆ.

[A species of scale insect found on the quince in Florida.] Prof. Comstock reports this species in Florida, and states in appearance the scale is indistinguishable from that of the greedy scale—(*A. rapax*).

REMEDIES.—Same as for San Jose Scale (*A. perniciosus*), Chap. XX.

CHAPTER LXV.

The Quince Curculio.   (Cal.)

(*Conotrachelus cratægi.*—Walsh.)

Order, COLEOPTERA ; Family, CURCULIONIDÆ.

[Living in quinces and pears, a whitish, footless grub, about four lines long, with a distinct tubercle on either side of each segment, and a reddish-brown head ; usually, but not always, deserting the fruit and entering the earth to pupate.]

The female curculio punctures the fruit and deposits an egg therein. This hatches out in a few days, and the grub works, for the most part, near the surface of the fruit, never, as far as known, entering the core. It acquires its full growth in about a month after leaving the egg, and then usually, but not always, deserts the fruit and burrows two or three inches into the earth. Here it forms a small cell, in which it remains unchanged until the following Spring, no matter whether it left the fruit as early as the first of August or as late as the first of October. In April of the following year, the larva assumes the pupa form, from which the beetle issues in the course of a week or two.

Fig. 103.

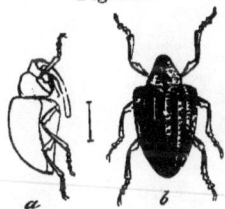

Fig. 103.—Quince Curculio, enlarged; *a*, side view : *b*, back view—color, ash-gray.

The beetle (Fig. 103) is about two lines long, of an ashen-gray color, mottled with pale-yellow, dusky and whitish, and at the base of the thorax is a somewhat triangular dusky spot. The body is broadest at the shoulders, and the wing cases are destitute of humps. The snout is longer than the thorax and is bent beneath the breast.

I have found but two specimens of this species. and these were found in 1882.

REMEDIES.—In the latter part of April, spray the foliage and fruit thoroughly with Nos. 5 or 7. Repeat in two weeks if necessary.

INSECTS INFESTING THE PEACH TREE.

CHAPTER LXVI.

The Peach Tree Borer

(*Ægeria exitiosa*.—Say.)

Order, LEPIDOPTERA ; Family, AEGERIDÆ.

[Boring into the trunks of peach, plum, cherry and similar trees, a pale yellow sixteen-legged larva.]

The peach tree borer is widely distributed over the greater part of the United States, and wherever found it is one of the most pernicious enemies of the peach tree.

The egg from which this borer hatches is deposited in the latter part of Summer upon the trunk of the tree, usually near the roots, but sometimes at the base of one of the lower limbs.

As soon as hatched, the young borer begins to eat its way downward through the bark and sapwood, continuing its course in this direction until the following Spring, when it turns about and directs its course upward. Before pupating, it forms a cocoon composed of its chips and castings, mixed with gum. The perfect insect issues during the latter part of July, or during the month of August. The presence of this borer may easily be detected by the mass of thick gum mixed with the castings of the larva, which accumulates around the opening of its burrow. It usually works in that part of the tree which is at or just beneath the surface of the ground, although it occasionally occurs in the crotches, or upon some other part of the tree.

Fig. 104.—Peach Tree Borer—color, yellowish.

Fig. 104.

The full grown larva (Fig. 104) measures from six to eight lines in length, and is pale yellow, and provided with sixteen legs; the head is reddish, marked with black.

Fig. 105.—Peach Tree Borer (moths)—colors, steel blue and yellow; 1, the female moth; 2, the male moth.

Fig. 105, 1.   Fig. 105, 2.

The perfect insects into which these borers are finally transformed somewhat resemble wasps, being provided with four nearly transparent wings. The female (Fig. 105, 1,) differs so much from the male (Fig. 105, 2,) as to cause her to be mistaken for a distinct species. Her body is of a steel blue color, with a dark orange colored band across the middle of the abdomen; the fore wings expand about one inch and six lines, and are of a steel blue color; the hind wings are transparent, but are veined and bordered with steel blue.

The male expands about one inch; the wings are transparent, and are bordered and veined with steel blue; the fore-wings crossed beyond the middle by a band of the same color; the body is also steel blue, and the edges of the collar and shoulders tufts, as well as two rings on the abdomen, and the brush at the end are pale yellow.

REMEDY.—Use No. 37 in July and August, or No. 98—the latter is preferable.

CHAPTER LXVII.

The Peach Moth. (Cal.)

(*Anarsia lineatella.*—Zeller.)

Order, LEPIDOPTERA ; Family, TINEIDÆ.

[A small reddish-pink larva, from four to five lines in length, boring into the fruit of the peach and apricot ; also, into the buds and the new growth of the peach.]

In 1882 the larva (Fig. 106, Plate 1) of this species was found boring into peaches (Fig. 107, Plate 1) and apricots in several of the fruit-growing districts of this State. The moth or perfect insect is small ; length, three lines ; spread of wings, about six lines ; color, dark-gray ; antennæ gray. ringed with brown ; wings gray, with brown streaks ; hind wings, smoky-gray ; cilia, gray. Larva length, nearly four lines ; color, reddish-pink ; head, brownish-black.

I found a larva in a peach early in June, 1882. It left the fruit on the 13th, and was changed to a pupa by the 16th. On the 25th of June, the perfect moth appeared. The larva in apricots matured to perfect insect in about the same length of time. I have also found the larva of this moth in peach buds, also in the end of the new growth. In the latter it can be easily detected, as the new leaves present a withered appearance. The larva passes its transformations in the debris in the crotches of the trees, or on the ground among fallen leaves, etc. I have found them more than half grown in the buds in January, but apparently in a semi-dormant state.

The variety found in the fruit of the apricot and peach is lighter in color than that found in the buds and new growth. The perfect insect (Fig. 106, Plate 1) that is reared from the fruit-eating larva, is lighter colored than those raised from the bud or new growth larva.

There are probably three broods each year. From the limited opportunities I have had to investigate the natural history and habits of this pest, I am inclined to think that a part at least of the so-called sap disease, or gum oozing from the buds is caused by the boring of the larva of *A. lineatella*, but will not say positively that such is the case. It is also reported as a destructive pest to the strawberry, by eating channels through the crown of the plant, and also burrowing into the runners.

Since writing the above. I have reared a number of the perfect insects. The larva leaves the fruit to pass its tranformations. When it selects a place it spins a few threads, then changes to pupa. It remains in the pupa state from seven to ten days, according to the temperature.

This season (1883), specimens of plums, prunes and nectarines have been received infested by the larva of this moth.

Peaches received in this city (Sacramento) in July last, had at least forty per cent. infested by this pest. The moth deposits the egg on the fruit, generally near the stem.

REMEDIES.—See No. 67, and use Nos. 5 or 7 as spray. For strawberries, see No. 68.

CHAPTER LXVIII.

The Peach-leaf Roller.   (Cal.)

(Cræsia persicana.—Fitch.)

SYNONYM.—*Ptycholoma persicana.*—Fitch.

Order, LEPIDOPTERA : Family, TORTRICIDÆ.

[The measurements of insects in this work are given in inches and lines. The above cut represents one inch divided into lines and fractious thereof.]

[Living singly in a nest of newly-expanded leaves on peach trees; a pale-green worm, with two white lines along the back, the head dull-yellowish.]

This leaf-roller assumes the pupa state in its nest. The fore-wings of the perfect moth expand about eight lines, and are of a yellowish color, varied with black, and marked with white spots.

REMEDY.—Use No. 24.

CHAPTER LXIX.

The Peach Aphis.   (Cal.)

(*Myzus persicæ.*—Sulzer.)

Order, HEMIPTERA ; }
Sub-order, HOMOPTERA : }   Family, APHIDDIÆ.

[Living on the underside of the leaves and on the new growth of peach, prune, and nectarine trees, causing the leaves to curl and thicken by puncturing them with their beaks and extracting the sap ; small, black or reddish-brown plant-lice.]

The wingless females are rusty red : the winged females are black or greenish-brown.

The winged males are a bright yellow, with a transverse brown streak on the thorax, and a few streaks of the same color on the abdomen.

REMEDIES.—When the tree is dormant, spray with No. 11 or 12 ; when in leaf, use Nos. 4, 5, or 7—5 or 7 preferable.

CHAPTER LXX.

The Indian Cetonia.

(*Cetonia inda.*—Linnæus.)

SYNONYM.—*Euryomia inda.*

Order, COLEOPTERA ; Family, CETONIDÆ.

[Eating into peaches, pears, and grapes ; a coppery-brown beetle, about six lines or half an inch long, sprinkled with brown dots, and thinly covered with yellowish hairs.]

Fig. 108.—Indian Cetonia—color, coppery-brown.   Fig. 108.

The larval and pupa stages of this insect are unknown, but it probably lives in the ground during the larval state, feeding upon the roots of plants.

The beetles (Fig. 108) make their appearance in the Autumn and again in the Spring, passing the Winter in some sheltered situation.

CHAPTER LXXI.

The Green Fruit-beetle. (Cal.)

(*Gymnetis nitida.*—Linnæus.)

SYNONYM.—*Allorhina nitida.*

Order, COLEOPTERA ; Family, CETONIDÆ.

[Feeding upon ripe peaches, pears, plums, figs, and melons ; a green beetle (Fig. 109*c*) measuring about one inch and two lines in length, having the wing-cases bordered with yellow ; or feeding upon the roots of strawberry plants, etc., a whitish six-legged larva.] (Fig. 109*a*.)

Fig. 109.

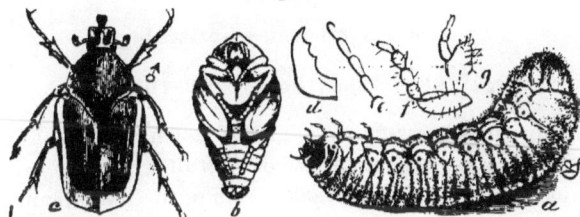

Fig. 109.—Green Fruit-beetle, Larva and Pupa ; *c*, the male beetle—colors, green and yellow ; *b*, the pupa—color, brown ; *a*, the larva, or grub—color, whitish ; *d*, its upper jaw, enlarged ; *e*, its antennæ, enlarged ; *f*, one of its legs, enlarged ; *g*, its upper lip and palpus.

This species is very scarce in this State, as but few specimens have been found. It is plentiful in Arizona Territory, and is very destructive to peaches and melons.

INSECTS INFESTING THE APRICOT.

CHAPTER LXXII.

The Red-bodied Saw Fly.   (Cal.)

(*Dolerus tejonicus.*—Norton.)

Order, HYMENOPTERA ; Family, TENTHREDINIDÆ.

[A four-winged saw-fly with dark wings and a reddish body, eating the leaves and shoots of young fruit trees.]

About the 15th of May, 1883, a young orchard planted in 1882, was attacked by this saw-fly (Fig. 110, Plate 1). The owner of the orchard writes on the 3d of June: "The flies injured my trees very much by eating the leaves and young shoots off entirely; many of the trees have sent out new growths, but many have not, and appear as though they were killed."

The following is the description of this new pest: Head, black ; antennæ black. nine-jointed ; thorax reddish, hinder portion and scutellum black : abdomen marked with black at the base ; legs. black ; wings clouded with smoky or blackish ; the veins and costa deep black ; length of body nearly half an inch : spread of wings, three fourths of one inch.

At a later date, June 26, the orchardist writes: "They are now feeding on the weeds around the orchard."

NOTE.—This saw-fly, if not identical with the European saw-fly, *Dosytheus lateritius*—Klug., very closely resembles that species. The larval history of this insect is unknown to us at the present time.

REMEDY.—Spraying the foliage with No. 5, or 7, or 65, will protect it from the ravages of this pest.

CHAPTER LXXIII.

The Apricot Leaf-roller. (Cal.)

(*Dichœlia Californiana.*—Walsingham.)

Order, LEPIDOPTERA ; Family, TORTRICIDÆ.

[A small caterpillar, living in a rolled leaf, and feeding upon the new leaves and growth of the apricot tree. It also feeds on the fruit after it sets forth from the bloom.]

This caterpillar appeared in several orchards in the Spring of 1882. In one instance it nearly destroyed the entire crop of apricots, by feeding upon the fruit. I did not succeed in rearing the moth last year, but this season I have been successful. It has attacked several orchards this Spring, 1883 ; in one case young trees planted last Spring from the dormant bud, were nearly denuded of their foliage by this pest. In other cases it has fed on the fruit. The caterpillar first attacks the fruit when about the size of a small marble—first fastening a leaf to it. At first it feeds upon the epidermis, or skin, but as the caterpillar grows larger it eats into the fruit. When not feeding, it returns to its nest. At other times it folds a leaf and feeds upon the new shoots and leaves near its nest—in some cases nearly cutting off the former.

The body of the caterpillar (Fig. 111, Plate 1) is yellowish-green, with a dorsal line of a darker shade ; head and cervical shield brownish-black, with a pale space between them ; spiracles, ringed with brown ; length, seven to eight lines.

The perfect insect (Fig. 111, Plate 1) emerged from the pupa case on the 6th, 7th, 11th, and 12th of June, after remaining in the pupa state about eleven days.

The moth is of a golden ochre-yellow color; the fore wings are crossed near the middle, by an oblique reddish-brown irregular band, on which are scattered some bluish scales; and there is a reddish-brown bar, tinged with yellow, across the outer end of the wings; hind wings, reddish-brown, marked with whitish on the front edge; legs, yellow; length, from front of head to apex of fore wings (when at rest), nearly four lines; spread of wings, about nine lines. There are two broods each year.

REMEDIES.—As directed in No. 24. Spray early in May, with Nos. 5 or 7, or 65.

———

CHAPTER LXXIV.

The Striped Bud-beetle.   (Cal.)

(*Disonycha limbicollis.*—Leconte.)

Order, COLEOPTERA ; Family, CHRYSOMELIDÆ.

[A five-striped beetle, similar in size and appearance to the striped cucumber beetle (*D. vittata*), feeding upon the buds of apricot trees.]

This species has appeared in immense numbers in several localities, and many persons supposed it to be the striped cucumber beetle. It hibernates in the perfect state (Fig. 112, Plate 1), and as soon as the warm weather begins it appears in the orchards and attacks the fruit-buds as soon as they begin to swell, seeming to have a preference for the apricot. After their appearance in the early Spring, at night and on cold days, they gather together in large numbers. As many as one or two gallons have been found on one tree, fence, frame of windmill, etc., apparently in a semi-dormant condition; but as soon as the weather became warm again they took to flight. This beetle is elongate-oval in shape; thorax, black, bordered with a narrow light-yellow margin; the wing-cases are light yellow, marked with five black lines. The difference between this species and the striped cucumber beetle can be readily distinguished. The larva probably lives in the roots of some kinds of plant.

REMEDIES.—Spray trees infested by this beetle with No. 5, or 7, or 65. If the trees are seriously infested, by stirring or mixing one pound of buhach in fifteen gallons of the mixture, it will effectually destroy the beetle. When gathered in large numbers in one place, as described above, the solution, mixed with buhach, should be used.

CHAPTER LXXIV½.

The Twelve-spotted Diabrotica. (Cal.)

(*Diabrotica 12-punctata.*—Olivier.)

Order, COLEOPTERA : Family, CHRYSOMELIDÆ.

[Feeding upon the buds and leaves of various kinds of plants, and also upon ripe or nearly ripe fruit; a yellow beetle (Fig. 112½) about three lines long, the head black, and the wing-cases marked with twelve black spots.]

Fig. 112½.

The early stages of this beetle have never been traced out, but it probably lives in the ground in the larva state, feeding upon the roots of plants.

Fig. 112½.—Twelve-spotted Diabrotica—colors, yellow and black.

REMEDIES.—The fruit in an orchard in this vicinity (Sacramento) was attacked by these pests in the month of August, 1883. The owner sprayed the trees with a solution composed of six pounds of buhach steeped in one gallon of alcohol, then diluted with twenty gallons of water; this destroyed the pests very effectually. I have succeeded in driving them off of the trees by spraying the latter with Remedy No. 5 or 7, one pound to each gallon of water used.

NOTE.— Since writing the article on the Horned Flower-beetle (page 259), I learn that these beetles sometimes burrow into ripe peaches, pears, and plums; and they are also charged with gnawing off the green grapes and letting them fall upon the ground.

When furnishing copy to the publisher, this Chapter was overlooked and not detected until too late.- M. C.

CHAP.

The Cherry-tree Borer (*Dicerca divaricata*).76

The Cherry Tortrix (*Loxotænia cerasivorana*) 76

The Cherry Worm 77

The following insects also infest the Cherry-tree :

The Woolly Aphis (*Schizoneura lanigera*).

The Lemon-peel Scale (*Aspidiotus nerii.*)

The Peach-tree Borer (*.Egeria exitiosa.*)

The Branch and Twig-burrower (*Polycaon confertus.*)

The Harvest-fly (*Cicada.*)

The Orchard Tent-caterpillar (*Clisiocampa Americana.*)

The Forest Tent-caterpillar (*Clisiocampa sylvatica.*)

The Red-humped Caterpillar (*Notodonta concinna.*)

The Canker Worms.

The Greater Leaf-roller (*Loxotænia rosaceana.*)

The Many-dotted Caterpillar (*Brachytænia malana.*)

The Turnus Butterfly (*Papilio turnus.*)

The Pear Slug (*Selandria cerasi.*)

The Red Spider (*Tetranychus telarius*).

The Yellow Mite.

The Rose Chafer (*Macrodactylus subspinosus.*)

The Angular-winged Katydid (*Microcentrum retinervis.*)

The Brown Strawberry-weevil (*Listronotus neeradicus.*)

The Negro-bug (*Corimelæna pulicaria.*)

The Plum Curculio (*Conotrachelus nenuphar*).

CHAPTER LXXV.

The Cherry-tree Borer. (Cal.)

(*Dicerca divaricata*—Say.)

Order, COLEOPTERA ; Family, BUPRESTIDÆ.

[Boring beneath the bark of cherry and peach trees; a yellowish footless grub, having the second segment greatly widened and flattened.]

This borer closely resembles the flat-headed apple-tree borer in all its stages, as well as in its habits.

The perfect beetle (Fig. 113) is from eight to eleven lines long, and is of a shining bronze or copper-color; the wing-cases are elongated, their tips separating quite widely from each other, and appearing as if broken squarely off at the apex.

Fig. 113.

Fig. 113.—Cherry-tree Borer—color, coppery-gray.

REMEDY.—Use No. 37.

CHAPTER LXXVI.

The Cherry-tree Tortrix.   (Cal.)

(*Loxotænia cerasivorana*—Fitch.)

Order, LEPIDOPTERA ;   Family, TORTRICIDÆ.

[Living on cherry trees, between two leaves, or in communities in a large nest formed by fastening the leaves and branches together with silken threads; a nearly naked, pale-yellow caterpillar, with the head and a spot on top of the fore and hind parts of the body, black.]

This caterpillar pupates within its nest, and a short time before the perfect moth issues, the pupa works itself part way out of the nest.

The moth (Fig. 114) expands from nine to thirteen lines, is of a pale ochre-yellow color, marked with pale leaden spots or bands; the hind wings, and the under side of all the wings are pale yellow.

Fig. 114.—Cherry Tortrix — colors, yellowish and brown.

REMEDIES.—When the tree is dormant, spray with No. 13—five pounds of the mixture to six gallons of water; or, No. 11 or 12. As soon as the fruit sets well from blossom, use Nos. 5 and 7.

Fig. 114.

CHAPTER LXXVII.

The Cherry Worm.   (Cal.)

Order, HYMENOPTERA ;   Family, TENTHREDINIDÆ.

[A small twenty-footed larva, eating into cherries.]

Specimens of cherries infested by a small, twenty-footed larva, have been received from at least three localities, situated about thirty miles from each other.

The larva (Fig. 115, Plate 1), when full grown, measures about three lines in length; color—body yellowish-white,

10

immaculate, anal shield a little darker than body; head small, round and pale-yellow; eyes black; twenty legs.

The egg is probably laid by the parent fly on the cherry when the latter is about the size of a pea; as soon as hatched the larva commences to feed upon the skin of the fruit and eats in toward the pit or stone. In the young fruit it eats into the pit, but when the cherry is more than half grown it seldom attacks the pit. When the larva is full grown it evidently leaves the fruit to prepare to go through its change (*metamorphosis*) in the ground or elsewhere. The specimens were received too late in the season to learn the natural history of this pest; and failing to rear the perfect insect, it is only by analogy that its history can be referred to, therefore its having twenty legs indicates that it is the larva of a saw-fly, and as the full grown larva is only three lines in length, the perfect insect or fly must be very small.

REMEDY.—The natural history of this insect being unknown to me at present, I can only recommend the picking off of the trees all infested fruit and boiling it, or otherwise making such use of it as will destroy the insect which it contains. It would be beneficial to the tree to spray it when dormant with No. 11 or 12. Or No. 13—five pounds to each six gallons of water used.

INSECTS INFESTING THE PLUM TREE.

————

CHAPTER LXXVIII.

The Plum Tree Aphis.   (Cal.)

(*Aphis pruni.*—Koch.)

SYNONYM.—*A. prunifolia.*—Fitch.

Order, HEMIPTERA ; } Family, APHIDIDÆ.
Sub-order, HOMOPTERA ; }

[The measurements of insects in this work are given in inches and lines. The above cut represents one inch divided into lines and fractions thereof.]

[Living on the under side of the leaves of the plum, which they puncture with their beaks and extract the sap; small greenish plant-lice, usually marked with black.]

The wingless lice are greenish-white ; or the head is black, the thorax green, with two transverse black lines, and the abdomen is green, dotted with black. and marked on the top with a large dark colored spot.

The winged lice have the head and thorax black, usually with a green ring around the neck; the abdomen is colored similar to that of the wingless lice, but is darker.

This species is very destructive to the plum and prune.

REMEDIES.—When the trees are in leaf, use No. 3 or 4, or No. 5 or 7; when dormant, spray with No. 11 or 12, as directed; or No. 13—five pounds to six gallons of water.

CHAPTER LXXIX.

The Plum Leaf-hopper.

(Bythoscopus clitellarius.—Say.)

Order, HEMIPTERA: }
Sub-order, HOMOPTERA; } Family, CECROPIDÆ.

[Puncturing the fruit-stems of plums and extracting the sap; a small cylindrical, slightly tapering leaf-hopper, about two and a half lines long: black or dark brown, with a bright, sulphur-yellow spot on the middle of its back and a pale yellow band in front of this; the head pale yellow, with two black dots on the forehead.—Fitch.]

REMEDIES.—When the tree is dormant, spray with No. 11 or 12, as directed; or with No. 13—five pounds to six gallons of water. For Summer wash, use Nos. 3, 4, 5, or 7; those mixed with sulphur are preferable.

CHAPTER LXXX.

The Plum Curculio.

(Conotrachelus nenuphar.—Herbst.)

Order, COLEOPTERA; Family, CURCULIONIDÆ.

[Living in plums, cherries, peaches, pears, nectarines, apricots, quinces and apples, a yellowish-white footless grub which undergoes its transformations in the earth.]

This is undoubtedly the worst enemy with which the fruit-grower has to contend—in fact its operations have become so extensive that the raising of plums has become almost entirely abandoned in several sections of this country.   This pest can be kept in check by following the proper course, but it requires constant watching; or, as Professor Riley remarks, "eternal vigilence is the price of fruit."

Fig. 116.—Plum, showing egg-puncture and crescent-mark of the plum Curculio; also, a curculio resting upon the plum.

Fig. 116.

The female curculio makes a small hole in the fruit (Fig. 116) with her snout, then turns around and deposits therein a single egg; after which she gnaws a crescent-shaped slit around and partially under the egg.   This precaution is probably taken in order to prevent the fruit from growing over and thus destroying the egg.   On account of this habit, the insect has been named the "Little Turk," the crescent being the national emblem of the Turkish Empire.

This crescent is a pretty sure indication that the fruit upon which it appears is infested with the curculio, although upon apples and similar fruits the growth of the fruit is so rapid as to obliterate the crescent in a short time.   Each female is supposed to have a stock of from fifty to one hundred eggs, and to deposit from five to ten a day.   While those which appear earlier begin this work by the middle of May, it is continued by others, which appear later, so that the period of egg-laying is extended to a period of about two months.

Fig. 117.—Larva of Plum Curculio, enlarged—color, yellowish-white.

Fig. 117.

The larva (Fig. 117) which hatches from the egg of the curculio is a small footless worm, somewhat resembling a maggot, except that it does not taper so much, and it has a distinct head.   It is of a glossy yellowish-white color, but partakes more or less of the color of the flesh of the fruit it infests.   There is a lighter line running along each side of the body, with a row of minute black bristles below, and a less distinct one above it.   The under part is reddish-brown, and the head is yellowish or pale brown.   When fully grown it measures about five lines in length. As soon as it reaches its full growth, the larva deserts the fruit—which usually falls to the ground before ripening—and enters the earth to the depth of a few inches, where it forms a small cell in which to pass the pupa state (Fig. 118). It remains in this state about three weeks when the change to the perfect state takes place.

Fig. 118.

Fig. 118.—Pupa of Plum Curculio, enlarged—color, yellowish-white.

Fig. 119.

Fig. 119.—Plum Curculio, enlarged—colors, brown, yellow, black and white.

The perfect beetle (Fig. 119) or curculio is about two lines long, and is of a dark brown color, varigated with white, yellow and black: the snout is rather longer than the thorax—the latter is uneven: the wing-cases have two black tubercles on them, one on the middle of each near the suture; behind these is a broad band of dull yellow and white; the thighs have two small teeth on the under side.

This insect lives not only in the fruits mentioned at the head of this article, but also in the black knot infesting plum and cherry trees. The perfect beetle feeds not only upon the fruit, but also upon the leaves, and even the bark of newly-formed twigs does not escape its attacks.

The number of broods which this insect produces in one year is not definitely known, but most authors regard it as being single-brooded: the perfect beetles hibernating beneath pieces of wood, etc., lying upon the ground.

I am not aware that this insect has been found in this State up to date; but as we have received so many injurious insects from the East, it is not at all improbable that the plum curculio will make its appearance among us. The greatest care should be taken to prevent its importation on nursery stock from infested districts.

REMEDIES.—Use Nos. 109 and 110.

CHAPTER LXXXI.

The Plum Gouger.

(*Anthonomus prunicida*—Walsh.)

SYNONYM.—*Coccotorus scutellaris*—Lec.

Order, COLEOPTERA; Family, CURCULIONIDÆ.

[Living in the pits of plums; a small milk-white footless grub with a yellowish-white head, passing through its transformations within the pit, and finally producing a brown snout-beetle, having the thorax pale-yellow.]

This insect, as its name indicates, seems to confine its attacks wholly to the plum. The female gnaws a hole into the fruit and deposits an egg therein; as soon as hatched, the young larva makes its way directly to the pit or stone, which it enters and feeds upon the kernel; after attaining its full growth, it cuts a round hole through the shell of the pit—which is now quite hard—and having thus prepared a place of exit, it casts off its skin and appears in the pupa form, from which the perfect insect issues in the course of a few weeks.

The larva of this species can easily be distinguished from that of the plum curculio by having the under part of its body white, this part being reddish-brown in the curculio.

The perfect insect (Fig. 120) is about one and a half lines long, exclusive of the snout, which is not much longer than the thorax; the latter is pale yellow, as are also the legs; the wing-cases are brown, with a dull grayish tint, and are destitute of tubercles.—Walsh and Riley.

Fig. 120.—Plum Gouger, enlarged—colors, yellow and grayish-brown.

Fig. 120.

In 1882 I received specimens of plums, in the pits of which was a small grub, but failed to rear the perfect insect, therefore cannot say if this species is found here.

REMEDIES.—Use Nos. 109 and 110.

CHAPTER LXXXII.

The Plum Moth.

(*Semasia pruniana*—Walsh.)

Order, LEPIDOPTERA; Family, TORTRICIDÆ.

[Living in plums, apples and crab-apples; a dingy white or brownish-yellow sixteen-legged worm, having a black head.]

It is not known with certainty whether this larva will attack sound fruit, or whether it only infests fruit which has been attacked by some other insect, but the latter is perhaps the case, and, if this view is correct, then this insect cannot be regarded as being very injurious to the orchard. When fully grown, this worm measures about three lines in length; it then deserts the fruit and spins a dark colored cocoon, which is fastened to some neighboring object.

The perfect moth has the fore-wings black and variously marked with red, blue and white, the latter forming seven short streaks along the front edge of the wing: the hind wings are grayish next the body, shading into black at the tips. This insect was bred by Mr. Walsh from the plum; the *black knot*: a gall produced by plant-lice on an elm leaf: and a gall made by a four-winged fly on an oak leaf, and Professor Riley has bred it from the apple, crab-apple and haws.

This insect is described in order that investigations may be made by those who have plums infested by a small whitish larva. I have been unable to procure specimens, but I know of at least six localities in which plums are infested by a small caterpillar.

INSECTS INFESTING THE PRUNE.

The following insects infest the prune, but are treated of elsewhere in this work :

The Robust Leaf-beetle (*Serica valida*).   The Peach Moth (*Anarsia lineatella*).
The Peach Aphis (*Myzus persicæ*).

Tree Cricket (Oecanthus).

I have received twigs of prune trees containing eggs similar to that of the gray tree cricket (*Oecanthus latipennis*—Fig. 121), but whether they belonged to this or to an allied species I am unable to say.

Fig. 121.—Eggs of Gray Tree Cricket ; *a*, the wood removed showing the eggs; *b*, punctures containing the eggs ; *c*, an egg highly magnified—color, white.

Fig. 121.

REMEDIES.—Use Nos. 25 and 28.

INSECTS INFESTING THE NECTARINE.

The following insects infest the nectarine, but are treated of in another part of this work :

The Peach Aphis (*Myzus persicæ*).   The Plum Curculio (*Conotrachelus nenuphar*).

INSECTS INFESTING THE PERSIMMON TREE.

CHAPTER LXXXIII.

The Persimmon Aphis.

(*Aphis diospyri*—Thomas.)

Order, HEMIPTERA ;   Family, APHIDIDÆ.
Sub-order, HOMOPTERA ;

The measurements of insects in this work are given in inches and lines. The above cut represents one inch divided into lines and fractions thrreof.

[Living upon the leaves of persimmon trees, which they puncture with their beaks and imbibe the sap ; small brown and black plant-lice.]

The wingless lice are purplish-brown : the head and thorax are dark ; the abdomen brownish with the extremity black. The winged lice are usually colored like the wingless ones.— Professor Thomas.

REMEDY.—Spray the foliage with either Nos. 3, 4, 5 or 7, as directed.

INSECTS INFESTING THE OLIVE.

CHAPTER LXXXIV.

The Black Scale. (Cal.)

(*Lecanium oleæ.*—Bernard.)

Order, HEMIPTERA ; } Family, COCCIDÆ.
Sub-order, HOMOPTERA ;}

[A dark brown hemispherical scale insect, or bark-louse, which infests all varieties of citrus trees, and nearly all varieties of deciduous fruit trees, and many shrubs, vines, etc.]

The black scale is more generally found in the orchards and gardens of California than any other species of the *Coccidæ*.

It infests the orange, lemon, lime, olive (Fig. 122, *L*,) apple, pear, peach, apricot, plum, prune, cherry and pomegranate trees. In the garden, it infests the honeysuckle, chrysanthemum, rose, oleander, and many other plants ; and this, or a closely allied species, infests the forest trees. The presence of this species can be readily detected by the appearance on the branches, foliage and fruit of a black smut, known to scientists as *Fumago salicina*, and the cause of its production is a question upon which authorities differ. I am convinced, from practical investigation, and also from information received from Mr. Alexander Craw, and Mr. Wolfskill, of Los Angeles, and the late A. B. Clark, of Orange, Los Angeles County, that the black smut is caused by a honeydew exuded by the females of the black scale insect, in the stage of their life between the time of the first formation of the calcareous secretion by which the insect is covered, and their reaching maturity or becoming fixed to any part of the plant.

Fig. 122.

Fig. 122.—Black Scale; *1*, an infested twig; *1a*, side view of one of the scales, enlarged—color, dark brown.

In relation to this smut or fungus, Professor Barlow writes: "The result of our examination of the diseased orange and olive leaves is briefly as follows: The disease, although first attracting the eye by the presence of the black fungus, is not caused by it, but rather by the attack of some insect which itself deposits some gummy substance on the leaf and bark, or so wounds the tree as to cause some sticky exudation on which the fungus especially thrives. It is not denied that the growth of the fungus greatly aggravates the trouble already existing by encasing the leaves, thus preventing the action of the sunlight. We only say that in seeking a remedy we are to

look further back than the fungus itself, to the insect, or whatever it may be, which has made the luxuriant growth of the
fungus possible."

The smut or fungus is found on the branches, foliage and
fruit of orange, lemon, lime and olive trees infested by the
black scale. I have also seen apricots and peaches, taken
from trees infested by this insect, so thoroughly covered by
this smut that it destroyed their market value for canning
purposes.

NATURAL HISTORY.

The black scale (Fig. 122, *1a*,) when full grown is of a dark
brown color, nearly hemispherical .in form, but is slightly
longer than broad; length, from two to two and a half lines;
width, about two thirds of the length; height, one and one half
lines; there are two ridges or bars across the body, apparently
dividing it into three parts, the middle being the largest; a
short ridge along the back joins the two cross ridges, forming
lines resembling the letter H; the edge of the covering of the
insect resting on the wood, foliage, etc., is margined, and has
a grooved or fluted appearance nearly one half the height of
the insect.

The eggs are oval in form; when first laid, whitish; before
hatching, a reddish-yellow. From seventy-five to one hundred
and seventy-five are deposited by each female of this species.

Fig. 123.

Fig. 123.—Larva of Black Scale, enlarged, back view—color, reddish-yellow.

The larva (Fig. 123) is one seventy-
fifth of an inch long; width, five eighths
of length; form, oval; antennæ, six or
seven jointed. From the time the secretion begins to form until the insect has
reached maturity, it assumes different
shades of color—first, greenish-brown;
half grown, reddish-brown, and at maturity, dark brown.

It is doubtful if there are more than one brood in each
year; the first brood are hatched, in this locality (Sacramento),
about the first of May, but do not attempt to leave from under

the scale until the twelfth; yet it is very common to find the females of this species depositing their eggs late in September, but whether they are of the Spring brood I am not prepared to say.

In relation to the length of time the lecaniums are capable of moving from one place to another, Mons. V. Signoret writes: "Before pregnancy, they have the power to move, if necessary."

REMEDIES.—Deciduous fruit trees : When the tree is dormant, spray with Nos. 11 or 12, as directed ; or No. 13—five pounds to six gallons of water. In Spring or Summer, when the young are hatched, spray thoroughly with Nos. 5 or 7. (No. 4 may be used and prove very effective, but the solution containing sulphur is preferable, as it destroys the black smut.) For citrus trees : (see No. 48), spray, etc., as directed in No. 49, and also as directed for olive trees. For olive trees : Use same as for citrus trees ; or No. 9, with one gallon of water added to every gallon of the solution. Example—To thirty gallons of No. 9, add thirty gallons of water—sixty in all.

CHAPTER LXXXV.

The Olive Worm.

(*Dacus oleæ.*)

Order. DIPTERA ; Family. ORTALIDÆ.

[Living in the olive berries (Figs. 126 and 127): small whitish footless maggots.]

Fig. 126.

Fig. 127

Fig. 126.—Olives infested by Olive Worms.
Fig. 127.—An olive cut open, showing work
of Olive Worms.

This pest is not found in California, so far
as I am aware, at present. The following is
taken from Figuier's account of this insect :

The parent (Fig. 124), which deposits the
eggs from which the maggots (Fig. 125, *left*,)
are produced, is about one half the size of the
common house-fly, and of an ashen-gray color; its head is
orange-yellow, with two black spots on the upper part of the
face; the eyes are green; the thorax is marked with four

light yellow spots, and the abdomen is brownish, spotted with black on the sides.

Fig. 124.   Fig. 125

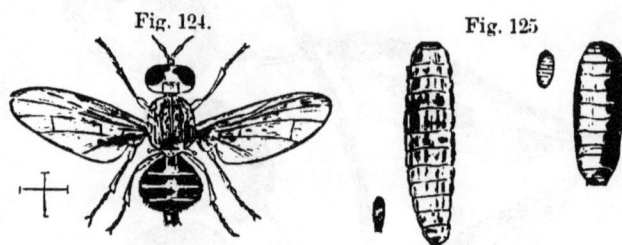

Fig. 124.—Olive Fly, enlarged—colors, gray, black and yellow.

Fig. 125.—Olive Worm and Pupa; at the left, the worm, natural size and enlarged—color, whitish; at the right, the pupa, natural size and enlarged—color, brown.

This fly punctures the skin of the olive and deposits therein a single egg; the maggot, which hatches from this, burrows into the berry until reaching the stone, which it leaves untouched. After attaining its full size, it forms a cell beneath the skin, and in this cell it assumes the pupa form (Fig. 125, *right*.)

REMEDY.—Spray with Nos. 4, 5 or 7.

A. N. Caudell.

Insects Infesting Orange, Lemon and Lime Trees.

CHAPTER LXXXVI.

The Red Scale. (Cal.)

(*Aspidiotus aurantii*—Maskell.)

Synonym.—*Aspidiotus citrii*—Comstock.

Order, HEMIPTERA ;
Sub-order, HOMOPTERA ;} Family, COCCIDÆ.

[A circular reddish scale insect, infesting the citrus trees, and has been found on grape vines and the foliage of walnut trees.]

The red scale (Fig. 128) infests some of the citrus groves of Southern California, and orange trees in Sacramento and Marysville. It has also been found on grape vines and on the foliage of walnut trees, but I do not think that any damage will be done to these plants by this pest. As the walnut sheds its foliage annually, the insects are likely to be destroyed; and those which I have examined on the grape vines in the month of September, and which appeared to be in a healthy

11

condition, were dead and shrunken when I examined the vines in the month of February following.

It is generally conceded that this species is an importation from Australia.

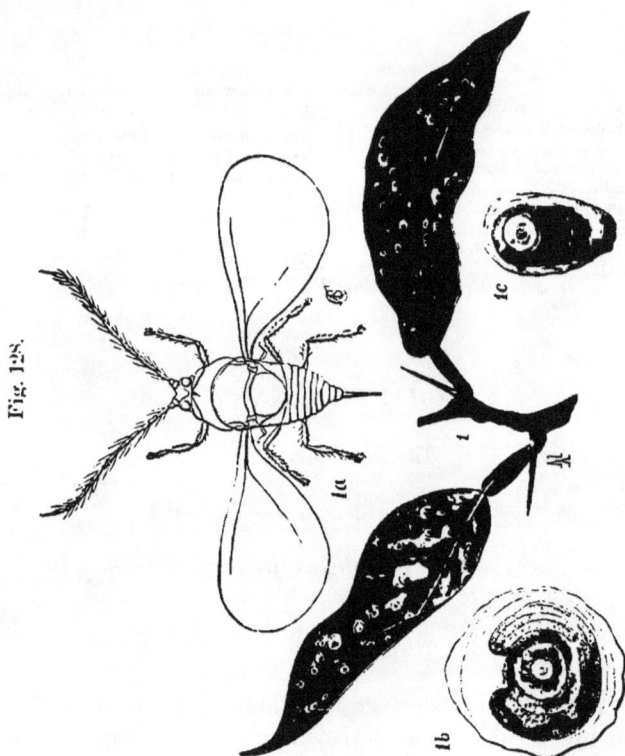

Fig. 128.—Red Scale : 1, a twig infested by these scales ; 1a, the male, highly magnified—colors, yellow and brown ; 1b, the female scale, greatly enlarged—colors, gray, yellowish or brown ; 1c, the male scale greatly enlarged—color, same as the female scale.

NATURAL HISTORY.

Female scale (Fig. 128, 1b), nearly transparent, circular, of a light-grayish color, and measures from one line to one and one quarter lines in diameter; exuviæ or cast skin in center, yellowish; second larval skin easily distinguished.

Male scale (Fig. 128, 1c) a little darker in color and smaller than the female scale; form, elongated; exuviæ nearest the anterior end.

Eggs.—It is thought by some writers that the females of this species are viviparous. I have watched the female insect ovipositing, and immediately examined the egg or sack under a microscope, using a high power, and could not detect any appendages; however, in twenty-four hours I noticed the presence of antennæ and legs. The insect produces from two to four of these eggs or sacks in twenty-four hours, and the number produced by each female is from twenty to forty-three; the latter is the highest number I have found.

In the month of September, 1882, I found a lemon at an orchard in Los Angeles County, on which the larvæ of thirty-nine male scale insects had located around the stem of the fruit, and as there was only one matured scale on the lemon, this was evidently the number produced by one female. Larva color, bright yellow; form, ovoid; length, one eightieth of an inch; antennæ, six-jointed; anal setæ, present.

Female (Fig. 129)—color, light or primrose-yellow when the scale is formed, but as it reaches maturity it becomes a brownish-yellow. The formation of the body is such that under the scale, when examined with a lens, its appearance is that of a broken ring, but when ovipositing the posterior end of the abdomen extends beyond the circular line of the body. The color of the natural insect is shown through the nearly transparent scale from which it derives its common name—Red Scale.

Fig. 129.

Male (Fig. 128, 1a)—color of body, amber-yellow, with dark marking on thorax; eyes, black.

Fig. 129.— Female Red Scale Insect, enlarged, ventral view—color, yellow.

The young larvæ can be found at all seasons of the year, and there are probably four or five broods in each year.

Fig. 130.   Fig. 131.

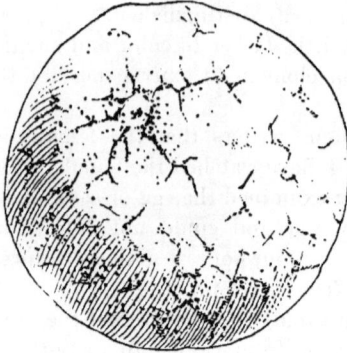

Fig. 130.—Orange infested by Red Scale.

Fig. 131.—Leaf infested by Red Scale; two of the scales at the left, enlarged—color, yellow or brown; the upper one the female; the lower one the male.

This species infests the small branches and foliage (Fig. 128, 1) and fruit (Fig. 130), but seems to prefer the fruit and foliage. (Fig. 131.)

REMEDIES.—See Nos. 48, 49, 50, 65, and 77. Spray.

CHAPTER LXXXVII.

The Red Scale of Florida.

(*Aspidiotus ficus*—Riley, MSS.; *Chrysomphalus ficus*—Riley. MSS. Ashmead.)

Order, HEMIPTERA;
Sub-order, HOMOPTERA;   } Family, COCCIDÆ.

[A species of scale insect infesting the branches, foliage and fruit of orange trees in Florada and the Island of Cuba.]

Fig. 132.—Red Scale of Florida; *2*, leaves infested by these scales; *2a*, the female scale, enlarged—colors, reddish-brown and brown; *2b*, the male scale, enlarged—colors, gray and brown: *2c*, young larva, highly magnified—color, yellow; *2d*, *2e* and *2f*, the scales in different stages of formation.

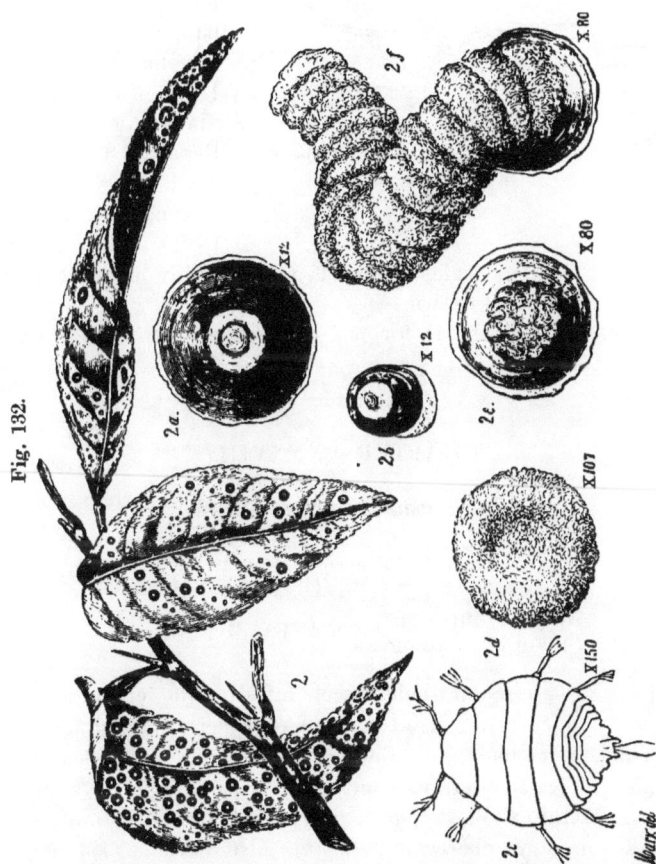

Fig. 132.

Professor Comstock describes this species as follows (Fig. 132): " *Female Scale.*—Color, the part of the scale covering the second skin is a light reddish-brown; the remainder of the scale is much darker, varying from a dark reddish-brown to black, excepting the thin part of the margin, which is gray; exuviæ nearly central, whitish in fresh specimens; form circular, one line in diameter. *Male Scale.*—The scale of the male is about one fourth as large as that of the female; the pos-

Fig. 133.

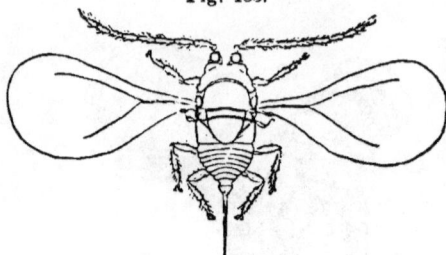

terior side is pro-
longed into a thin
flap, which is gray
in color (Male, Fig.
133). (See United
States Agricultural
Report, 1880; and
A s h m e a d   i n
"Orange Insects,"
1880.)

Fig. 133—Red Scale of Florida (male, highly magnified)—
colors, yellow and brown.

REMEDIES.—Same as for red scale (*A. aurantii*). Nos. 48.
49. 50. 65, 77 or 44.

CHAPTER LXXXVIII.

The Lemon-peel Scale.   (Cal.)

(*Aspidiotus nerii.*—Bouche.)

Order, HEMIPTERA;
Sub-order, HOMOPTERA; } Family, COCCIDÆ.

[A whitish circular scale insect. infesting the lemon. plum,
cherry and currant; also the oleander, acacia, magnolia, etc.]

This species has been known to scientists as the "Oleander
Scale" (Fig. 134), from which it derives its specific name,
nerii. Within the last four or five years it has been found on
the lemon, plum, cherry and currant; also on the acacia, magnolia. etc. It seems to prefer the fruit of the lemon, and in
many cases infests the skin or peel to such an extent as to
reduce its market value. California cannot claim a sole pro-
prietary right to this pest, as lemons imported from Europe
are often offered for sale in our market which are seriously
infested by *A. nerii*.

Fig. 134.—Lemon-peel Scale; 1, leaves and twigs infested
by these scales: 1*a*, the male scale insect greatly magnified—

colors, yellow and brown ; 1*b*, scale of male greatly enlarged—
color, white ; 1*c*, scale of female highly magnified—color, whitish or gray.

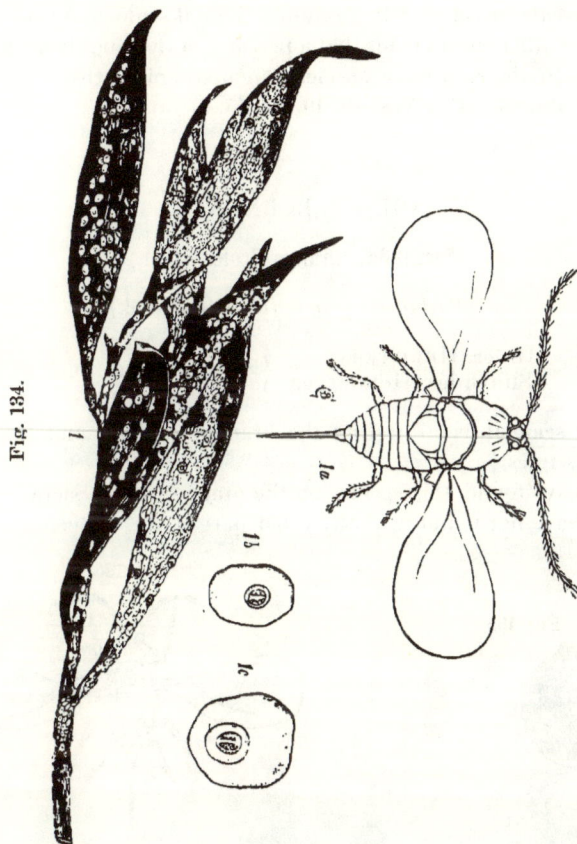

Fig. 134.

NATURAL HISTORY.

The female scale (Fig. 134, 1*c*,) is of a whitish color, and
nearly circular, measures one line in diameter ; exuviæ or cast
skin, yellowish, and near the center. Male scale (Fig. 134,
1*b*,) white, smaller, and not as circular as that of the female.
Egg, light yellow. Larva, yellowish white ; length, one eighty

fifth of an inch. Female, light yellow, with darker blotches; body, circular; abdominal segments appear as a pointed projection at one part of the circle. Male insect (Fig. 134, 1*a*.) winged; body yellowish, with dark markings. The lemonpeel scale insect closely resembles the red scale, and it is only by the difference in color that a person not thoroughly acquainted with the respective species can distinguish them.

REMEDIES.—Use Nos. 48, 49, 50, 65, 77 or 44.

CHAPTER LXXXIX.

Pergande's Orange Scale.   (Cal.)

(*Parlatoria pergandii.*—Comstock)

Order, HEMIPTERA :   }
Sub-order, HOMOPTERA : }   Family, COCCIDÆ.

[A scale insect infesting the branches, foliage and fruit of citrus trees.]

I have found this species on the orange tree in Sacramento, but have not found it in any other part of the State.

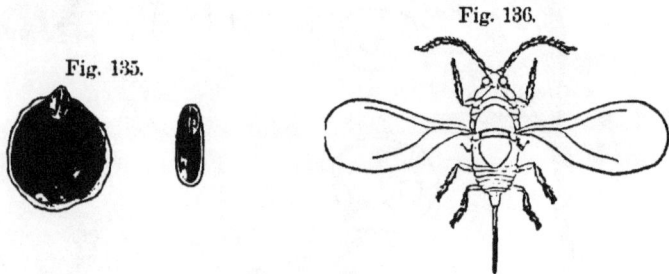

Fig. 135.

Fig. 136.

Fig. 135.—Pergande's Orange Scale; at the left, the female scale enlarged—color, dirty gray; at the right, the male scale enlarged—color, gray.

Fig. 136.—Pergande's Orange Scale; male, highly magnified—color, purplish.

The female scale (Fig. 135, left,) is somewhat elongated in form, but nearly circular, the exuviæ at one side of the center—color, grayish; exuviæ yellow, and generally oval in shape.

Male scale (Fig. 135, right,). The scale of the male is elongated and narrow—color, dirty white; exuviæ at the anterior end. Female—color, purplish, with posterior end of body yellowish, and is nearly as broad as long. Eggs—color purplish, elongated; from nine to twenty found under each female scale. Larva length, nearly one nineteenth of an inch—color purplish. Male (Fig. 136)—color dark purplish.

REMEDIES.—Nos. 48, 49, 50, 65, 77 or 44.

CHAPTER XC.

The Citrus Leaf and Fruit Scale.

(*Mytilaspis citricola*—Packard.)

SYNONYM.—*Aspidiotus citricola*.—Packard.

Order, HEMIPTERA ; } Family, COCCIDÆ.
Sub-order, HOMOPTERA ; }

[An elongated slightly curved scale insect, infesting citrus trees.]

This species of scale insect has not been found on any of the citrus trees in this State, so far as I know, but it will be strange if it is not found in the near future. It is not a rare occurrence to find it on oranges, etc., which are imported from Europe, Australia and Tahiti, and offered for sale on fruit stands throughout the State.

The scale of this species (Fig. 137, *1a, 1b*) is similar in form and appearance to that of the oyster shell bark-louse (*M. pomorum*, Chap. XIX), excepting that it may be a little wider at the posterior end. Length of female scale about one and one half lines. The male scale (Fig. 137, *1c*) is similar to other species of *Mytilaspis* in having a hinge-like joint, posterior to the middle of the scale, so that by lifting the posterior part up, the perfect insect (Fig. 138*a*) can emerge.

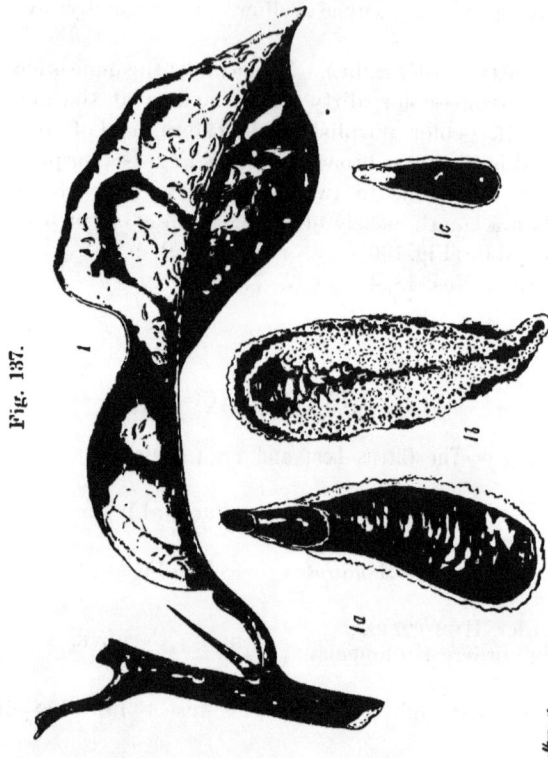

Fig. 137.—Citrus Leaf and Fruit Scale; *1*, a leaf infested by these scales; *1a*, the female scale greatly enlarged, back view —color, brown; *1b*, the same, ventral view, showing eggs—color of each, white; *1c*, the male scale enlarged, back view—color, brown.

Fig. 138.—Citrus Leaf and Fruit Scale, enlarged: *c*, the male scale — color, brown; *a*, the male scale insect—color, red; *b*, the larva—color, yellowish.

REMEDIES.—Nos. 44, 48, 49, 50, 65, and 77.

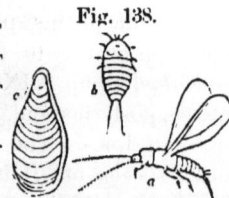

CHAPTER XCI.

The Soft Orange Scale. (Cal.)

(*Lecanium hesperidum.*—Linnæus.)

Order, HEMIPTERA ; } Family, COCCIDÆ.
Sub-order, HOMOPTERA ; }

[An oval flattened scale insect, infesting citrus trees, especially the orange.]

The soft orange scale is found in California in nearly every locality where citrus trees are grown; it infests the wood, foliage and fruit. This, or a closely allied species, is found on plants in hot-houses.

Fig. 139.

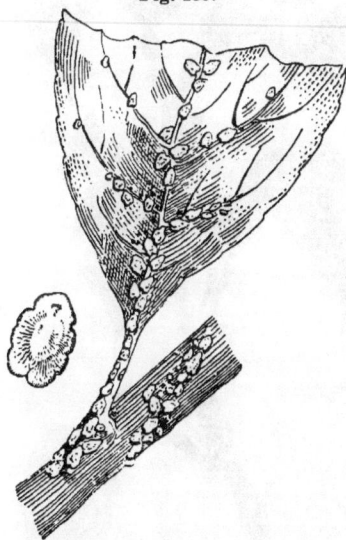

Fig. 139. — Soft Orange Scales; at the left, one of the scales enlarged—color, yellowish.

In the Treatise on Injurious Insects, 1881, I described the male of this species as winged. Prof. Comstock, in his Entomological Report of 1880, writes: "The male of this species has never been found, although it has been studied from the time of Linnæus down."

In September, 1880, I prepared a dry mounting of a specimen of *Lecanium hesperidum* for microscopic use at the State Fair of that year; early in the week a small insect was noticed coming from under a specimen beneath the glass, and finally released itself. It proved to be a male scale insect.

NATURAL HISTORY.

Female (Fig. 139)—a broad oval scale, measuring from one and one quarter to one and one half lines in length, widest at the posterior end—color, dark brown on top, and a lighter brown surrounding the margin. Two indentations on the margin on each side, and a large indentation on the posterior end. It has powers of locomotion similar to those of other *Lecaniums*. I have not found the egg of this species, but have found large numbers of the young larvæ—as many as forty-five under one specimen. The young larvæ appear about the first of May in this vicinity (Sacramento). Larva length, one eighty fifth of an inch—color, dark or dirty yellow; antennæ, six jointed (some specimens appear to have seven joints); two anal setæ.

DESCRIPTION. — Length of body, one seventy-second of an inch; from front of head to apex of wing, one twenty-fourth of an inch; posterior stylets, one forty-fifth of an inch, or one half the length of body — color, body, immaculate golden-yellow; eyes, dark or black; antennæ (from the peculiar position in which they are placed I can only count seven joints), golden-yellow and hairy; legs, golden-yellow.

As it did not agree with the description of any of the male scale insects I had read of, or specimens of males of *aurantii, perniciosus, persea, rapax, rosea* or *purchasi* in my possession, I could only imagine that it was the

Fig. 140.

male of *L. hesperidum* (be what it may, it come from under the *L. hesperidum* scale), and fortunately I preserved the mounting.

Fig. 140.—Soft Orange Scale—color, brown.

REMEDIES.—No. 4 or 9, one gallon of water added to each gallon of the mixture; No. 5 or 7, four pounds of mixture to each five gallons of water. Apply all the above at a temperature of one hundred and thirty degrees Fahrenheit. (Spray.)

CHAPTER XCII.

The Cottony Cushion Scale. (Cal.)

(*Icerya purchasi.*—Maskell.)

Order, HEMIPTERA ;
Sub-order. HOMOPTERA ; } Family, COCCIDÆ.

[A white cushion-like scale insect. feeding upon citrus trees. deciduous fruit trees, forest trees, and on some varieties of vegetables.]

Fig. 141.

Fig. 142.

Fig. 141.—Portion of a branch infested by Cottony Cushion Scales.

Fig. 142.—Cottony Cushion Scales, natural size—colors, orange, red, whitish and pale yellowish.

This species of scale insect (Figs. 141 and 142) I consider the most dangerous of any that infests fruit and other trees in California, as it may be said to be a general feeder: it is found on all varieties of citrus trees, deciduous fruit trees, on many varieties of ornamental trees. forest trees and shrubs; also on some varieties of vegetables. The apparent color of this scale insect at first sight is white, with a dark colored head. On examination, it is found that the part indicated by the dark color is the insect, and the white portion a bag or case spun by the insect to conceal her eggs when deposited.

Fig. 143.—Cottony Cushion Scale—color, yellowish-white.

Fig. 143.

The females (Fig. 143), after ovipositing (the egg case included), differ in size, some measuring six lines in length; but the general length is from three to four lines; width, one and one half to three lines, and slightly tapering toward the posterior end. Each female deposits from two hundred to five hundred eggs. In one instance I counted seven hundred and three. The eggs are oblong-ovate in form, and of a pale red color.

Fig. 144.—Larva of Cottony Cushion Scale—color, red.

Fig. 144.

Larva (Fig. 144)—color, body red; antennæ six jointed, clubbed at the apex, on which are six long hairs—color, smoky black; legs, smoky black (the joints of the antennæ and legs are lighter in color than the balance); there are six long anal hairs; the margin of the body and back is also dotted with hairs; length of body, one thirty fifth of an inch.

Fig. 145.

Fig. 145.—Female of Cottony Cushion Scale Insect—color, reddish-brown.

Mature female (Fig. 145) before spinning egg-case. The female insect during her growth assumes a variety of colors; principally yellowish-red, with irregular blotches of white, green and yellow. At full growth, and before spinning egg-case, she is ovoid in form; the hairs on the anal margin and sides are used as spinerets, exuding a cottony-like secretion of which the egg-case is formed. During her growth, and before beginning to spin her egg-case, the females exude a honeydew, which forms a black smut on the branches and foliage, as described in the chapter on the Black Scale.

Male insect (Fig. 146, Plate 3), winged—color, thorax and

body dark brown, abdomen red ; antennæ dark colored, with light brown hairs extending from each joint ; wings brown, irridescent.

REMEDIES.—Nos. 96, 44, 77 or 50—one part to twelve of water. If No. 50 is used, wash thoroughly with No. 5 or 7 after the leaves drop off. Simple remedies are of no avail in fighting this pest.

INSECTS INFESTING THE FIG.

The Green Fruit-beetle (*Cymnetis nitida*), and the Cottony Cushion-scale (*I. purchasi*) infest the fig, but are treated of in another part of this work.

INSECTS INFESTING THE GRAPE.

CHAPTER CXIII.

The Grape-root Borer.

(*Egeria polistiformis.*—Harris.)

Order, LEPIDOPTERA; Family, ÆGERIDÆ.

[Feeding upon the bark and sapwood of the roots of the grapevine; a whitish sixteen-legged larva.]

The female moth (Fig. 146*b*) usually deposits her eggs upon the vine close to the ground, and the young borer, as soon as issuing from the egg, begins to excavate a burrow through and just beneath the bark, proceeding directly to the roots, devouring the sapwood and leaving the heart untouched. When fully grown (Fig. 147) it measures about one inch and six lines in length. Before pupating, it forms an oval pod-like cocoon near the infested roots. The moth, or perfect insect, issues in July or early in August. Male insect (Fig. 146*a*)—The forewings of the moth expand a little over one inch, and are of a brownish-black color, with one or two nearly transparent spots at the base of each; the hind wings are transparent, with the veins and outer border brownish-black. The body is black, and variously marked with yellow; in some the basal part of the abdomen is black, with the remainder dull yellowish; in others the abdomen is wholly black, with the exception of one or two yellow rings.—[Condensed from Riley.

Fig. 146.

Fig. 147.

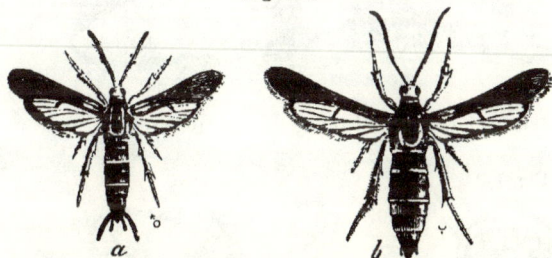 

Fig. 146.—Grape-root Borer Moths—colors, dark brown and orange: *a*, the male moth; *b*, the female.

Fig. 147.—Grape-root Borer—color, yellowish.

REMEDY.—Use No. 37 or 98. The latter is preferable.

CHAPTER CXIV.

The Broad-necked Prionus. (Cal.)

(Prionus laticollis.—Drury.)

Order, COLEOPTERA; Family, CERAMBYCIDÆ.

[Living upon and usually hollowing out the roots of apple trees, grape and hop vines; a large white nearly footless

12

larva or grub, having the first segment of its body larger than
any of the others.]

This larva (Fig. 148) is of a creamy-white color, with a pale
bluish line along the back; the first segment is as long or
longer than the next three combined; the body tapers gradually from the third segment backward; the head is brown, and
the legs are minute. It is supposed to pass three years in the
larva state. When fully grown it measures about three inches
in length; it then deserts the roots and forms a smooth cavity
in the earth, wherein to undergo its transformations. The
perfect beetle appears in about three weeks after the change
to the pupa form (Fig. 149) has taken place.

Fig. 148.

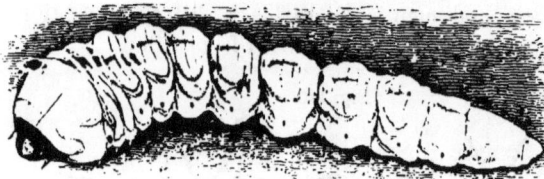

Fig. 148.—Larva of Broad-necked Prionus—color, yellowish-white.

Fig. 149.   Fig. 150.

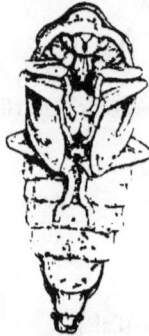   

Fig. 149.—Pupa of Broad-necked Prionus—color, yellowish-
white.

Fig. 150.—Broad-necked Prionus—color, brownish-black.

These beetles (Fig. 150) vary in length from a little over an inch to an inch and eight lines, and are of a brownish black color; on each side of the thorax are three teeth, the middle one the most prominent.

There is another beetle very closely related to the above species, which, like the latter, also infest the roots of grape and hop vines in its larva state. This is known as the tile-horned prionus (*Prionus imbricornis*). It differs from the broad-necked species in having from sixteen to nineteen joints in each antenna, whereas those of the broad-necked species have only twelve joints.

REMEDY.—Use No. 99.

___

CHAPTER CXV.

The Grape-root Louse. (Cal.)

(*Phylloxera vastatrix.*—Planchon.)

Order, HEMIPTERA ;   } Family, APHIDIDÆ.
Sub-order, HOMOPTERA ; }

[A minute yellow louse, feeding upon the roots of the grape vine.]

Fig. 151.

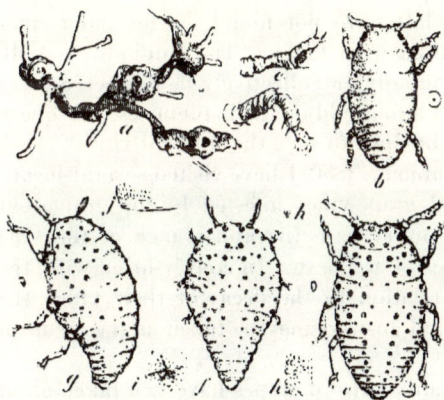

Fig. 151.—Grape Root Louse (root-inhabiting form); *a*, an infested root ; *b*, hibernating louse enlarged — color, yellow: *c*, its antenna enlarged: *d*, one of its legs enlarged: *e, f,* and *g,* the lice enlarged—color, yellow: *i*, a tubercle enlarged ; *h* and *j,* granulations of the skin, enlarged: *k,* the simple eyes, enlarged. I will not give an extended descrip-

tion of this pest, nor an account of the damage done to vine-growers by its ravages, as the reports of the State Board of Viticultural Commissioners, the report of Charles A. Wetmore, Chief Executive Viticultural Officer; papers by Prof. E. W. Hilgard and others, of this State, and Prof. C. V. Riley's Missouri Reports, have given full information of its natural history, habits, etc.

Unfortunately the presence of the grape-root louse (Fig. 151) in some of the vineyards of this State is established beyond a doubt, and that a large acreage of vineyard property has been destroyed by the ravages of this pest many of the vine-growers can offer substantial verifications.

Various opinions have been expressed as to how this insect spreads from one vine-growing district to another. A asserts that the winged individuals fly from one vineyard to another, as other species of *Aphididæ* spread on their respective food plants. B considers that the insect is spread by transporting from infested vineyards rooted vines and cuttings on which the Winter egg is deposited. C claims that the grape-root louse was brought on his premises by boxes returned from market in which grapes had been shipped. D insists that he has not had any boxes returned to his premises, that the cuttings he used were grown in his own vineyard, yet on a patch of ground five rods square the vines are infested by the *Phylloxera*, although the latter are not found on his older vines from which the cuttings were taken. The opinions of A, B, C, and D, may individually or collectively be correct, and as the pest has secured a foothold on their premises, the query arises: "What can be done to save the infested vines?"

Since the first of January, 1880, I have visited several localities in which I found grape-vines infested by the grape-root louse, and in nearly every case the appearance of the vine indicated the presence of the pest. In conversation with the vineyard owners, in relation to the decay of their vines, the following questions and answers may be taken as the result of such investigations:

Question—How many crops of grapes have you taken off of these vines? Answer (from different growers)—From ten to fifteen crops have been taken from this vineyard.

Q.—Were the vines pruned each year? A.—Yes.

Q.—What was done with the cuttings or prunings? A.— Some of them were used for cuttings for planting, the balance were hauled off the ground and burned.

Q.—Has the ground been thoroughly cultivated each year? A.—I believe so.

Q.—And of course you gathered the crop of grapes each year? A.—That is what I planted the vines for.

Q.—Have you at any time during the last eight years used any kind of fertilizers on your vineyard? A.—No.

Q.—How did the yield of the vines planted fourteen years ago or more, compare last year with those of four or five years ago? A.—Five years ago I had a much larger yield on the same vines.

Q.—Did you notice any difference in the yield of the crops of 1881 and 1882 of the vines planted, say seven years ago? A.—The crops of 1882 were heavier: but that may be attributed to a favorable season, as much at least as to the increased age of the vine.

My reasons for asking questions as indicated above, were this : In 1880 I had formed an opinion that a remedy for the grape-root louse could be found by fertilization, and so stated at a meeting of the State Horticultural Society—that "The presence of the grape-root louse and the serious damage done by it to the vine might, to a great extent, be attributed to the weakly condition of the vine at the time it was attacked by the pest, and which was probably caused by the vine being deprived of some nutriment that it required, and which did not then exist in the soil in which it was growing." At that time some of the vine-growers took issue with me on that opinion. In the month of September following the meeting of the State Horticultural Society referred to, a meeting was held of the " Phylloxera Congress," at Paris. France, at which the President of the Congress, M. Henricy, expressed the following opinion : "The phylloxera were but the result of the long use and fatigue of the vine, and that they might get rid of them by restoring health and vigor to the soil and plant."

Fig. 152.—Grape-root Louse, highly magnified (the root-inhabiting form); *a*, a healthy root; *b*, a root on which the lice are working; *c*, a root which they have deserted; *d, d, d, d*, the lice on large roots; *e*, the female pupa, back view—color, yellowish; *f*, the same, ventral view; *g*, the winged female, back view—color, yellow; *h*, the same, ventral view; *i*, her antennæ, more highly magnified; *j*, side view of wingless female, laying eggs; *k*, a section of an infested root, showing effects of their work.

Fig. 152.

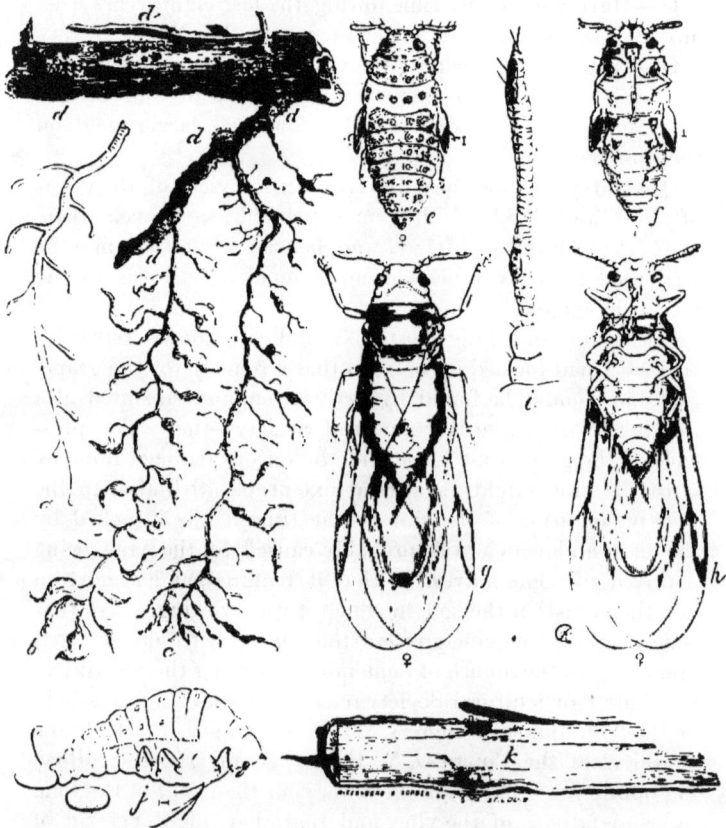

In discussing the merits of fertilization, vine-growers have used the following argument: "If fertilizers are used to produce a vigorous growth of the vine, so as to offset the damage done by the attack of the grape-root louse, the value of the grapes would be destroyed to a great extent for making fine wines."

Fig. 153.

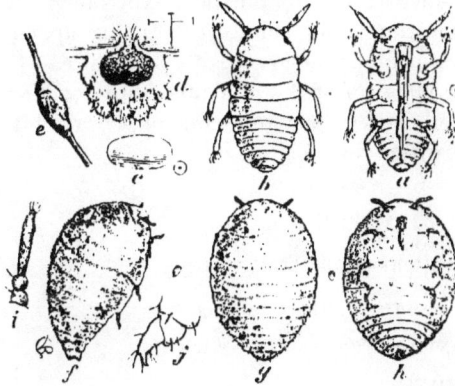

Fig. 153.— Grape-root Louse (g a l l - inhabiting form); *d,* a gall in the leaf cut open; *e,* a gall in the tendril; *c,* an egg, highly magnified—color, pale y e l l o w; *b,* back view of n e w l y hatched louse, enlarged; *a,* ventral view of the same, enlarged — color, yellow; *f, g,* and *h,* adult female lice—side view, back view, and ventral view, enlarged; *i,* an antenna—color, orange-yellow, greatly magnified; *j,* a foot, greatly magnified.

To this argument I only answer: Taking into consideration the general richness of the lands yet to be planted in this State, the visitation of this pest may be considered a benefit rather than a misfortune to the vine-growers who grow grapes for wine-making purposes. In regard to grapes grown for table use and raisin making, the objection to the use of fertilizers is untenable, as I know a very extensive grower of these varieties who has been very successful in maturing heavy crops of a choice quality by the use of fertilizers. A vine-grower residing within forty miles of Sacramento had thirty-nine acres, principally table grapes, which netted him a revenue of nearly six thousand dollars per year. The vines commenced to decay to such an extent that thirty-four acres were dug out and peach and apricot trees were planted in their place, and these are in full bearing at the present time—June, 1883. It was ascertained afterwards, by examining some of the roots left in the ground, that the grape-root louse was the cause of the decay of the vines. The few acres remaining were found to be infested. Every alternate row was taken up, and fertilizers used. The improvement in the appearance and bearing of the vines in two years was such that the grower expressed the opinion that had he known at the time his vines began to decay, what he has since learned by practical experiments, he might have saved his thirty-four acres of vines dug up, and also the loss of three or four years' crop.

REMEDIES.—If the vines infested are planted close together, say less than ten feet between the rows, dig out every third row, and use as a fertilizer No. 35.

CHAPTER CXVI.

The Cottony Grape Scale. (Cal.)

(*Lecanium vitis.*—Fitch.)

SYNONYM.—*Coccus vitis.*—Kirby.

Order, HEMIPTERA: Family, COCCIDÆ.
Sub-order, HOMOPTERA ;

[Infesting the grape vine: a scale insect which exudes a white cottony-like secretion in which its eggs are concealed.]

Fig. 155.

Fig. 154.

Fig. 154.—Cottony Grape Scale—color of the scales, reddish-brown ; of the cottony masses, white.

Fig. 155.—Cottony Maple Scale (*Pulvinaria innumeribilis, syn. Lecanium accricola, L. maclura, L. accricorticis*)—colors, brown and white.

This species is closely allied to the *Lecanium accricola* (Figs. 154 and 155), which infests the maple, and has been found in several parts of this State. It infests the foliage and branches of the grapevine.

NATURAL HISTORY.

The female is oval in form, and measures about two lines in length ; color, dark-brown ; near the posterior end are ridges or carinæ, and the lines separating them are darker than the other parts ; anal indentation similar to *L. hesperidum* or soft orange scale. The eggs are white when first laid, but change to a yellowish tinge before hatching ; oval in form, and about one ninetieth of an inch in length.

In one of the cottony-like masses attached to a female of this species, I have found nearly three hundred eggs. Larva—color, yellowish-white ; form, ovoid ; length, one ninetieth of an inch. Similar in appearance to the larva of the soft orange scale, except in color.

REMEDIES.—When the vine is dormant, use Nos. 14, 12, or 13, after scraping off the loose bark and burning it. In Summer use No. 4, or No. 3 if No. 4 cannot be had.

CHAPTER CXVII.

The Grape Scale.

(*Aspidiotus uvæ.*—Comstock.)

Order, HEMIPTERA ;   }
Sub-order, HOMOPTERA ; } Family. COCCIDÆ.

[A small circular scale, found on the trunk of the vine and wood left from growth of the previous year.]

I have only had one specimen of this species sent me, which

was found on imported vines. Professor Comstock reports it from Indiana, and describes it as infesting the lower part of the grapevines from the ground to the shoots of the second year's growth. In comparison with the red scale it is smaller, and the color is lighter, being a yellowish-brown.

REMEDIES.—When the vine is dormant, use Nos. 11, 12, or 13. In Summer use No. 1.

CHAPTER XCVIII.

Decaying Wood Borers.   (Cal.)

Order, COLEOPTERA; Family, PTINIDÆ.

[Grubs from three to five lines in length, found in decaying wood of fruit trees and grape-vines.]

Between the months of October, 1882, and March, 1883, several specimens of grubs were received that were found in decaying wood of fruit trees and grape-vines.

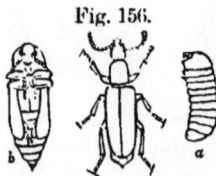

Fig. 156.

Fig. 156.—Decaying Wood Borers (*Bitoma opaculus.—*Lec.): *b*, the pupa—color, yellowish; *a*, the larva—color, white; in the middle, the beetle—color brown. Lives in dead grape canes.

I kept the specimens of larvæ (Fig. 156*a*) until they assumed the pupa form (Fig. 156*b*), but did not succeed in rearing the perfect insect (Fig. 156), as the insects died when casting the pupa skins; however, they were sufficiently matured that they could be identified as to family and genus. The specimens belonged to the genus *Lyctus* of Latreille (in Crotch's Check List, the *Bitoma* of Hubner). Those insects whose grubs are found boring in decayed wood are not injurious to trees of a healthy growth.

REMEDY.—Cut out and burn all decayed wood on vines.

CHAPTER XCIX.

The Achemon Sphinx.   (Cal.)

(*Philampelis achemon.*—Drury.)

Order, LEPIDOPTERA :   Family, SPHINGIDÆ.

[Feeding upon the leaves of the grapevine, a large green, yellowish or brown worm, with six narrow cream-colored spots on each side of the body.]

Fig. 157,

Fig. 157—Caterpillar of Achemon sphinx—colors, yellow and brown.

When fully grown (Fig. 157) it measures about three inches and six lines in length : it then deserts the vine and enters the

Fig. 158.

earth, where it forms a smooth cell in which to undergo its transformation.   (Fig. 158.)

Fig. 158.—Pupa of Achemon sphinx—color, brown.

The young worm is green, with a long slender horn on the hind end of the body : after casting its skin several times the horn disappears, and in its place is a polished black tubercle : the body is now a pale straw-color on the back, the sides brown and marked with six narrow scalloped cream-colored spots.

The fore-wings of the moth (Fig. 159) expand from three to four inches, are of a dark olive-gray color, marked with three dark olive spots : the basal part of the hind wings is roseate, followed by a dark stripe, next to which is the olive border. The body is fawn color, with two triangular olive colored spots on the thorax.

Fig. 159.—Achemon Sphinx—colors, gray, brown and pink.

Fig. 159.

REMEDIES.—Nos. 14, 100 and 101. (See letter of Mr. Blower, No. 33.)

CHAPTER C.

The Satellite Sphinx. (Cal.)

(*Philampelis satellitia.*—Linn.)

SYNONYM.—*P. pandorus.*—Hubner.

Order, LEPIDOPTERA; Family, SPHINGIDÆ.

[Feeding upon the leaves of the grapevine: a large brown worm, having five or six large cream-colored spots on each side of the body.]

The egg from which this worm hatches is deposited singly upon a leaf. When first hatched, and for some time afterwards, the worm is green, with a tinge of pink along the sides, and on the hind part of the body is a long curved horn. After the worm has cast its skin about three times, the horn disappears, and in its place is a polished, black tubercle; the color of the body is now a rich velvety brown, with five or six cream-yellow spots upon each side. When fully grown (Fig-

160) it measures three inches or more in length; it then enters the earth, and forms a smooth cell in which to undergo its transformations.

Fig. 160.

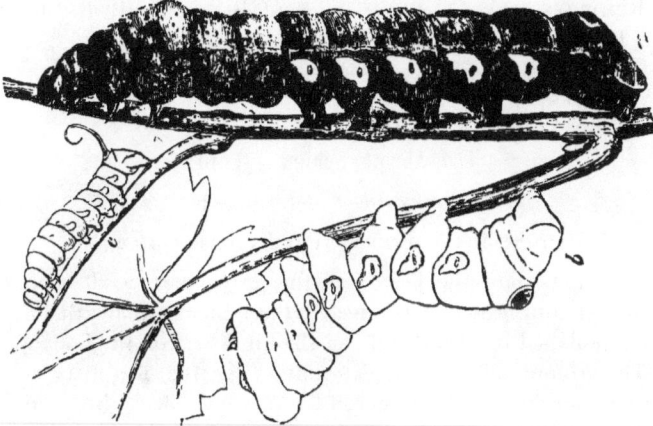

Fig. 160.—Caterpillars of Satellite Sphinx—colors, green or brownish and yellowish-white; *c*, the young caterpillar; *b*, the same nearly grown, at rest; *a*, the full grown caterpillar.

Fig. 161.

The moth (Fig. 161) into which this worm is finally transformed, expands from four to five inches; the fore-wings are pale olive-gray, marked with darker olive-green patches, one of which is at the tip, one near the outer hind angle, and a third on the hind margin near the body.

Fig. 161.—Satellite Sphinx—colors, gray and dark olive-green.

REMEDIES.—Use Nos. 14, 100, and 101. (See R. B. Blowers' letter, No. 33).

CHAPTER CI.

The Abbott Sphinx. (Cal.)

(Thyreus Abbotii—Swainson.)

Order, LEPIDOPTERA ; Family, SPHINGIDÆ.

[Feeding upon the leaves of the grapevine ; a checkered yellowish and brown worm, nearly three inches in length, having a polished black tubercle on the hind part of the body.]

The ground color of this caterpillar (Fig. 162) is a dirty yellowish, marked with numerous transverse black lines and black-brown patches, forming a line along each side of the body. The head is slightly roughened and dark, and is marked with a light streak, and there is another streak in the middle, which sometimes forms the letter X. When about to pupate, it deserts the vine and spins a cocoon beneath the leaves, etc.

Fig. 162.

Fig. 162.— Abbot Sphinx; upper figure, the caterpillar —colors, yellowish and black : lower figure, the moth—colors, grayish-brown and pale yellow.

The perfect moth(Fig. 162) has the outer edge of the wings deeply scalloped : the fore-wings are dull grayish-brown, variegated with dark brown; the hind wings are pale yellow, with a dark brown hind border.

REMEDIES.—Use Nos. 14, 100, and 101.

CHAPTER CII.

The Hog Caterpillar.

(*Chærocampa pampinatrix*—Abbott and Smith.)

SYNONYM—*Darapsa myron.*—Cramer.

Order, LEPIDOPTERA; Family, SPHINGIDÆ.

The measurements of insects in this work are given in inches and lines. The above cut represents one inch divided into lines and fractions thereof.

[Feeding upon the leaves of the grape-vines; a brown or green worm, with a light stripe on each side of the body and a curved horn on the posterior end.]

The eggs of this species are perfectly round, of a uniform yellowish-green color, and are glued, singly, to the underside of the leaves.

Fig. 163.—Hog Caterpillar—color, green and yellowish.

Fig. 163.

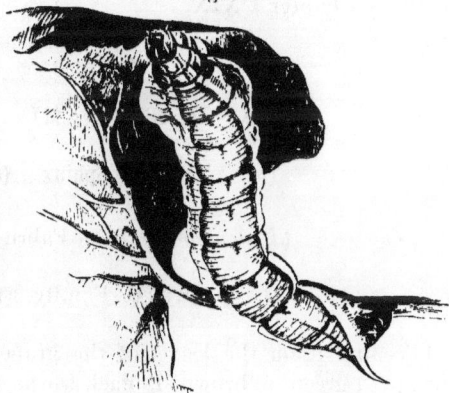

The full grown worm (Fig. 163) measures about one inch and six lines in length; is of a pale green or brown color, marked with a yellow stripe on each side of the body, and with a row of dark spots on the back. It receives its name of " Hog Caterpillar " from the forepart of the body being greatly swollen, thence tapering to the head, giving to this part a vague resemblance to the head of a fat hog. When about to

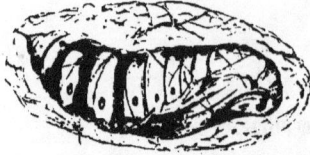

Fig. 164.

pupate it deserts the vine and spins an imperfect cocoon beneath the leaves, etc. (Pupa, Fig. 164.)

Fig. 164.—Pupa of Hog Caterpillar in its cocoon — color, brown.

The perfect moth (Fig. 165) expands about two inches and three lines, and has the body and fore-wings of a lilac gray

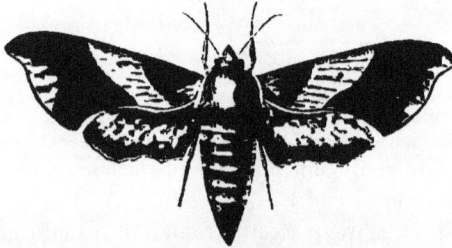

Fig. 165.

color, marked and shaded with olive green, while the hind wings are of a deep rust color, with a shade of gray near their inner angle.—Riley.

Fig. 165.—Moth of Hog Caterpillar—colors, gray, olive-green and rust-red.

Use the same remedies as recommended for the Achemon Sphinx—Chapter CXIX.

CHAPTER CIII.

The White-lined Sphinx.   (Cal.)

(*Deilephila lineata*—Fabricius.)

Order, LEPIDOPTERA: Family, SPHINGIDÆ.

[Feeding upon the leaves of the grape, apple, melon and turnip; a green or brownish-black worm, having a horn upon the hind end of the body.]

Fig. 166.

Fig. 166.—Caterpillar of White-lined Sphinx—colors, green and yellow.

This worm appears in two entirely different colorings; in the more common form (Fig. 166) it is yellowish-green, with a row of oval spots along each side of the back, which are usually connected by a yellow line; these spots consist of two curved black spots enclosing a yellow and a crimson spot.

Fig. 167.

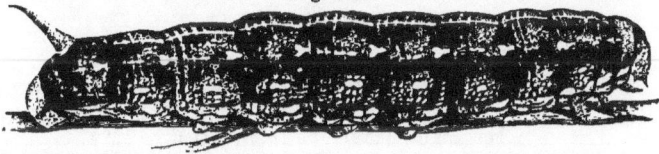

Fig. 167.—Caterpillar of White-lined Sphinx—colors, black and yellow.

In the other form (Fig. 167) the body is black, usually marked on the back with a yellow line, and with a number of yellow spots on each side of the body. When fully grown it measures from three to four inches in length; it then creeps beneath some leaves or other rubbish, sometimes even entering the earth, and forms a smooth cell in which to undergo its transformations.

The fore-wings of the moth (Fig. 168) expand about three inches, are of a dark olive-green color, with a whitish line extending from the base to the tip of each wing; the hind wings are black, with a rose-colored band extending lengthwise through the middle. This moth is quite common in this State, and may frequently be seen in the evening twilight hovering over the flowers like a humming-bird.

13

Fig. 168.

Fig. 168.—White-lined Sphinx—colors, olive-green, white, black and rose-colored.

REMEDIES.—Nos. 14, 100 and 101.

CHAPTER CIV.

The Eight-spotted Forester.   (Cal.)

(*Alypia octomaculata.*—Fabricius.)

Order, LEPIDOPTERA; Family, ZYGÆNIDÆ.

[Feeding upon the leaves of the grapevine; a bluish-white worm, dotted with black, and having on the middle of each segment a broad transverse orange band, on each side of which are four black rings.]

Fig. 169.—Eight-spotted Forester and Caterpillar: *c*, the moth—colors, blue-black and yellow; *a*, the caterpillar—colors, bluish-white, black and orange; *b*, one segment of its body, enlarged.

Fig. 196.

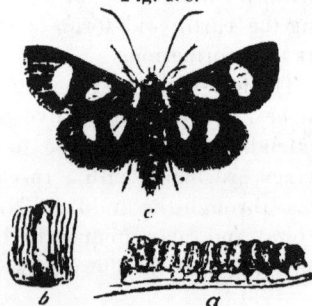

This worm when fully grown (Fig. 169*a*) measures one inch and four lines in length; it pupates within a slight cocoon spun beneath some leaves, or just beneath the surface of the ground. The fore-wings of the moth (Fig. 169c) are of a deep blue-black color, and each are marked with two light yellow spots; the hind wings are also blue-black, and are each marked with two white spots. The larvæ are found in the latter part of May. One specimen was received about May 27th, and one on the 3d of June, 1882; both specimens were nearly full grown.

REMEDIES.—Where vines are seriously infested, spray with No. 5 or 7, with an equal number of gallons of No. 9 added; also No. 101. The larva will creep under chips, etc., laid on the ground under the vine, to pupate, and can be gathered in the Autumn.

CHAPTER CV.

The Beautiful Wood Nymph.

(*Eudryas grata.*—Fabricius.)

Order, LEPIDOPTERA ; Family, ZYGÆNIDÆ.

[Feeding upon the leaves of the grapevine ; a bluish-white caterpillar dotted with black, and marked on the middle of each segment with a transverse orange band, on each side of which are three black rings.]

Fig. 170.—Caterpillar and eggs of the Beautiful Wood Nymph ; *a*, the caterpillar—colors, bluish, black and orange ; *b*, a segment of its body enlarged ; *d*, back view of hump on eleventh segment enlarged : *c*, back view

Fig. 170.

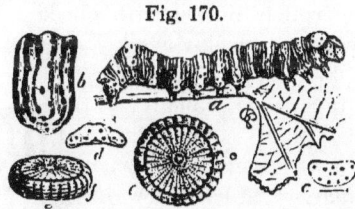

of the top of first segment enlarged ; *f*, side view of an egg enlarged (natural size indicated beneath) ; *e*, an egg as seen from above enlarged (natural size indicated at the right)— colors, yellowish and black.

The full grown larva (Fig. 170) measures about one inch and six lines in length ; it then deserts the vines and burrows a short distance into a piece of soft wood, or enters the earth and forms a cell in which to pass the pupa state.

Fig. 171.

Fig. 171.—Beautiful Wood Nymph (female moth)—colors, white, yellow, brown and olive-green.

The fore wings of the moth (Fig. 171) expand about one inch and six lines, and are of a white color, broadly bordered and marked with reddish-brown and olive-green; the hind wings are pale yellow, broadly marked with pale brown on the hind border.—Riley.

I have not found this species in this State, although moths of a similar description have been reported.

REMEDIES.—Same as recommended for the Eight-spotted Forester, Chapter CIV.

CHAPTER CVI.

The Grapevine Epimenis.   (Cal.)

(*Psychomorpha epimenis.*—Drury.)

Order, LEPIDOPTERA; Family, ZYGÆNIDÆ.

[Feeding upon the leaves of the grapevine: a bluish-white caterpillar, dotted with black and marked on each segment with four black rings.]

Fig. 172.—Caterpillar of Grapevine Epimenis: *a,* the caterpillar—colors, white and black; *b,* segment of its body, enlarged; *c,* back view of the top of the eleventh segment,

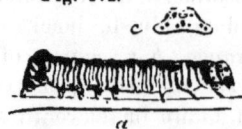

Fig. 172.

enlarged. This caterpillar (Fig. 172) lives in a sort of nest formed by fastening several leaves together with silken threads. When fully grown, it measures about one inch in length; it then deserts the vine and bores into wood or other sufficiently soft substance, and forms a cell in which to undergo its transformations.

Fig. 173. — Grapevine Epimenis (male moth)—colors, black, white, and red.

Fig. 173.

The fore-wings of the moth (Fig. 173) expand a little over an inch, are of a velvety black color, marked with blue; and a little beyond the middle of each is a large yellowish spot, which on the hind wings is of a deep orange color. I have found the nest and caterpillar as described early in May, but failed to rear the perfect insect.

REMEDIES.—Use No. 24, and the same as recommended for the Eight-spotted Forester. Chap. CIV.

CHAPTER CVII.

The American Procris.

(*Procris Americana.*—Bois.)

Order, LEPIDOPTERA : Family, ZYGÆNIDÆ.

[Feeding in companies upon the leaves of the grapevine : a small yellowish caterpillar, usually having small tufts of black hairs on the body.]

Fig. 174. — American Procris Caterpillars on a leaf—colors, yellow and black.

Fig. 174.

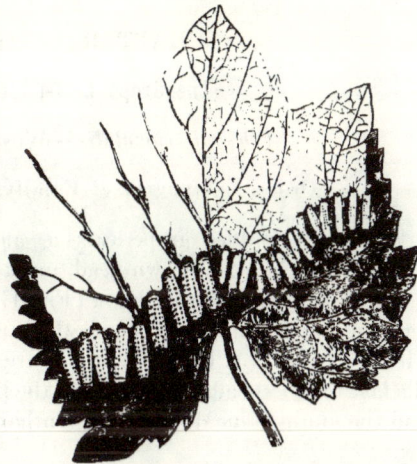

These caterpillars (Fig. 174) arrange themselves side by side, their heads all pointing in one direction. When young, they eat only the surface of the leaves, but when they get stronger, they devour all the leaf excepting the larger veins.

The full grown caterpillar (Fig. 175a) measures about six lines in length, is of a pale yellow, with a row of black prickly tufts on each segment: the first segment is black, with a yellow edge, and the head is brown. When about to pupate they desert the plants and spin in some sheltered place their tough, flattened, whitish cocoons (Fig. 175e). The perfect moth (Fig. 175, d and e,) expands a little under an inch, and is wholly black except the deep orange collar. I have not found this species in this State, although its presence here has been reported. The above description is taken from Riley's Second Missouri Report.

Fig. 175.

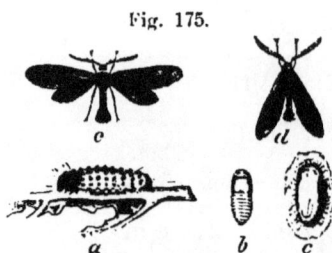

Fig. 175.—American Procris: c, the moth, with its wings expanded: d, the same, with its wings closed —colors, black and orange; a, the caterpillar — colors, yellow and black: b, the pupa—color, brown; e, the cocoon—color, whitish.

REMEDIES.—Use No. 5 or 7, mixed with equal quantity of No. 9; on non-bearing or young vines, use No. 10 or No. 103.

CHAPTER CVIII.

The Grape Leaf-folder.

(*Desmia maculalis.*—Westwood.)

Order, LEPIDOPTERA; Family, PYRALIDÆ.

[Living in a folded grape leaf; a pale-green larva or caterpillar, with a reddish-brown head, marked with darker spots.]

When fully grown this larva (Fig. 176, 1,) measures about ten lines in length. It assumes the pupa form (Fig. 176, 3,) within its nest. At least two broods are produced in one year: the last brood spends the Winter in the pupa or chrysalis state, and the moths issue during the month of May of the following year.

The perfect insect or moth (Fig. 176, *4* and *5*,) is of a black color, the fore-wings marked with two white spots and the hind wings with only one white spot—but this is sometimes divided into two spots; all of the wings are bordered with white.

Fig. 176.

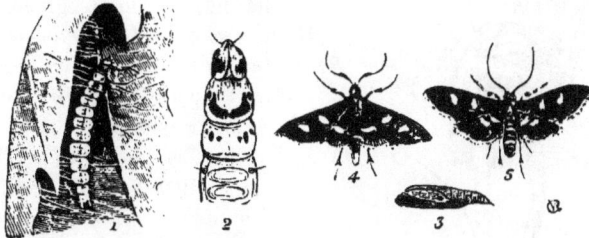

Fig. 176.—Grape Leaf-folder; *1*, the caterpillar—color, green or yellowish; *2*, the head and fore part of its body, enlarged; *3*, the pupa—color, brown; *4*, the male moth—colors, black and white; *5*, the female moth—colors, black and white.

REMEDY.—Use No. 24; where seriously infested use No. 5 or 7, with an equal quantity of No. 9.

CHAPTER CIX.

The Grapevine Plume. (Cal.)

(*Pterophorus periscelidactylus.*—Fitch.)

Order, LEPIDOPTERA; Family, PTEROPHORIDÆ.

[Living in a rolled grape leaf, upon which it feeds; a small pale greenish caterpillar, about six lines long, with numerous spreading clusters of whitish hairs.]

This caterpillar (Fig. 177*a*) usually fastens the opposite edges of a leaf together with silken threads, but it sometimes fastens several leaves together, forming a large cavity in which it resides. When about to pupate—which is about the tenth of May—it suspends itself by the hind feet. The pupa (Fig. 177*b*) is angular, and the anterior end appears as if it had been obliquely cut off; on the middle of the back are two angular projections.

Fig. 177.

Fig. 177.—Grapevine Plume: *d*, the moth—colors, pale yellow, white, and brown; *a*. the caterpillars in their nests — color, greenish-yellow, with white hairs; *b*, the pupa — color, green or brown; *c*. one of the horns on the back of the pupa, enlarged; *e*, one of the leg-bearing segments of the larva, enlarged.

The perfect moth (Fig. 177*d*) expands a little over nine lines, is of a tawny yellow color, and each fore-wing is marked with about five white spots; these wings are cleft or cut nearly to the middle; the hind wings are cleft twice, the forward cleft reaching nearly to the middle, and the second cleft reaches nearly to the base of the wing; they are rusty brown at the base, with the remaining part tawny yellow. I have found the perfect insect; also, a rolled leaf on vines.

REMEDY.—Use No. 24. Spray in the latter part of April, or early in May, with No. 5 or 7.

CHAPTER CX.

The Grapevine, or Steel-blue Flea-beetle. (Cal.)

(*Haltica chalybea.*—Illiger.)

Order, COLEOPTERA; Family, CHRYSOMELIDÆ.

[Feeding upon the buds or leaves of the grape and willow; an elongate brownish six-legged larva, marked with black dots and with a black head; finally changing to a small blue-black beetle, about two lines in length.]

Fig. 178.

Fig. 178.—Grape-vine Flea-beetle and Larvæ; *a*, several larvæ on a leaf; *b*, a larva enlarged—colors, brown and black; *c*, the cocoon; *d*, the beetle enlarged—color, steel-blue.

These beetles have the hind thighs greatly enlarged, which enables them to leap to a considerable distance, like a flea; it is from this peculiarity that they take the name of "flea-beetles." They spend the Winter in some sheltered situation, and come forth early in the following Spring and feed upon the buds of grapevines, usually hollowing them out. As soon as the leaves are expanded the females deposit their eggs upon them; in a few days the young are hatched, and immediately begin to feed upon the leaves. When fully grown (Fig. 178*b*) they descend to the ground, which they enter, and form small cells (Fig. 178*c*) in which to pass the pupa state. These insects are very destructive to grapevines, and every effort should be made to exterminate them.

REMEDIES.—No. 19, 20, 21 or 102. On young non-bearing grapevines, etc., use No. 103, or No. 10 may be used instead.

CHAPTER CXI.

The Imported Grape Flea-beetle.   (Cal.)

(*Adoxis vitis.*—Linnaeus.)

Order, COLEOPTERA : Family, CHRYSOMELIDÆ.

[A small dark-brown beetle, feeding upon the leaves of the grapevine.]

This species was reported in 1882, from at least six of the vine-growing districts, as damaging the foliage of grapevines, and is reported this Spring (1883) to be destroying the vines infested. This is an enemy of the grapevine, which must be eradicated. Their mode of attack on the vines is similar to that of the steel-blue or grape flea-beetle (*Haltica chalybea*), and they frequently damage young vines to such an extent that they die. One vineyard in this vicinity (Sacramento) has been damaged seriously this Spring. The perfect insect (Fig. 179, Plate 3,) is black, with a sub-metallic luster of a greenish hue, but after emerging from the pupa state is of a reddish-brown color. It is ovate and convex, and measures from two lines to two and one quarter lines in length : antennæ, eleven jointed. This insect, although known to be widely destructive on the Pacific Coast and in the mountainous regions of the Atlantic States, had not been reported as being very injurious in this State until 1882. In Europe it is well known as a dangerous enemy to the grapevines, especially in France, where it is called *C. gribourier*, and where much has been written in regard to its destructive habits.

REMEDIES.—Nos. 19, 20, 21 and 102. On young non-bearing grapevines Nos. 103 and 10 may be used.

CHAPTER CXII.

The Rose Chafer. (Cal.)

(Macrodactylus subspinosus.—Fabr.)

Order, COLEOPTERA; Family, SCARABÆIDÆ.

[Feeding upon the leaves of the grape, rose, apple, cherry, etc.; a slender brownish-yellow beetle, about four lines long.]

This insect is a very general feeder, and at times becomes very injurious. During its larva stage it lives in the ground, feeding upon the roots of various kinds of plants.

The eggs laid by each female are about thirty in number, and are deposited from one to four inches below the surface of the earth.

The larvæ lie upon their sides, their bodies being curved so that the head and tail nearly meet each other. They are of a yellowish-white color, with the head pale reddish. They pass the Winter in their earthen cells, and assume the pupa form the following Spring, there being but one annual brood.

Fig. 180.

Fig. 180.—Rose Chafer—color, brownish-yellow.

The beetles, or chafers (Fig. 180), as they are commonly called, measure about four lines in length; the thorax is produced into a small point on each side of the middle; the head and thorax are black, and the wing-cases brown, but the entire insect is colored with yellowish scales which give it a brownish-yellow appearance.

REMEDIES.—Nos. 19, 20, 21, 102; or No. 103 or No. 10 on non-bearing vines or plants.

CHAPTER CXIII.

The Spotted Pelidnota.

(Pelidnota punctata.—Linnæus.)

Order, COLEOPTERA ; Family, SCARABÆIDÆ.

[The measurements of insects in this work are given in inches and lines. The above cut represents one inch divided into lines and fractions thereof.]

[Feeding upon the leaves of the grape vine : a robust clay-yellow beetle, about one inch long, marked with a black spot on each side of the thorax, and with three black spots on each wing-cover.]

Fig. 181.—Spotted Pelidnota, Larva and Pupa ; *c*, the beetle—colors, yellowish and black ; *a*, the grub—color, white ; *d*, the tip of its body : *e*, one of its antennæ, enlarged ; *f*, one of its legs, enlarged ; *b*, the pupa in its earthen cell—color, brown.

Fig. 181.

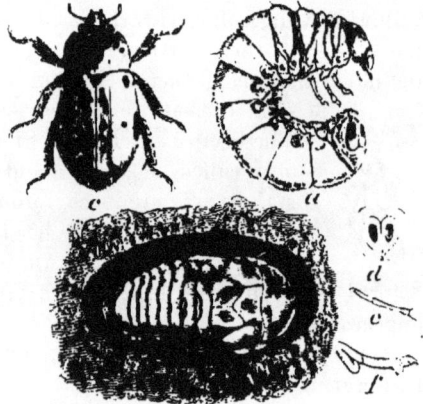

These beetles (Fig. 181*c*) sometimes occur in destructive numbers upon the leaves of the cultivated grapevine, but only in limited localities.

The larvæ or grubs (Fig. 181*a*) closely resemble the white grubs, but differ in having a heart-shaped swelling at the hind end of the body.

REMEDIES.—Use Nos. 19, 20, 21, 102 ; use Nos. 103 and 10 on non-bearing trees and vines.

CHAPTER CXIV.

The California Grapevine Hopper.   (Cal.)

(*Erythroneura comes.*—Say.)

Order, HEMIPTERA :   ⎱ Family, CERCOPIDÆ.
Sub-order, HOMOPTERA ⎰

[Infesting the leaves of the grapevine, which they puncture
with their beaks and imbibe the sap, causing the leaves to turn
yellow prematurely ; a yellowish-white vine hopper marked
with orange-red.]

The perfect insects (Fig. 182, Plate 3,) are yellowish-white,
with oblique confluent orange-red bands on the wing-cases,
and a short oblique line on the middle of the outer margin of
each ; on the thorax are usually three red stripes, the middle
one forked anteriorly and confluent with two red stripes on
the crown of the head ; length, about one line and a half. The
scutellum is large, triangular and marked with three orange
spots which form a triangle ; the two anterior spots are sometimes black. At the outer fore-angle of the wing-cases is a
black dot, and there is another black dot at the outer hind
angle ; the posterior tibiæ are densely spined, and there is a
row of spines on the anterior tibiæ.

These insects hibernate in the perfect state around the vines,
etc., and come forth in the following Spring and feed upon the
new growth of vegetation. The females deposit their eggs
upon the leaves of the grapevine as soon as the latter are
expanded.   The young vine-hoppers closely resemble the
adults in form, but are destitute of wings. They cast their
skins several times before arriving at maturity, and these
white cast off skins may frequently be found adhering to the
leaves or scattered upon the ground beneath them. I have
first observed the perfect insect in the month of July.

These insects puncture the leaves of the vine and extract
the sap to such an extent as to cause the leaves to turn yellow
prematurely, preventing the canes from ripening, and when the
grapes, picked off of the invested vines, are used for shipping
to the Eastern States they are liable to mildew.

REMEDIES.—Nos. 32 and 33.   (See Nos. 19, 20, 21 and 103.)

CHAPTER CXV.

The Grapevine Aphis.

(*Siphonophora viticola.*—Thomas.)

Order, HEMIPTERA ;   } Family, APHIDIDÆ.
Sub-order, HOMOPTERA ; }

[Living upon the leaves and twigs of grapevines, which they puncture with their beaks and extract the sap; small brownish plant lice.]

The wingless females are dusky brown ; the legs and honey tubes are black ; the latter are about one fourth as long as the body. The winged lice are colored similar to the wingless ones, but are darker.—Professor Thomas.

REMEDIES.—Use No. 5 or 7 ; No. 4 will also be effectual, or No. 64.

CHAPTER CXVI.

The False Chinch Bug.   (Cal.)

(*Nysius destructor.*—Say)

Order, HEMIPTERA ;   } Family, LYGÆIDÆ.
Sub-order, HOMOPTERA ; }

[A small grayish-brown bug, feeding on the foliage of the grapevine, etc.]

In 1882 specimens of the larva and pupa (Fig. 183*b*) of this species were sent me, accompanied by a note stating that these insects were destroying the foliage of grapevines (Fig. 183*a*), and in two or three weeks later specimens of the perfect insect were received. The first letter received with these specimens stated that vines on about five rods square of ground were attacked ; ten days later nearly two acres were infested before the insects were checked from spreading.

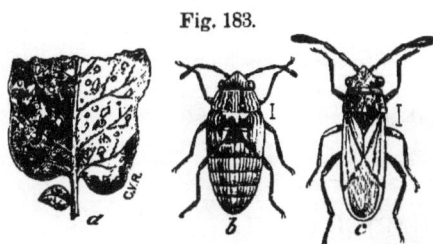

Fig. 183.

Fig. 183. — False Chinch Bug ; *a*, a leaf, showing its work ; *b*, a pupa— colors, dingy yellow, red, and brown ; *c*, the adult — colors, grayish-brown and black.

The perfect insect (Fig. 183*c*) is of a grayish-brown color ; the wing covers (hemelytra) are nearly transparent, and generally colorless ; the legs are yellowish, inclining to brown ; length, one and one half lines—in some specimens a little over. The pupa is nearly of the same color as the perfect insect, excepting that the longitudinal lines are brighter ; in color, red and brown ; wing-pads visible. The larva is of a brownish-yellow color.

This plant bug is reported as feeding on the potato, cabbage, etc., but has only been reported on grapevines in this State.

REMEDIES.—Use Nos. 36 and 51 ; or No. 65. (See No. 20.)

CHAPTER CXVII

The Yellow Mite. (Cal.)

(*Tetranychus Sp.?*)

Class, ARACHNIDA ; } Family, TROMBIDIDÆ.
Sub-class, ACARINA ;

[A small yellow mite infesting the branches and foliage of deciduous fruit trees, grapevines and nursery stock.]

As this species was only brought to notice last year (1882), I cannot give any particulars as to its natural history. The color is immaculate primrose-yellow, excepting that the male (Fig. 184, Plate 3,) has two bright minute vermillion-red spots on the anterior portion of the body (or on the shoulders). The female (Fig. 185, Plate 3,) is about one seventieth of an inch in length ; form, ovate ; the male is smaller.

They are very destructive to the foliage of fruit trees, grape-vines and nursery stock, and appear to spin more than the red mite. Mr. W. B. West, of Stockton, has found them on forest trees. Mr. Williams, of Fresno, reports them on weeds growing on uncultivated land.

REMEDY.—Use No. 34.

CHAPTER CXVIII.

The Angular-winged Katydid.   (Cal.)

(*Microcentrum retinervis*—Burmeister.)

Order, ORTHOPTERA; Family, LOCUSTIDÆ.

[Feeding upon the leaves of the orange, apple, cherry and grapevine: a large green grasshopper.]

The following account of the manner in which this insect deposits her eggs is condensed from the excellent history of this species given by Professor Riley :

"Selecting a twig about the size of a common goose-quill, this provident mother prepares it for the reception of her eggs by biting and roughening the bark with her jaws for a distance of two or three inches. When this operation is accomplished to her satisfaction, she commences at one end of the roughened portion of the twig and, after fretting it anew with her jaws, and feeling it over and over again with her palpi as if to assure herself that all is as it should be, she curls the abdomen under until the lower edge of the curved ovipositor is brought between the jaws and palpi, by which it is grasped and guided to the right position. It is then gently worked up and down for from four to six minutes, while a viscid fluid is given out apparently from the ovipositor. Finally the egg gradually rises and adheres to the roughened bark ; the insect now rests for a few minutes, soon to resume her efforts and repeat the like performance in every particular, except that the egg is placed to one side, and a little above the first."

Fig. 186.

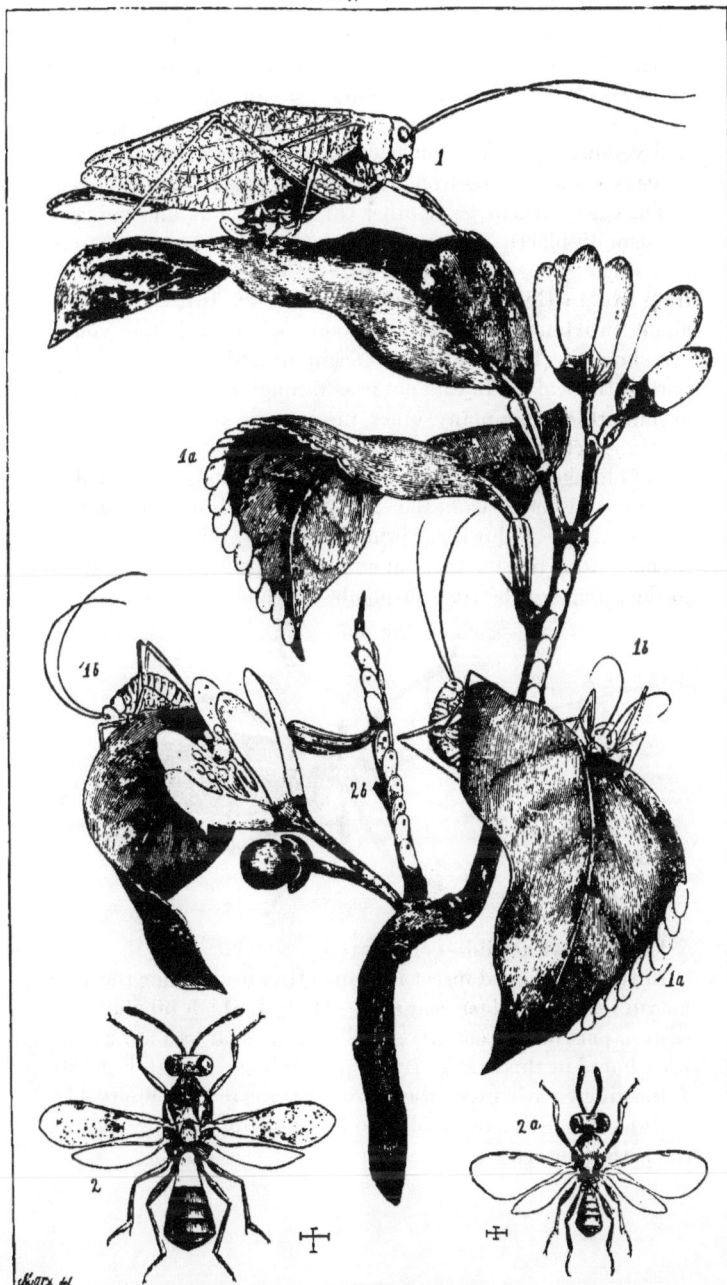

Fig. 186.—Angular-winged Katydid; *1*, the adult—color,
green; *1a*, the eggs—color, brownish or slate color; *1b*, the
young—colors, green and yellow; *2*, an egg parasite (the
Eupelurus mirabilis of Walsh) female, enlarged—colors, black
and yellowish; *2a*, the male, enlarged—color, metallic-green;
2b, eggs, showing holes from which these parasites had emerged.

The eggs overlap each other (Fig. 186, *2b*,) at one end and
are usually placed in two rows, side by side, but more rarely
in a single row. The number of eggs laid at one time varies
from two to thirty, and each female (Fig. 161, *1*.) deposits
about two hundred eggs. As soon as hatched, the young
grasshoppers (Fig. 186, *1b, 1b*,) begin to feed upon the pulpy
part of the leaf. They do not pass through a quiet pupa state,
as butterflies and many other insects are known to do, but
continue active from the time they leave the egg until they
die of old age or some other cause. Two broods are probably
produced in one season, the eggs of the last brood not hatching until the following Spring. Should this species ever
become numerous, a great amount of damage would be done
to the foliage of the trees or plants on which they feed.

Fig. 187.

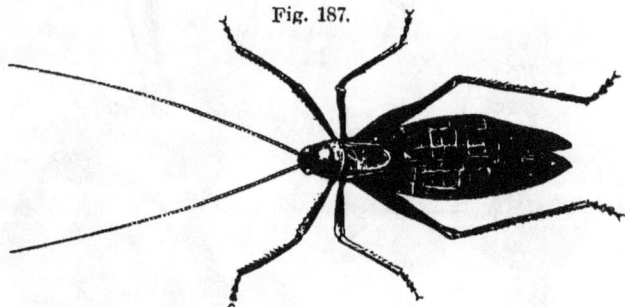

Fig. 187.—Katydid—color, green.

NOTE.—The above insect is frequently mistaken for the true
katydid (*Platyphyllum concavum*—Harris) which inhabits the
eastern part of this country, but has not, so far as I am aware,
been found in this State. By comparing Fig. 187 with Fig. 186,
1, the difference between these two species is readily observable.

REMEDIES.—No. 18; and also by capturing and destroying
the perfect insects.

CHAPTER CXIX.

The Snowy Tree-cricket.

(*Œcanthus niveus.*—Harris.)

Order, ORTHOPTERA : Family, GRYLLIDÆ.

[Puncturing the tender twigs of the grapevine, apple and peach trees, raspberry and blackberry bushes, and depositing her eggs therein ; a greenish-white cricket.]

Fig. 188.

Fig. 188.—Snowy Tree-cricket, female—color, greenish white.

In depositing her eggs, the female cricket (Fig. 188) first uses her jaws for the purpose of slightly tearing away the outer bark ; the ovipositor is next inserted into the twig, and an egg thrust into the puncture thus made.

Fig. 189.—Eggs of Snowy Tree-cricket ; *a*, the egg punctures in a twig ; *b*, the eggs exposed — color, pale yellow ; *c*, an egg highly magnified ; *d*, upper end of the same, still more highly magnified.

Fig. 189.

These eggs (Fig. 189) are usually placed diagonally across the central pith. In this way the female proceeds until her stock of over two hundred eggs is exhausted. These punctures are frequently made within four lines of each other, and extend in an irregular row a distance of one foot and six lines or over.—Riley.

Fig. 190.

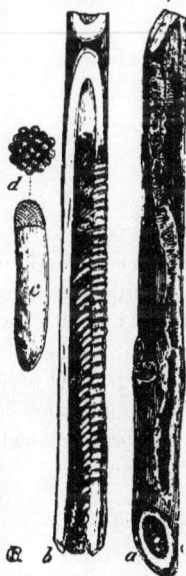

Fig. 190.—Snowy Tree-cricket, male—color, greenish white.

When first hatched, the young cricket feeds upon plant-lice, eggs, etc., and has even been known to attack and devour one of its own kindred. After acquiring wings they sometimes do considerable mischief by gnawing off the stalks of green grapes, permitting the latter to fall to the ground. The eggs are usually deposited in the latter part of Summer or early in the Autumn, and these do not hatch out until the following Spring.

REMEDY.—Use No. 28.

———

CHAPTER CXX.

The Grape Curculio.

(*Caliodes inaqualis.*—Say.)

Order, COLEOPTERA; Family, CURCULIONIDÆ.

[Living in grapes, a whitish or bluish footless grub about two lines long, with a brownish head; when fully grown deserting the fruit and entering the earth to pupate.]

Fig. 191.

Fig.1 b

Fig.1

Fig.1 a

Fig. 191.—Grape Curculio enlarged—color, grayish-black: *a*, one of its fore legs; *b*, its larva—color, yellow-ish-white.

The female curlio (Fig. 191, 1,) excavates a small cavity in the grape (Fig. 192*a*). and then deposits therein a single egg of a bright yellow color. The grub (Fig. 191*b*) which hatches from this egg feeds upon the pulp or flesh of the grape, and rarely upon the seeds. When fully grown (Fig. 192*b*) it is about two lines long, and each segment of its body bears on each side a large fleshy tubercle.

Fig. 192.

Fig. 192.—Grape Curculio; *a*, an infested grape; *b*, the larva enlarged—color, white.

When about to assume the pupa form the larva deserts the fruit, which sometimes drops from

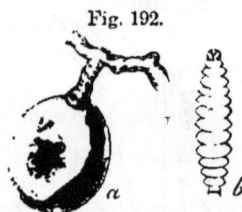

the vines previous to this, and enters the earth to the depth of a few inches, where it forms a small cavity in which it shortly afterwards assumes the pupa form. The beetle issues in the Fall, and passes the Winter in some sheltered place. The perfect beetle is nearly hemispherical in outline, and of a black color, but is covered with short appressed scale-like white hairs, which give it a grayish tinge; the legs are reddish, and on the upper and outer edge of each fore and middle shank (tibia) is a rectangular tooth. The body measures about one line in length. I have not found this species in this State.

CHAPTER CXXI.

The Grape-seed Maggot. (Cal.)

(*Isosoma vitis.*—Saunders.)

Order, HYMENOPTERA; Family, CHALCIDIDÆ.

[Living within the seeds of grapes; a minute footless maggot.]

This maggot (Fig. 193) burrows into the grape while the covering of the seeds are still soft and tender; it makes its way directly to the seed, which it enters and feeds upon the kernel. It undergoes its transformations within the seed, and the fly, when about to issue, gnaws its way out.

Fig. 193. — Grape-seed Maggot — color, whitish.

Fig. 193.

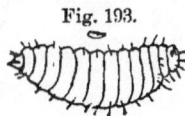

The perfect fly is black, and the forewings expand about one line. Only one brood is usually produced in a year, and these pass the Winter in the larva state.

Mr. Charles A. Wetmore, Chief Executive Viticultural Officer, called my attention to the larva found in the seed of the California wild grape, from which I bred the fly, and found it to be the *Isosoma vitis.* I do not know of any cultivated grape being infested by this pest, but close attention should be given to examinations lest it gains a foothold.

REMEDIES.—Should this pest spread to the cultivated varieties, all grapes should be picked off the vines at the end of the season, and if not fit for any use, destroyed; also, all seed from wine presses, etc., should be destroyed.

CHAPTER CXXII.

The Grape Leaf-roller.   (Cal.)

(*Tortrix Sp?*)

Order, LEPIDOPTERA; Family, TORTRICIDÆ.

[Living in a rolled grape leaf: a small greenish caterpillar.]

Last season (1882) specimens of grape leaves rolled up and fastened with silken threads, as in Figs. 194 and 195, were

Fig. 194.   Fig. 195.

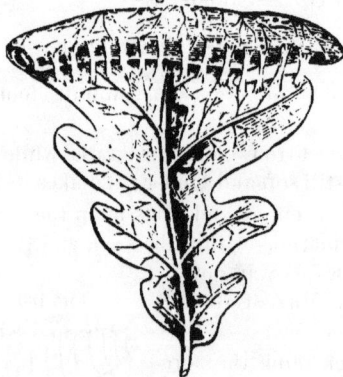

sent me from Fresno County. As I did not succeed in breeding the perfect insect, I am unable to give the name of the species to which this leaf-roller belongs; but, judging from the account given by Miss Ormerod and others of the oak leaf-roller (*Tortrix viridana*—Stephens), it is evident that our species is closely related to it.

REMEDY.—Use No. 24.

CHAPTER CXXII½.

The Red-shouldered Grapevine Borer.

(*Sinoxylon basillare.*—Say.)

Order, COLEOPTERA; Family, PTINIDÆ.

[Living in the canes of grapevines; a small yellowish six-legged larva or grub, finally changing into a black beetle which is usually marked with a red spot at the base of each wing-cover.]

This grub (Fig. 195½*a*) lives in the canes of grapevines, boring the wood beneath the bark, and also the heart-wood, in various directions. It also lives in the trunks of apple, peach and hickory trees.

The pupa form (Fig. 195½*b*) is assumed in the burrow.

The beetle (Fig. 195½*c*) measures about two and a half lines in length, is of a black color, and there is usually a large reddish spot at the base of each wing-cover; the thorax is armed with short spines in front, and the wing-cases are armed with several small teeth at the apex or tip.

Fig. 195½.

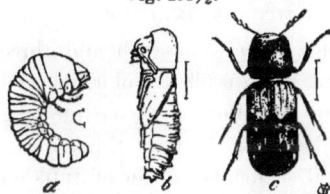

Fig. 195½. — Red-shouldered Grapevine Borer, enlarged; *a*, the larva—color, yellowish; *b*, the pupa — color, yellowish; *c*, t h e beetle—color, b l a c k and reddish.

REMEDIES.—Use Nos. 26 and 27.

INSECTS INFESTING THE RASPBERRY.

CHAPTER CXXIII.

The Raspberry-root Borer.   (Cal.)

(*Egeria rubi.*—Riley.)

Order, LEPIDOPTERA;   Family, ÆGERIDÆ.

[Living in the stems and roots of raspberry and blackberry
bushes ; a pale yellow sixteen-legged larva.]

Fig. 196.

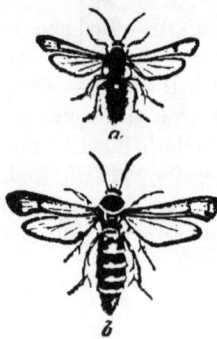

a.

b.

Fig. 196. — Raspberry - root Borer—
colors, brown, black, and yellow : *a*, the
male moth ; *b*, the female.

The perfect insect (Fig. 196) expands
from one inch to one inch and three
lines, and is of a black color, marked
with gray, as follows : A narrow ring
around the neck, the hind third of the
abdominal segments ; a row of tufts on
the back, and another row along each
side of the abdomen, besides a few
streaks on the thorax.

The eggs from which these borers hatch are deposited upon

the bushes at a distance of from four to six inches from the ground. As soon as hatched, the young borer enters the stem to the pith, and then directs its course downward to the roots, which it reaches at the approach of Winter. Here it remains until the following Spring, when it directs it course upward, burrowing out a different stem than the one by which it had entered the roots. After attaining its full growth it prepares a place of exit, and soon afterward assumes the pupa form.

I have only found one specimen of the larva of this species and have had specimens of roots sent me from which the borer had matured.

REMEDY.—Use No. 37 in the latter part of June and early in July: or No. 5 or 7.

CHAPTER CXXIV.

The Raspberry-root Gall-fly.

(*Rhodites radicum.*—Osten Sacken.)

Order, HYMENOPTERA ; Family, CYNIPIDÆ.

[Living in a large swelling on the roots of raspberry bushes ; small whitish footless grubs.]

If one of these galls were to be cut open, it would be found to be composed of a yellowish pithy substance, and scattered through it are a number of small cells, each containing a small white larva or grub. These soon change to pupæ, which in turn produce the perfect flies; and the latter gnaw their way out of the gall, leaving small holes to mark their places of exit.—Condensed from Saunders.

CHAPTER CXXV.

The Raspberry Borer.

(*Agrilus ruficollis.*—Fabricius.)

Order, COLEOPTERA ; Family, BUPRESTIDÆ.

[Living in the stems of raspberry and blackberry bushes, producing a gall-like swelling; a pale yellow footless grub, which is finally transformed into a slender blackish beetle, with the head and thorax of a brilliant copper color.]

The location of these borers in the canes may easily be detected by the gall-like swelling in the canes; this swelling is about an inch long, and is very rough as compared with the rest of the cane. Occasionally several larvæ will be found inhabiting the same gall, thus lengthening the latter and causing it to assume a very irregular shape.

Fig. 197.—Raspberry Borer, enlarged; *b*, the larva—color, yellowish; *a*, the hind part of its
body still more enlarged; *c*, the
beetle — colors, black and cop-
pery-red.

When fully grown, the larva
(Fig. 197*b*) measures about seven
lines in length, and is of a yel-

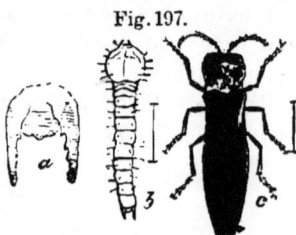

Fig. 197.

lowish-white color; the first segment behind the head is greatly dilated on each side, and the last segment is armed with two dark brown horns, each with three teeth on the inner edge. It assumes the pupa form within the gall in April or May, and the perfect beetle (Fig. 197*c*) issues in the course of a few weeks.

This species has been reported to me as infesting the black-berry in this State, but I have not seen any specimens.

REMEDIES.—Use No. 27, 28, 37, or 67.

CHAPTER CXVI.

The Raspberry Spanworm.

(*Aplodes rubivora.*—Riley.)

Order, LEPIDOPTERA; Family, PHALÆNIDÆ.

[Feeding upon the leaves and fruit of the raspberry and blackberry; a yellowish ten-legged span-worm, thinly covered with small spines, on which are usually affixed small pieces of leaves.]

Fig. 198.

Fig. 198. — Rasp-
berry Spanworm and
Moth; *a*, the worm—
color. yellowish-
gray; *b*, a segment
of its body greatly
enlarged; *c*, the
moth—colors, green
and white; *d*, two of
its wings enlarged.

Before pupating,
this worm (Fig.198*a*)
spins a loose cocoon
in which to undergo
its transforma-
tions.

The perfect moth (Fig. 198*c*) expands about six lines, and
is of a delicate green color, and all of the wings are crossed by
two curved light colored lines.—Riley.

REMEDIES.—When the caterpillars appear, spray with No.
83 or No. 85.

CHAPTER CXXVII.

The Raspberry Leaf-roller.

(*Exartema permundana.*—Clemens.)

SYNONYM.—*Eccopsis permundana.*—Clemens.

Order, LEPIDOPTERA; Family, TORTRICIDÆ.

[Living within a rolled leaf on raspberry, strawberry and
various other plants; a greenish worm with a black head.]

Besides rolling up and devouring the leaves, this leaf-roller
sometimes fastens several blossoms together with silken
threads, afterwards feeding upon them. It has nearly the
same habits as the Greater Leaf-roller (Chapter XL.)

Fig. 199.—Raspberry Leaf-roller, natural size and enlarged—colors, yellowish
or brownish.

Fig. 199.

The perfect moth (Fig. 199) has the
fore-wings yellowish, varied with brown
streaks and patches. The caterpillar
appears about the time the berry is in
bloom. I have found the moth of this
species, and also the nest of the larva, but have never found
the larva.

REMEDIES.—(See Remedies, Chapter XL.)

CHAPTER CXXVIII.

The Raspberry Aphis.

(*Siphonophora rubi.*—Kaltenbach.)

Order, HEMIPTERA ;
Sub-order, HOMOPTERA : } Family, APHIDIDÆ.

[Living on the stems and leaves of blackberry and raspberry bushes, which they puncture with their beaks and
extract the sap; small, greenish plant-lice.]

The wingless and winged lice of this species are almost
entirely of a greenish color.—Prof. Thomas.

REMEDY.—Use No. 3, 4, 5, or 7.

CHAPTER CXXIX.

The Negro Bug.

(*Corimelæna pulicaria.*—Germar.)

Order, HEMIPTERA ;
Sub-order, HOMOPTERA : } Family, SCUTELLARIDÆ.

[Living upon the stems and fruit of the strawberry, raspberry, cherry, and quince; a small black bug, with a white
stripe on each side of the wing-covers.]

" These bugs puncture the young twigs and fruit and imbibe the sap; but the injury they occasion in this direction is as nothing compared to the effect which their presence has upon the fruit, as they exhale an offensive oder which renders the fruit upon which they congregate wholly unfit to be eaten; besides this, their small size renders their detection very difficult, so that a person is likely to get one or more of them into his mouth, along with the fruit, without being aware of it."— Riley.

Fig. 200.—Negro Bug, natural size and magnified—colors, black and white.

Fig. 200.

The young bugs closely resemble the adults (Fig. 200), but are of a more brownish color and are entirely destitute of wings.

I have a specimen found on the cherry that agrees with the above description, excepting stripes on the wing-case. It is probably a closely allied species, but I have found the genuine negro bug in Southern California, on purslane.

INSECTS INFESTING THE BLACKBERRY.

CHAPTER CXXX.

The Blackberry Cane-Borer.

(*Oberea tripunctata.*—Fabricius.)

Order, COLEOPTERA ; Family, CERAMBYCIDÆ.

[Living within the stems of blackberry and raspberry bushes ; a yellow footless grub, transforming into a slender black long-horned beetle, having the top of the thorax yellowish, and usually marked with two or three black dots.]

Fig. 201.—Blackberry Cane-borer—colors, black and yellowish.

Fig. 201.

The perfect beetles (Fig. 201) appear in May or June. The female beetle, after gnawing two rings around the growing cane, punctures the latter between the girdled rings, and deposits therein a single egg ; the grub which is produced from this egg burrows into the central pith, where it lives until reaching its full growth, when it forms a cell in its burrow, and soon afterward assumes the pupa form : in due time the pupa is transformed into the perfect insect, or beetle, which gnaws its way out of the burrow.

REMEDIES.—Use Nos. 60 and 66.

CHAPTER CXXXI.

The Blackberry Aphis. (Cal.)

(*Sipha rubifolii.*—Thomas.)

Order, HEMIPTERA ; } Family, APHIDIDÆ.
Sub-order, HOMOPTERA ; }

The measurements of insects in this work are given in inches and lines. The above cut represents one inch divided into lines and fractions thereof.

[Living on the underside of the leaves of blackberry bushes, which they puncture with their beaks and extract the sap; small, black and green or pale green plant-lice.

The wingless lice are pale green, marked with a darker green, the head tinged with yellowish. The winged lice have the head and thorax black, and the abdomen colored as in the wingless lice.--Professor Thomas.

REMEDIES.—Use No. 4, 5, or 7 ; No. 3 will do if No. 4, 5, or 7 cannot be obtained.

INSECTS INFESTING THE CURRANT.

CHAPTER CXXXII.

The Currant Borer. (Cal.)

(*Egeria tipuliformis.*—Linn.)

Order, LEPIDOPTERA : Family, ÆGERIDÆ.

[Eating out the central pith of currant and gooseberry bushes ; a whitish sixteen-legged worm.]

Among the numerous insect enemies of our small fruits. none are more widely distributed nor better known than the currant borers.

The eggs from which these borers (Fig. 202) hatch are deposited near the buds, only one egg usually being consigned to a single plant. As soon as hatched, the young borer penetrates the stem to the pith, which it devours, forming a burrow several inches in length in the interior of the stem. As it increases in size it enlarges the hole communicating with its burrow, so as to admit of the more easy passage of its castings, which it pushes out of this opening. It reaches its full growth in the following Spring, and then measures a little over six lines in length.

Fig. 202.—Currant Borer; upper figure, the larva—color, white; lower figure, the pupa—color, white.

Fig. 202.

When about to assume pupa form (Fig. 202) this borer takes up a position near the opening of its burrow, and closes the burrow above and below it with its castings, and then spins around its body a silken cocoon. Soon after completing this task the borer is changed to a pupa, from which the perfect insect is evolved in the course of a week or so. The latter usually makes its appearance in April or May, and soon afterward deposits its eggs.

The fore-wings of the perfect moth (Fig. 203) expand about ten lines; they are transparent, with the veins and margins black, and crossed near the middle with a black band; at the tip of each wing is usually a black spot, which is more or less tinged with copper color; the body is blue-black, with the edges of the collar and shoulder tuft, and three rings on the abdomen, golden yellow.

Fig. 203.—Currant Borer (moth)—colors, blue-black and yellow.

Fig. 203.

REMEDIES.—Use Nos. 60 and 66. In the latter part of April or early in May, spray with No. 5 or 7; use No. 3 or 4 if No. 5 or 7 cannot be obtained.

CHAPTER CXXXIII.

The American Currant-borer.

(*Psenocerus supernotatus.*—Say.)

Order, COLEOPTERA; Family, CERAMBYCIDÆ.

[Living within the stems of currant bushes; a whitish footless larva with a brown head.]

"This larva feeds upon the pith until reaching its full growth, when it gnaws a hole to the bark; it then retreats to

15

a short distance below it and forms a cell in which to undergo its transformations. It remains in its cell unchanged throughout the Winter, and is changed to a pupa in the following Spring, the beetle issuing in April.

Fig. 204.—American Currant-borer. enlarged—colors, brownish and white.

This beetle (Fig. 204) is a little over two lines long, and is of a black color, with the margins of the thorax and wing-covers pale brown. Near the middle of each wing-cover is a crescent-shaped white spot, and a short distance toward the base of the wing-cover from this are two yellowish or ash-gray spots.—Fitch.

REMEDIES.—Use Nos. 60 and 66.

Fig. 204.

CHAPTER CXXXIV.

The Currant Mite.

(*Tyroglyphus ribis.*—Fitch.)

Class, ARACHNIDA; Family, ACARIDÆ.

[Living in the burrows of the currant-borers; minute eight legged mites of a white color, and measuring scarcely one hundredth of an inch in length.]

In specimens of currant stocks infested by borers sent me by Mr. W. H. Jessup, of Haywards, in September, 1880, he called my attention to nests of eggs in the debris left by the borer from which hatched mites. These are not true insects. but belong to the same class as the spiders, and are characterized by having in the adult state four pair of legs, whereas no insect in the perfect state has more than three pairs. They belong to the same family as the itch-mites—minute creatures which live beneath the skin of man, producing that loathsome disease commonly known as the itch.

The habits and economy of the currant mite have never

been studied up, and hence it is impossible to say whether it should be regarded as a beneficial or as an injurious insect.

Another species belonging to the same genus—the phylloxera mite, *tyroglyphus phylloxeræ*—Planchon and Riley—feeds in the early part of its existence upon partly decomposed vegetable and animal matter, while later in life it preys upon the living phylloxeræ or grape-root lice; and so we may conclude that the currant mite is a friend and not a foe; or at least that it is not destructive to living plants.

CHAPTER CXXXV.

The Imported Currant-worm.

(*Nematus ventricosus.*—Klug.)

Order, HYMENOPTERA; Family, TENTHREDINIDÆ.

[Feeding upon the leaves of currant and gooseberry bushes: a naked green twenty-legged worm, marked with black spots.]

Fig. 205.

Fig. 205.—Eggs of Imported Currant Saw-flies; 1, the eggs; 2 and 3, holes made in the leaf by the young worms.

The eggs (Fig. 205) from which these worms hatch are laid along the larger veins on the underside of a leaf. In the course of eight or ten days these hatch into pale worms, having a large whitish head with a black spot on each side. After casting the skin the color becomes green, and the body is marked with numerous polished black spots (Fig. 206); the head is also black. After the last molt the black spots disappear, and the head becomes greenish, with a dark spot on each side.

Fig. 206.—Imported Currant Worms; *a, a* and *a*, the worms of different sizes—colors, green and black; *b*, a segment of a worm's body enlarged.

Fig. 206.

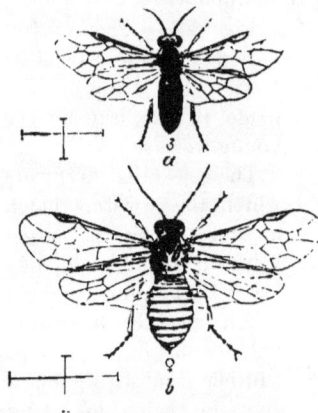

When fully grown they measure about nine lines in length. They then desert the plants and hide themselves beneath the leaves, or enter the earth and spin their tough brownish cocoons. At least two broods are produced in one season; the last brood passing the Winter in their cocoons.

Fig. 207.

Fig. 207.—Imported Currant Saw-flies—colors, yellow and black; *a*, the male enlarged; *b*, the female enlarged.

The female fly (Fig. 207*b*) is of a bright honey-yellow—color, the top of the head and thorax marked with black; the wings expand about six lines, and are transparent with black veins.

The male fly (Fig. 207*a*) differs so much from the female as to be easily mistaken for a distinct species. It is black, with the top of the thorax marked with yellow, and the tip and the under side of the abdomen are marked with yellow.

Fig. 208.

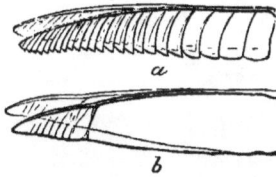

Fig. 208.—Ovipositors of Saw-flies highly magnified; *a*, ovipositor of willow saw-fly (*N. salispomum*); *b*, ovipositor of imported currant saw-fly.

REMEDIES.—When the larva is noticed on the foliage, use No. 64, No. 5 or No. 7. See also No. 80.

CHAPTER CXXXVI.

The Native Currant Worm.

(*Pristophora grossulariæ.*—Walsh.)

Order, HYMENOPTERA; Family, TENTHREDINIDÆ.

[Feeding upon the leaves of currant and gooseberry bushes; a naked green twenty-legged larva or worm.]

Fig. 209.

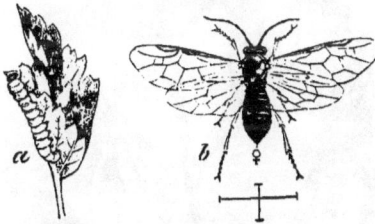

Fig. 209.—Native Currant Worm and Fly; *a*, the worm—color, green; *b*, the fly—colors, black and yellowish.

When fully grown (Fig. 209*a*) this worm measures about six lines in length; it then spins a tough cocoon in some sheltered place, usually among the leaves of the plant it infests.

Two broods are usually produced in one season, and the second brood are changed to flies (Fig. 209*b*) in the Fall. The latter deposit their eggs upon the twigs, and these eggs do not hatch until the following Spring.

The wings of the perfect fly expand from four to six lines; the body is black, and the thorax is marked with pale yellowish.—Riley, ninth Missouri Report.

REMEDIES.—When the larvæ appear on the foliage, use No. 64, No. 5 or No. 7. See also No. 80.

CHAPTER CXXXVII.

The Currant Span-worm.

(*Eufitchia ribearia.*—Fitch.)

SYNONYM.—*Abraxas,* or *Ellopia ribearia.*

Order, LEPIDOPTERA; Family, PHALÆNIDÆ.

[The measurements of insects in this work are given in inches and lines. The above cut represents one inch divided into lines and fractions thereof.]

[Feeding upon the leaves of currant and gooseberry bushes;
a whitish ten-legged span-worm, marked with yellow stripes
and numerous black spots.]

Fig. 210.—Currant
Span-worms and Pupa;
a and *b*, the worms in
different positions—col-
ors, yellow and black; *c*,
the pupa—color, brown.

Fig. 210.

This span-worm (Fig.
210, *a* and *b*), when fully
grown, measures a little
over an inch in length,
is of a whitish color, spotted with black, and
marked with a yellow
stripe on the back and a
similar one on each side
of the body: the under
side of the body is also
white, and has a yellow
line in the middle. When
about to pupate, it deserts the bushes and creeps beneath the
fallen leaves, etc., or enters the earth and forms a small cell
in which to undergo its transformations. (Fig. 210*c*.)

Fig. 211.

Fig. 211.—Currant Span-worm Moth (female)—colors, yellow and leaden.

The wings of the perfect moth (Fig. 211) expands about an inch, are of a pale yellowish color, marked with several dusky spots.

This moth appears in June or July, and deposits her eggs upon the stalks of the currant and gooseberry bushes; these eggs do not hatch until the following Spring, there being but one brood of . these insects produced in one year.

Remedies.—When the larvæ appear on the foliage, use No. 64 or No. 5, or No. 7. See, also, No. 80.

CHAPTER CXXXVIII.

The Progne Butterfly. (Cal.)

(*Grapta progne.*—Fabricius.)

Order, Lepidoptera; Family, Nymphalidæ.

[Feeding upon the leaves of the currant and gooseberry bushes; a grayish sixteen-legged caterpillar, thinly covered with whitish spines tipped with black.]

When fully grown, this caterpillar measures about one inch and three lines in length; the body is marked with alternate black and white bands, and on each side are two rows of yellow spots. When about to pupate, it suspends itself by the hind feet.

The pupa or chrysalis is pale brown, faintly clouded on the sides with olive-brown, and the abdomen is broadly striped with the same color on the back and sides; across the middle of the back is a deep depression, on each side of which are two silvery spots.

The fore-wings of the butterfly expand about one inch and ten lines, are of a reddish-brown color, marked with black spots; the outer edges of the wings are scolloped; the under side of the wings is blackish-gray, paler at the tips, and on each hind wing is a silvery character, resembling the letter L.

REMEDIES.—When the caterpillars appear on the foliage. use No. 64, or No. 5, or No. 7. See No. 80.

CHAPTER CXXXIX.

The Currant Aphis. (Cal.)

(*Myzus ribis.*—Linnæus.)

Order, HEMIPTERA ;   Family, APHIDIDÆ. •
Sub-order, HOMOPTERA ;

[Living on the under side of the leaves of currant and gooseberry bushes, which they puncture with their beaks and extract the sap; small green or yellowish plant-lice.]—Thomas.

REMEDIES.—Use No. 64, or No. 5, or No. 7 ; Nos. 3 or 4 will give good results.

INSECTS INFESTING THE GOOSEBERRY.

CHAPTER CXL.

The Frosted Leaf-hopper. (Cal.)

*(Pœciloptera pruinosa.—*Say.)

Order, HEMIPTERA ;
Sub-order, HOMOPTERA ; } Family, FULGORIDÆ.

[Puncturing and sucking the juices of the leaves and tender
shoots of gooseberry bushes and the rhubarb plant ; a small
four-winged moth-like insect of a dusky bluish color, covered
with a white mealy powder.]

This insect (Fig. 212) measures about four lines in length,
and when at rest holds its wings slanting over the back like a
steep roof.

Fig. 212.   Fig. 212.   Fig. 213.

—Frosted
Leaf - hop-
per enlarg-
ed—color, dull leaden or
pale green.

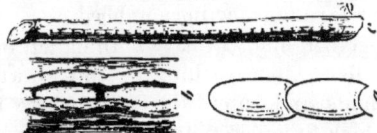

Fig. 213.—Eggs of Frosted Leaf-hopper ; *a*, the eggs highly
magnified—color, yellow ; *b*, the same in the twig, highly magnified ; *c*, the same in a twig, natural size.

The female deposits her eggs (Fig. 213) in a continuous slit in a twig of a tree or shrub, placing them upon one side, end to end. The young leaf-hopper is of the same general form as the adult, but is destitute of wings. It is covered with a cottony matter which envelopes the entire body, excepting the head.

I have found these insects on rhubarb in May.

REMEDIES.—Use No. 28, No. 5 or No. 7.

CHAPTER CXLI.

The Gooseberry Fruit-worm. (Cal.)

(*Pempelia grossulariæ.*—Packard.)

Order, LEPIDOPTERA: Family, PYRALIDÆ.

[Eating out the interior of currants and gooseberries; a pale green worm.]

Fig. 214.—Gooseberry Fruit-worms—color, green.

Fig. 214.

The caterpillar (Fig. 214) after eating out the interior of one berry will fasten a neighboring one to it with silken threads, and in this way whole bunches of currants or gooseberries are sometimes webbed together, and will have nothing left of them except the empty hulls. After reaching its full growth this worm deserts the plants and enters the earth, where it forms a small cell in which to undergo its transformations. Only one brood is produced in a season, and these pass the Winter in the pupa state.

Fig. 215.

Fig. 215.—Gooseberry Fruit-worm Moth and Cocoon: at the left the cocoon—color, grayish-brown: at the right the moth—color, gray.

The fore-wings of the perfect moth (Fig. 215) expand nearly an inch: are of a pale gray color, and marked with transverse white and dark lines. I have found the caterpillar and pupa of this insect, but failed to raise the perfect moth.

REMEDY.—Use No. 28. Pick off and destroy all infested fruit, in addition to gathering the larvæ.

CHAPTER CXLII.

The Gooseberry Midge. (Cal.)

(*Cecidomyia grossulariæ*—Fitch.)

Order, DIPTERA; Family, TIPULIDÆ.

[Living in gooseberries, causing them to turn prematurely red and to decay; a bright yellow footless maggot.]

The midge which produce these maggots quite closely resembles a mosquito; it is only about one line to the end of the closed wings, and is of a pale yellow color, the eyes black, the legs yellow and dusky, and the wings are transparent and faintly tinged with dusky.

This midge or fly punctures the young gooseberries and deposits therein one or more eggs. The maggots hatched from these eggs are entirely destitute of feet, and are of a bright yellow color. They assume the pupa form within the berries; the latter usually become prematurely red, and drop to the ground sometime previous to this.

The only remedy which suggests itself is to gather up and destroy the fallen fruit shortly after it falls, or before the maggots have completed their transformations and escaped.

Remedies as above, and when the fruit is fully formed use No. 5 or 7, and repeat in two weeks.

INSECTS INFESTING THE STRAWBERRY.

CHAPTER CXLIII.

The Strawberry Crown-borer.

(Analcis fragariæ—Riley.)

Order, COLEOPTERA; Family, CURCULIONIDÆ.

[Living in the roots of strawberry plants; a whitish curved grub or larva, about two lines long, with the head yellow; finally changing into a chestnut-brown snout-beetle.]

Fig. 216.

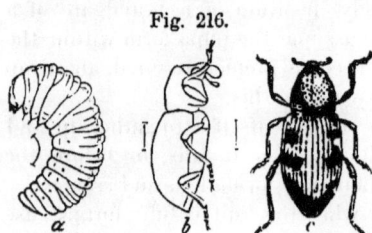

Fig. 216.—Strawberry Crown-borer; *a*, the larva, enlarged—color, white; *b*, the beetle, side view, enlarged; *c*, the same, back view, enlarged—color, chestnut-brown.

The egg from which this grub (Fig. 216*a*) hatches is probably laid upon the crown of the plant, and as soon as hatched the grub burrows downward into the pith. Here it remains until it has acquired its full growth, working in the thick, bulbous root: it undergoes its transformations to the pupa and perfect beetle state within the roots, and the beetle (Fig. 216c) makes its appearance above ground in the month of August.—Riley.

I have not found this species in this State, although I hesitate to say that it is not here. I have found grubs in strawberry plants which I supposed to belong to this species, but I did not succeed in rearing the perfect insect.

REMEDY.—Use No. 104.

CHAPTER CXLIV.

The Brown Strawberry Weevil. (Cal.)

(*Listronotus nevadicus.*—LeConte.)

Order, COLEOPTERA; Family, CURCULIONIDÆ.

[A small snout-beetle feeding on strawberry and cherry leaves, also eating into the crown of strawberry plants.]

This species was found, last Summer, eating the leaves and also the crown of the strawberry plants; it also eats cherry leaves. The fruit grower sending the specimens writes:

"I found this weevil on a strawberry patch; it eats the leaves, making them appear like lace work, and it also eats into the heart of the plant. I have also found it eating cherry leaves. In the daytime it hides in crevices in the bark of trees or in the ground under the strawberry plants; it is a night feeder."

This beetle (Fig. 127, Plate 3,) measures from two and a quarter to two and one half lines in length; is of a reddish-brown color, mottled with yellow, whitish or light markings on each side of the thorax and the outer base of the wing-cases, also at the apex of the wing cases; inner edges of the latter bordered with black. From the fact that the larvæ of this group of insects usually live on the roots of plants, it is probable that the larvæ of this species live in the roots of strawberry plants.

REMEDY.—Use No. 105.

The Strawberry Worm.

*(Emphytus maculatus—*Norton.)

Order, HYMENOPTERA : Family, TENTHREDINIDÆ.

[Feeding upon the leaves of strawberry plants; a naked dirty-yellowish worm, which is provided with twenty-two legs.]

The eggs (Fig. 218, 9,) from which these worms hatch are deposited in the stems of the plants; this operation is performed by the female fly by means of a saw-like instrument with which all the flies belonging to this family are provided. When fully grown the worms (Fig. 218, 4,) measures about six lines in length; they then enter the earth and form small cells in which to undergo their transformations. Two broods are usually produced in one season, and the last brood remain in their cells unchanged until the following Spring.

Fig. 218.

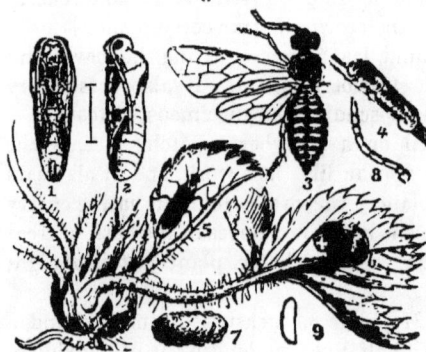

Fig. 218.—Strawberry Worm, Pupa, Fly, etc.; *1*, the pupa, enlarged, ventral view; *2*, the same, side view— color, greenish-white; *3*, the fly, enlarged (the wings on one side not represented)—colors, black and dirty white; *8*, her antennæ, greatly enlarged; *5*, the fly with its wings closed; *6*, the worm at rest; *4*, the same extended—color, dirty yellow; *7*, the cocoon; *9*, the egg, magnified—color, white.

The perfect flies (Fig. 218, *3* and *5*,) are deep black, with two rows of dirty white spots on the abdomen; the wings are transparent, and expand about six lines.—Riley's Ninth Missouri Report.

REMEDY.—If the plant is bearing fruit, use No. 19, or spray with No. 83 or 85.

CHAPTER CXLVI.

The Strawberry Leaf-roller.

(*Anchylopera fragaria*—Riley.)

SYNONYM.—*Phoxopteris fragaria.*—W. and R.

Order, LEPIDOPTERA ; Family, TORTRICIDÆ.

[Rolling up and feeding upon the leaves of the strawberry ; a yellowish or greenish brown caterpillar about four lines long, with a yellowish-brown head having a black spot on each side.]

This caterpillar (Fig 219*a*) feeds upon the upper surface of the leaf, giving the latter the appearance of having been scorched. It pupates within its nest, and the moth (Fig. 219*c*) which issues a few weeks later has the head, thorax and fore-wings reddish-brown, the latter streaked and spotted with black and white. It expands about six lines.

Fig. 219.—Strawberry Leaf-roller ; *a*, the cater-pillar—color, dull olive-green ; *b*, head and first four segments of its body, enlarged ; *d*, two last segments of its body, enlarged ; *c*, the moth enlarged—colors, reddish-brown, black and white.

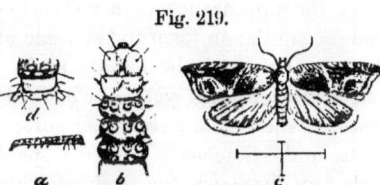

Fig. 219.

There are two broods of these leaf-rollers each season, the last brood passing the Winter in the pupa state.

I have found a small caterpillar on the strawberry leaves similar to that described here, but failed to rear the perfect insect, and therefore cannot say to what species it belonged.

REMEDY.—Use No. 24.

INSECTS INFESTING THE WALNUT TREE.

CHAPTER CXLVII.

The Walnut Scale. (Cal.)

(*Aspidiotus juglans-regiæ*.—Comstock.)

Order, HEMIPTERA ;
Sub-order, HOMOPTERA ; } Family, COCCIDÆ.

[A grayish-brown scale, infesting the walnut tree.]

Mr. Alexander Craw, of Los Angeles, sent me specimens of this scale in 1881. I have examined several trees infested by this species of scale, and find that they spread rapidly on the bark of the limbs and branches.

NATURAL HISTORY.

The female scale measures one line in diameter—color, grayish-brown; form—circular, or slightly elliptical; exuviæ or larval skin on one side of the center, giving the shell in some cases the appearance of a cockle shell. The scale of the male is similar in form to the male of the red scale, and is of the same color as the female scale, which is grayish-brown. The female is pale yellow, with golden colored spots; when matured she is of a rich amber color. This species infests the limbs and branches. I have a specimen of a limb over two inches in diameter, the bark of which is entirely covered by this pest.

REMEDIES.—Use No. 11, 12 or 13, when the tree is dormant ; or No. 4 in Summer.

INSECTS INFESTING THE CHESTNUT TREE.

CHAPTER CXLVIII.

The Brown Chestnut Bud Beetle. (Cal.)

(*Pityophthorus pubipennis.*—Leconte.)

Order, COLEOPTERA ; Family, SCOLYTIDÆ.

[A small dark-brown beetle eating into the base of the buds of chestnut trees, especially the variety known as the Italian chestnut.]

Last year (1882), specimens were received of branches (Fig. 220, Plate 3,) of chestnut trees, the buds of which were infested by a small beetle. The beetle (Fig. 221, Plate 3,) measured from three fourths of a line to one line in length, and is not half as wide as long—color, dark-brown ; and as its specific name, *pubipennis*, indicates, the wing-cases are covered with a fine hairy-like down.

The perfect insect infests the bud at the base of the outer side, and eats into the heart, entirely destroying it, as only the outer shell remains. This beetle is probably a native species, as Dr. Leconte found it twenty years ago in this State, living under the bark of oak trees. It is also found feeding on the oaks in the vicinity where it is found feeding in the buds of chestnut trees. This species hibernates in the perfect state, as I found the beetles in buds last March in a semi-dormant condition. I am not aware that the larvæ have been found upon the chestnut.

REMEDIES.—Use No. 67, No. 11 or 12 ; or No. 13—five pounds of the mixture to each six gallons of water used. These solutions should be used when the tree is dormant.

16

INSECTS INFESTING THE ALMOND TREE.

CHAPTER CXLIX.

The Red Spider. (Cal.)

(*Tetranychus telarius.*—Linn.)

Class, ARACHNIDA ; ⎫
Sub-class, ACARINA ; ⎭ Family, TROMBIDIDÆ.

[A small red mite, infesting the trunk, limbs, and foliage of fruit trees and nursery stock, and also garden and hot-house plants.]

The red mite (Fig. 222), commonly called the "red spider," may be said to be a universal pest of the orchard, garden and hot-house, and has been allowed to spread to an alarming extent in orchards in this State. Although a feeder on nearly all varieties of deciduous fruit trees, and also on citrus trees, the almond seem to be the worst infested by this species.

Fig. 222.—Red Spider, adult—color, reddish.

The perfect mites have eight legs; form, ovate; length, one sixtieth of an inch; color, various, sometimes brick-red, rust-red, or greenish, dotted with red; head, yellowish.

The egg when first exposed is colorless, but changes to a bright red; in form it is round, and measures one one-hundred-and-fortieth of an inch in diameter.

I have no reason to change my opinion expressed in 1881, in my "Insects Injurious," etc., that the female

Fig. 222.

does not deposit the eggs, but at maturity fastens herself to the bark or leaf and dies; the covering is removed from the body by the males, exposing from fifteen to twenty whitish eggs. I have counted one hundred and twenty-seven female bodies fastened by the anterior part to the bark, and have also noticed the male insects removing the skin of the dead females, and in this way exposing the eggs.

Fig. 223.

Fig. 223.—Larva of Red Spider, enlarged —color, bright red.

The young mite, when first hatched from the egg, has only six legs (Fig. 223); in a few days a film seems to cover the body, and from this emerges the perfect mite, having eight legs. (Fig. 222.)

The injury done by this species is principally on the fruit buds, blossoms, and on the leaves; they eat the epidermis or skin of the latter, and also cover them with a fine web, causing them to wither and fall off. The trunk, limbs and branches of trees seriously infested by this mite appear of a reddish color; this is from being covered by a multitude of the eggs of these mites. (Fig. 224.)

Fig. 224.—Portion of a twig infested by Red Spiders.

Fig. 224.

In some cases where not seriously infested, it has the appearance of being covered with iron rust.

When this species first attack a tree the mites are generally found around the bud and new foliage.

The red spider is also occasionally found on trees infested by scale insects.

REMEDY.—Use No. 34.

CHAPTER CL.

The Almond Aphis.   (Cal.)

(*Aphis amygdali.*—Blanch.)

Order, HEMIPTERA ;   } Family, APHIDIDÆ.
Sub-order, HOMOPTERA ; }

[Living upon the under side of the leaves of almond and peach trees, causing them to become crisp and wrinkled, by puncturing them with their beaks and extracting the sap; small green plant-lice.]   (Fig. 225.)

Fig. 225.

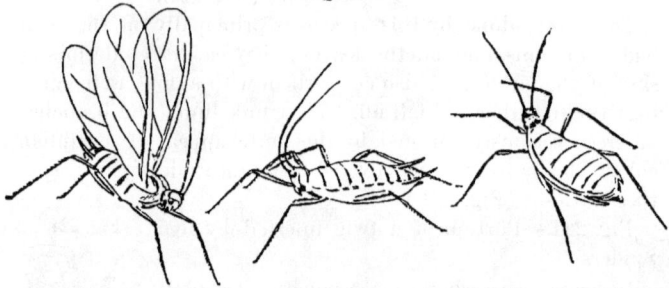

Fig. 225.—Plant Lice (*Aphis*) highly magnified—color, green.

REMEDIES.—Use No. 5 or 7 ; No. 3 or 4 will give good results if No. 5 or 7 cannot be had.

INSECTS INFESTING THE FILBERT

CHAPTER CLI.

The Filbert Scale. (Cal.)

(Lecanium hemisphæricum.—Targioni.)

Order, HEMIPTERA ;
Sub-order, HOMOPTERA ; } Family, COCCIDÆ.

[A dark-brown hemispherical scale insect, or bark-louse, feeding upon the filbert and orange trees, and on the oleander and other ornamental trees.]

Fig. 226.

I take the liberty of designating this species as the "Filbert Scale" so as to distinguish it from the soft orange scale (*L. hesperidum*) on account of the similiarity of their technical names.

Fig. 226.—Filbert Scale ; *3*, leaves infested by this scale ; *3a*, the adult female, enlarged—color, brown.

The filbert scale (Fig. 226. *3a*,) when full grown is of a dark brown color, hemispherical in form, and measuring one and three-quarters lines in length; width, one and a half lines ;

height, varying, but about one line. The size of the branch on which the insect locates sometimes alters or changes its form; if the branch is small, the flattened edge of the insect will spread downward, clasping the branch or twig, thus giving the insect an elongated and narrowed appearance, different from those located on a flat surface or leaf.

NATURAL HISTORY.

Eggs—length, one eighty fifth of an inch; width, one half of the length; color, pinkish. From seventy to one hundred are laid by each female. Larva—color, reddish; length, one seventy fifth of an inch; antennæ, seven jointed; anal setæ present.

The females of this species (Fig. 226, 3a—enlarged), like other Lecaniums, have the power of locomotion until they become fixed to the plant. When young they are of a reddish color, changing to a light brown, and finally becoming a dark brown color when they reach maturity. They infest the filbert and orange, also the oleander and other ornamental trees. They can be readily distinguished from the black scale (*L. oleæ*) by the absence of the carina or ridge, forming the letter H on the back.

REMEDIES.—Use the same as for the black scale on deciduous trees. (Chapter LXXXIV.)

INSECTS INFESTING THE EUCALYPTUS TREE.

The following insects infest the Eucalyptus Tree, and are treated of in another part of this work:

The San Jose Scale (*Aspidiotus perniciosus*).   The Greedy Scale (*Aspidiotus rapax*).

INSECTS INFESTING THE LOCUST TREE.

The measurements of insects in this work are given in inches and lines. The above cut represents one inch divided into lines and fractions thereof.

CHAPTER CLII.

The Locust Carpenter Moth, or the Legged Locust Borer.   (Cal.)

(*Xyleutus robiniæ.*—Peck.)

Order, LEPIDOPTERA; Family, BOMBYCIDÆ.

[Living in the trunks and larger limbs of locust, willow and oak trees; a greenish-white, sixteen legged worm.]

The common names given to this borer, "Legged, etc.," are to distinguish it from the grub of *Clytus robiniæ*, which also infests locust trees and is destitute of legs.

Fig. 227.

Fig. 227.—Legged Locust Borer; the full grown caterpillar —color, greenish-white, with a tinge of pink or yellow.

This borer (Fig 227) keeps its burrow open, and not packed full of woody fibres as many other borers are known to do.

When about to pupate it spins a cocoon in the lower part of its burrow, and before the perfect moth emerges the pupa works itself up to and partially out of the opening of its burrow, and while in this position the moth (Fig. 228) makes its escape.

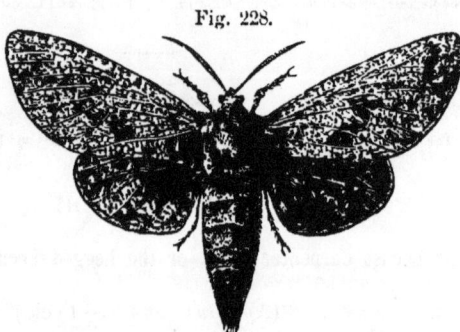

Fig. 228.

Fig. 228.—Locust Carpenter Moth (female)—colors, gray and black.

The sexes differ widely from each other, both in size and color. The female is the largest, expanding from two inches to two inches and six lines; the wings are gray, dotted with black and marked with a network of black lines.

The male moth expands only about an inch and six lines; the wings are darker than those of the female, and there is a large yellow spot on each hind wing. This species is very destructive to locust trees.

REMEDY.—Use No. 37.

CHAPTER CLII.

The Legless Locust Borer.   (Cal.)

(Clytus robinia.—Forster.)

Order, COLEOPTERA: Family, CERAMBYCIDÆ.

[Boring into the trunks and branches of the locust tree; a nearly footless whitish grub, about one inch long and as thick as a goose-quill.]

When young this borer lives in the sap-wood, but as it grows longer it burrows into the solid wood, perforating it in every direction. At first it casts its chips out of the opening of its burrow, which finally becomes packed full of the coarse and fibrous parts of the wood. It assumes the pupa form in its burrow, and the beetle issues late in July or early in August.

Fig. 229.—Legless Locust Borer (the beetle)— colors, yellow and black.

This beetle (Fig. 229) is of a black color and is marked with several transverse and oblique yellow lines; it is sometimes very destructive to locust trees.

REMEDY.—Use No. 37.

INSECTS INFESTING THE WILLOW.

CHAPTER CLIV.

Weeping Willow Borer. (Cal.)

(*Sciapteron robiniæ.*—Edwards.)

Order, LEPIDOPTERA; Family, ÆGERIDÆ.

[A yellowish-white sixteen-legged larva, about one inch in length, boring into the wood of the weeping willow, locust and cottonwood.]

In September, 1882, I was requested to examine a number of weeping willows planted for ornamental purposes near a residence. I found them so infested by the larva (Fig. 230, Plate 3,) of a moth that I recommended they be cut down, which was immediately done. The wood of the stems was thoroughly burrowed. I found a number of larvæ, and one pupa. On the 30th of September the perfect insect emerged from the pupa. At the present time, June 9th, one of the larvæ found is spinning a cocoon; it is evidently about to undergo its changes. The perfect insect (Fig. 231, Plate 3,) is about nine lines long; spread of wings, about one inch and two lines; color, head dark-brown; palpi, dark-orange; antennæ, yellowish-red; thorax, dark brown, with a narrow yellow marginal line around the upper surface; first segment of abdomen, dark brown; second segment, yellow; third and fourth segments, dark brown; all the segments posterior to the fourth are yellow; anal segment slightly tufted; fore-wings, brick red, with black veins; hind-wings clear, transparent, the inner edge bordered with a narrow brown band, yellow at the base; legs yellowish red. Pupa—color, dark amber. Larva—length, one inch; color, pale yellowish; second segment, yellowish; head yellowish-red; mouth part dark brown; about six fine yellowish hairs on each segment, growing on as many wart-like spots; spiracles small, round, and of a light-brown color; true legs light-brown; pro-legs represented by eight rings of hooks. This species is found in localities one hundred and fifty miles apart. This species also infest the locust tree.

NOTE.—Since writing the above the perfect insect has emerged from the pupa of the above mentioned larva. It proved to be a specimen of the *Sciapteron robiniæ* of Edwards. (July 13.)

REMEDY.—Use No. 37.

CHAPTER CLV.

The Willow Scale.   (Cal.)

(*Aspidiotus convexus.*—Comstock.)

Order, HEMIPTERA ;   )
Sub-order, HOMOPTERA ;)   Family, COCCIDÆ.

[A species of scale insect infesting willows.]

This species resembles the greedy scale (*A. rapax*) in appearance, and it is only by scientists that a distinction can be made. I think it beyond question that this species will spread on cultivated trees, especially the apple and pear.

Near the city of Los Angeles, the orchards of McKinlay Brothers and Mr. Kieser were protected by over two miles of wind breaks made by planting willows, which formed an excellent protection from the winds and supplied a large quantity of firewood annually. Two years ago it was noticed that the fruit—apples and pears—was seriously infested by the *A. convexus.* It was found that the willows were also infested by the same species. I advised the digging out of the willows, which was done, and the wind breaks replaced by planting eucalyptus trees. Last year there was no scale on the fruit.

REMEDIES.—Same as for San Jose scale, Chapter XX.

———

CHAPTER CLVI.

The San Bernardino Willow Scale.   (Cal.)

(*Chionaspis ortholobis.*—Comstock.)

Order, HEMIPTERA ;   }
Sub-order, HOMOPTERA ;   } Family, COCCIDÆ.

[A scale insect found on the willows at San Bernardino, Cal.]

Prof. Comstock found this species on willows, and described it as follows: "Color white, about one line in length. The body of the female is dark purple; eggs dark purple. This species infests chiefly the bark of the small whip-like limbs which spring from the trunk of the trees."

REMEDIES.—Same as for San Jose scale, Chapter XX.

———

INSECTS INFESTING THE POPLAR AND COTTONWOOD.

———

The following insects infest the poplar and cottonwood, and are treated of in another part of this work :

The San Jose Scale (*Aspidiotus perniciosus*).   The Legged Locust-Borer (*Xylentus robiniæ*).

INSECTS INFESTING THE ELM.

The Semicolon Butterfly (*Grapta interroga tionis*), also infests the elm.

CHAPTER CLVII.

The Elm and Locust Scale   (Cal.)

(*Lecanium Sp.?*)

Order, HEMIPTERA ;
Sub-order, HOMOPTERA ;   } Family, COCCIDÆ.

[A dark brown scale insect infesting the elm and locust trees.]

Fig. 230.

Fig. 230.—Elm and Locust Scales—color, brownish-black.

This species is found on the elm trees, especially the cork elm, and also on the locust trees—varieties, honey locust and Chinese locust; they infest the branches (Fig. 230) and leaves.

The female scale is of a dark-brown color, ovate in form, and measures three lines in length, two and one quarter lines in width, and nearly two lines in height. This scale differs from the black scale (Chapter LXXXIV) and filbert scale, in the form being conical (Fig. 230).

Fig. 231.

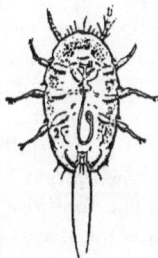

Fig. 231.—Larva of Elm and Locust Scale enlarged—color, reddish-yellow.

Larva (Fig. 231)—length, one ninetieth of an inch; form, elongate-ovoid. Eggs—color yellowish-white, from two hundred to three hundred being produced by each female.

REMEDIES.—Same as for black scale. Chapter LXXXIV.

INSECTS INFESTING THE OAK.

CHAPTER CLVIII.

The Acorn Moth.   (Cal.)

(*Holocera glandulella.*—Riley.)

Order, LEPIDOPTERA : Family, TINEIDÆ.

[Living in acorns; a yellowish or grayish-white sixteen-legged caterpillar, from three to six lines long.]

This caterpillar (Fig. 232*a*) is supposed to infest only those acorns which have been infested by some other insect, such as the larva of a weevil. It assumes the pupa form in its burrow.

Fig. 232. — Acorn Moth; *f*, the moth enlarged—color, ash-gray; *b*, an acorn, showing hole where the caterpillar entered; *a*, the caterpillar in an acorn—color, grayish-white ;

Fig. 232.

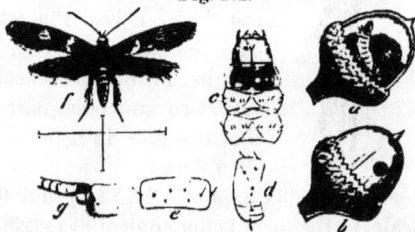

c, back view of a segment of its body, enlarged ; *d*, side view of same, enlarged ; *e*, top of the head and first three segments of the caterpillar's body, enlarged—color of head and first segment, light brown.

The fore-wings of the moth (Fig. 232*f*) expand from six to nine lines, are of an ashen-gray color, marked near the middle with two dark spots, and with a pale transverse stripe across the basal third of the wing. The moths issue during the Summer season, or from April to September.

NOTE.—The above account of this insect is given because several persons have insisted that it is the codlin moth.

INSECTS INFESTING THE PINE.

CHAPTER CLIX.

The Pine Weevil. (Cal.)

(*Pissodes strobi.*—Peck.)

Order, COLEOPTERA : Family, CURCULIONIDÆ.

[Living in the terminal shoots of pine trees ; a footless grub' which is finally changed into a brownish beetle, marked with two large whitish spots behind the middle of the wing-cases.]

Fig. 233.

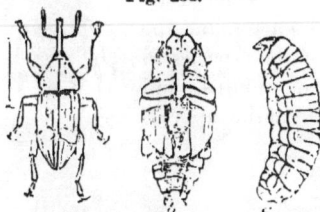

Fig. 233.—Pine Weevil, enlarged ; at the left, the weevil —colors, brown and white ; *b*, the pupa, ventral view—color, white ; *a*, the larva—color, white, the head reddish.

The larvæ (Fig. 233*a*) or grubs of this weevil are sometimes very injurious to pine trees, by destroying the terminal shoots ; as many as forty have been found in one shoot, which they had perforated in various directions. They assume the pupa form (Fig. 233*b*) within their burrows, first gnawing a passage to the outside for the egress of the perfect beetles (Fig. 233, left). In the vicinity of Sacramento these insects have been found on pine trees that had been planted for ornamental purposes.

REMEDIES.—Nos. 26 and 27.

CHAPTER CLX.

The Pales Weevil.   (Cal.)

(*Hylobius pales.*—Herbst.)

Order, COLEOPTERA;   Family, CURCULIONIDÆ.

[Living beneath the bark of the pine tree; a whitish footless grub, finally transforming into a dark brown or black snout-beetle, about four lines long, and marked with numerous whitish spots.]

Before pupating the larva gnaws a passage to the outside of the bark, but leaves a thin covering to its burrow in which it assumes the pupa form.

Fig. 234.—Pales Weevil—colors, black, brown and white.

Fig. 234.

In due time it is changed to a beetle (Fig. 234) which gnaws through the thin covering of its burrow and makes its escape. I have found the grub, pupa and perfect insect of this species in pine slabs shipped from the mountains, and have found a specimen having a similar appearance on pear trees.

REMEDIES.—Use No. 26 or 27.

CHAPTER CLXI.

The Norfolk Island Pine Scale.   (Cal.)

(*Chleria araucaria.*—Comstock.)

Order, HEMIPTERA;   }
Sub-order, HOMOPTERA;   }   Family, COCCIDÆ.

[Infesting the Norfolk Island pine tree; a white, nearly circular scale insect.]

This species is found only upon the Norfolk Island pine, and occurs in the Counties of Sacramento and Santa Barbara.

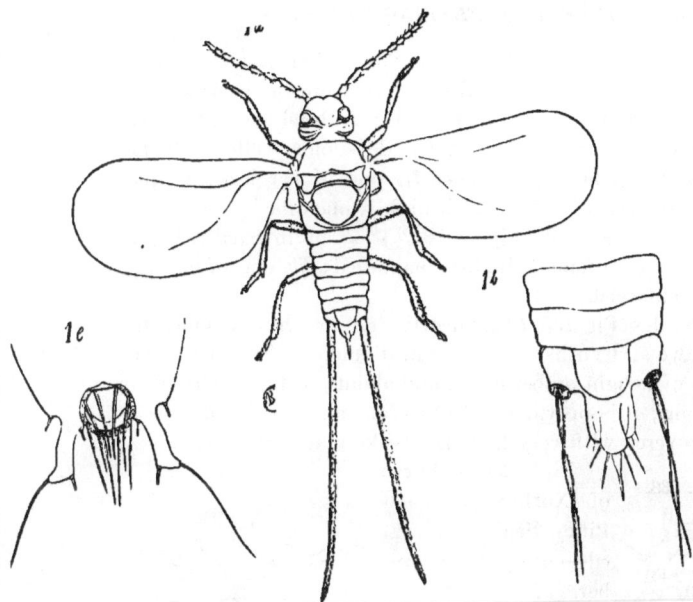

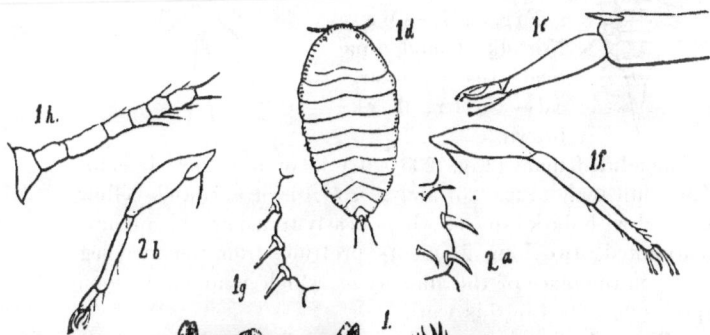

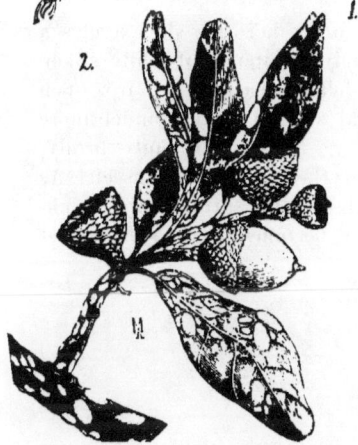

Fig. 235.—Norfolk Island Pine Scale (Figs. *1 to 1h*); *1*, an infested twig; *1a*, the male—colors, white and brown; *1b*, the hind end of the body of the male; *1c*, hind end of the body of the female; *1d*, the adult female—color, yellow; *1h*, her antennæ; *1g*, her spinnerets; *1f*, one of her legs; *1e*, portion of the leg of a male; all highly magnified. Oak Scale (*Rhizococcus quercus*) (Figs. *2 to 2b*); *2*, an infested twig; *2a* spinnerets of a female, highly magnified; *2b*, one of her legs, greatly enlarged.

The scales (Fig. 235, *1*,) are nearly circular, white, and measure about two sixteenths of an inch in diameter. The larva (Fig. 236) is of a light amber color, and about one-hundredth of an inch long; it is provided with two long anal setæ, and appears to be covered with very fine armor-like plates or scales.

Fig. 236.

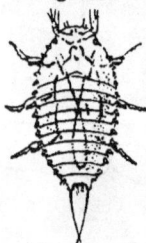

Fig. 236.—Larva of Norfolk Island Pine Scale, enlarged—color, light amber.

Fig. 237.—Male Norfolk Island Pine Scale Insect; enlarged—color, dark-brown.

Fig. 237.

The adult female (Fig. 235, *1d*,) is of a yellowish color. The adult male (Figs. 237 and 235, *1a*,) is of a whitish-yellow, marked with dark brown; wings nearly transparent; antennæ ten jointed; two long filaments protrude from the anal segment, in the place of the anal stylet, which is found in some species described in this work.

" When the female is ready to lay her eggs she excretes a cocoon-like covering to the body, composed of white silken threads. The sac is dense like felt but easily torn; it is open on the middle line of the ventral surface, or very much more delicate on that part. It adheres to the tree quite firmly, remaining where excreted after the death of the insect. As the eggs are laid, the body of the female shrinks away, making room for them, and finally it becomes a very small pellet in the anterior end of the sac, the remainder of the space being filled with eggs. These are light yellow in color. When the male larva is ready to undergo his metamorphosis he secretes a covering to his body resembling the sac excreted by the female, except that it is very much smaller, measuring only one five hundredth of an inch in length."—Comstock.

REMEDIES.—Use Nos. 4, 5 and 7.

—

INSECTS INFESTING THE JUNIPER

——— ———

——— ———

CHAPTER CLXII.

The Juniper Scale.

(*Diaspis carueli.*—Comstock.)

Order, HEMIPTERA ;   }
Sub-order, HOMOPTERA ;} Family, COCCIDÆ.

[A white scale insect infesting various species of juniper, and some other shrubs of the pine family.]

Professor Comstock describes this species as being very common in Washington, D. C.

The scale of the female (Fig. 237½, 2a,) is circular, snowy-white, with the exuviæ central or nearly so ; diameter of scale, from one half to three fourths of a line. The females are of a yellow color, nearly circular in outline, but a little elongated posteriorly.

Fig. 237¼.

Fig 237½.

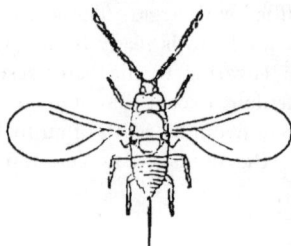

Fig. 237¼.—Juniper Scale; 2, an infested branch; 2a, the female scale—color, phite; 2b, the male scale—color, white.
Fig. 237½.—Male Juniper Scale Insect, highly magnified—color, orange-yellow.

REMEDIES.—Use same as for the Red Scale. Chap. LXXXVI.

INSECTS INFESTING THE RED BAY TREE.

CHAPTER CLXIII.

The Red Bay Scale.

(*Aspidiotus perseæ.*—Comstock.)

Order, HEMIPTERA;
Sub-order, HOMOPTERA; } Family. COCCIDÆ.

[A circular scale infesting the foliage of the red bay tree, and also the olive tree.]

The red bay scale insect infests the foliage of the red bay tree, and also the foliage of trees planted in the vicinity of trees infested by this species. I have also found this, or a closely allied species, on a privet hedge.

The scale of the female is circular, and from three fourths of a line to one line in diameter; the outer margin is dark reddish-brown with a yellowish tinge; the central part is a darker brown; the exuviæ is nearly in the center, and forms a nipple-like prominence.

Fig. 238.—Male Red Bay Scale Insect, enlarged, ventral view—color, yellowish.

Fig. 238.

The male scale (Fig. 238) is of a dirty white color, elongated; the exuviæ or larval skin nearly central, and of a golden-yellow color. The eggs are yellowish, and of an elongate-oval shape; from twenty to thirty-five are deposited by each female. The larvæ are yellow.

The female is orange-yellow; form, similar to the red scale of Florida (*A. ficus*).

REMEDIES.—Same as for the red scale, Chapter LXXXVI.

INSECTS INFESTING THE AZALEA.

CHAPTER CLXIV.

The Azalea Bark Louse.

(*Eriococcus azalea.*—Comstock.)

Order, HEMIPTERA : ⎫ Family. COCCIDÆ.
Sub-order, HOMOPTERA ;⎭

[A bark-louse or scale insect, enclosed in a white felt-like sac, feeding on the azalea.]

This species can be readily distinguished from other species of scale insects or bark-lice described in this work, by being enclosed in a felt-like sac. In form the sac or covering is hemispherical, but more pointed at one end than at the other, and snow-white in color.

The female insect is dark purplish in color and oval in form, the posterior end being the narrowest : the eggs and young larvæ are of a reddish-purple or carmine color.

REMEDIES.—In conservatories use No. 88, or No. 5 or 7.

INSECTS INFESTING THE OLEANDER.

CHAPTER CLXV.

The Oleander Aphis.

(*Aphis nerii.*—Fonscol.)

Order, HEMIPTERA ;
Sub-order, HOMOPTERA ; } Family, APHIDIDÆ.

[The measurements of insects in this work are given in inches and lines. The above cut represents one inch divided into lines and fractions thereof.]

[Living upon the leaves and tender twigs of the oleander, which they puncture with their beaks and extract the sap; small yellowish plant-lice, sometimes marked with brown.]

The wingless lice are yellow. The winged lice are also yellow, having the thorax marked with brown.

REMEDIES.—No. 3 or No. 4; but No. 5 or No. 7 are preferable. No. 64.

INSECTS INFESTING THE ROSE BUSH.

CHAPTER CLXVI.

The Rose Scale, or White Scale. (Cal.)

(*Diaspis rosæ*.—Sandberg.)

Order, HEMIPTERA :
Sub-order, HOMOPTERA : } Family, COCCIDÆ.

[A white scale-insect infesting the rose, raspberry, blackberry, currant, etc.]

The rose or white scale insect (Fig. 239) infests several
varieties of rose bushes, and also the raspberry, blackberry,
currant. etc.

Fig. 239.

Fig. 239. — R o s e
Scales — color. white :
at the top are two
scales. magnified. the
one to the right being
a side view and the one
to the left being a top

Fig. 240.

view.   Fig. 240.—Female Rose Scale-insect. enlarged.

Female (Fig. 240) form. elongated ; abdomen. distinctly
segmented : color. dark red.

Male (Fig. 242). winged—color, light amber. with dark
irregular markings ; wings white. the veins slightly colored ;
anal stylet half the length of the body ; legs. yellowish.

Fig. 241

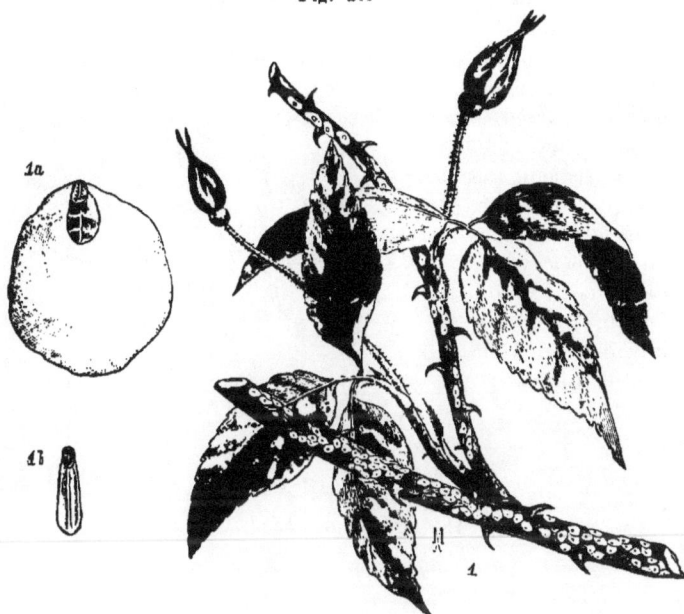

Fig. 241.—Rose Scale; *1*, a portion of an infested bush; *1a*, the female scale. greatly enlarged—color, white; *1b*, the male scale, enlarged—color, white.

It is generally found on the stems and branches of the plants (Fig. 241) and its presence can be easily detected, as its color contrasts strongly with the color of the bark of the plant which it infests. I have recently received specimens of currant roots taken from below the surface of the ground which are covered by this species of scale-insect. Query— Does this species infest the roots of plants? Will fruit-growers please investigate?

The scale of the male (Fig. 241, *1b*,) is elongated and meas-ures three fourths of a line in length; color, white; exuviæ near the anterior end.

Eggs—color, red; form, oval: length, one one-hundredth of an inch—from twenty to fifty under each scale.

Larva—length, one eighty-fifth of an inch, and about two

thirds as wide; color, reddish; antennae, six-jointed; two anal setae, but very fine.

Fig. 242.—Rose Scale-insect (male, greatly enlarged)— color, yellow.

The scale of the female insect (Fig. 241, *la.*) is circular, or nearly so; color, snowy white; length, from one line to one and one

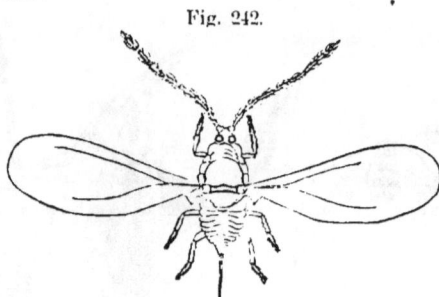

Fig. 242.

quarter lines; exuviæ or cast skin to one side of the center.

REMEDIES.—No. 3 or 4; but No. 5, 6, or 7 are preferable.

CHAPTER CLXVII.

The Rose Aphis.   (Cal.)

(*Siphonophora rosæ*.—Reaumur.)

Order, HEMIPTERA;
Sub-order, HOMOPTERA;   } Family, APHIDIDÆ.

[Living on the stems and leaves of the rosebush, which they puncture with their beaks and extract the sap; small green or reddish plant-lice, usually marked with black or brown.]

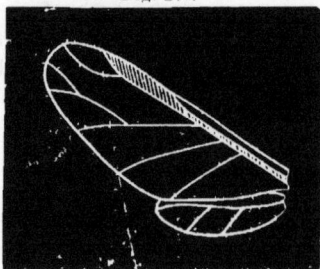

Fig. 243.

Fig. 243.—Wing of Aphis, showing venation.

The wingless lice are green in color, excepting one variety, which is reddish. The winged lice (Fig. 243, wing,) are green, the head and thorax brown or black, the abdomen marked with brown or black.

REMEDIES.—Use No. 64; No. 3 or 4 is very effective, but 5 or 7 is better. No. 83 or No. 85 are excellent.

CHAPTER CLXVIII.

The Rose Slug-worm. (Cal.)

(*Selandria rosæ.*—Harris.)

Order, HYMENOPTERA : Family, TENTHREDINIDÆ.

[Feeding upon the leaves of the rosebush; a naked green worm, provided with twenty-two legs.]

Fig. 244.—Rose Slug-worm—color, green.

Fig. 244.

These slug-worms (Fig. 244) eat only the upper surface of the leaves, leaving the remainder untouched, thus giving the leaves the appearance of having been scorched. These worms have the head yellowish, with a black spot on each side, and on the edge of the first segment are two triple-pointed warts. When fully grown they desert the plants and burrow a short distance into the earth, where each one forms a small cell in which it spins a tough elliptical cocoon. Two broods are usually produced in one year, the last brood passing the Winter in their cocoons.

Fig. 245.—Rose Saw-fly—color, black.

Fig. 245.

The perfect fly (Fig. 245) has four smoky wings, which expand about five lines : the body is of a uniform black color. The female fly deposits her eggs singly in punctures made in a leaf by means of a saw-like instrument with which her abdomen is armed.

REMEDIES.—Use No. 64, 5, 7 or 6 ; the latter is excellent, but might mark the foliage.

CHAPTER CLXIX.

The Horned Flower-beetle. (Cal.)

(*Notoxus monodon.*—Fabricius.)

Order, COLEOPTERA : Family, ANTHICIDÆ.

[Feeding upon the petals of the garden rose; a grayish-brown beetle, having a brown band across the middle of the wing-cases: two or more brown spots at the base of the same, and an obscure band at the tip: the thorax projects in the form of a long horn over the head.]

This beetle (Fig. 246, Plate 3,) measures about a line and a half in length. Where it lives during its larval stage is not known, but the perfect insect passes the Winter beneath pieces of wood, etc., that lie upon the ground.

REMEDIES.—Spray with No. 5, 7 or 64: or dust after sunset with No. 19.

INSECTS INFESTING THE FLOWER GARDEN.

CHAPTER CLXX.

The Verbena Aphis. (Cal.)

(*Siphonophora Verbenæ.*—Thomas.)

Order, HEMIPTERA; ⎰ Family, APHIDIDÆ.
Sub-Order, HOMOPTERA ⎱

[Living upon the leaves of the garden verbena, which they puncture with their beaks and imbibe the sap; small green or yellowish plant-lice.—Thomas.]

REMEDIES.—Use No. 3 or 4, one pound to each two gallons of water used; or No. 83 or No. 85.

CHAPTER CLXXI.

The Carnation Aphis. (Cal.)

(*Rhopalosiphum dianthi.*—Schrank.)

Order, HEMIPTERA; ⎰ Family, APHIDIDÆ.
Sub-order, HOMOPTERA ⎱

[Living upon the stems and leaves of the carnation pink and German ivy, which they puncture with their beaks and imbibe the sap: small green plant-lice, sometimes marked with black.]

The wingless lice are yellowish-green, striped with darker

green. The winged lice have the head and thorax black, the abdomen dark olive-green, with darker transverse lines.

REMEDIES.—No. 3 or 4, one pound to each two gallons of water used; or No. 83 or No. 85.

CHAPTER CLXXII.

The Tulip Aphis.   (Cal.)

(Rhopalosiphum tulipa.—Fonscol.)

Order, HEMIPTERA;   } Family, APHIDIDÆ.
Sub-order, HOMOPTERA ;}

[Living upon the leaves and stems of the garden tulip, which they puncture with their beaks and imbibe the sap; small green plant-lice, sometimes marked with black.]

The wingless lice have the head and thorax blackish-green, the abdomen dark green, with darker transverse lines and a row of black dots along each side.

REMEDIES.—No. 3 or 4, one pound to each two gallons of water used; or No. 83 or No. 85.

CHAPTER CLXXIII.

The Snowball Aphis.   (Cal.)

(Aphis viburni.—Scopoli.)

Order, HEMIPTERA;   } Family, APHIDIDÆ.
Sub-order, HOMOPTERA ;}

[Living upon the leaves and stems of the garden snowball and the high-bush cranberry, which they puncture with their beaks and imbibe the sap; small brown or blackish plant-lice.—Thomas.]

REMEDIES.—No. 3 or 4, one pound to each two gallons of water used; or No. 83 or No. 85.

INSECTS INFESTING THE CONSERVATORY.

CHAPTER CLXXIV.

The Common Mealy-bug. (Cal.)

(Dactylopius adonidum.—Linnæus.)

Order, HEMIPTERA ;
Sub-order, HOMOPTERA ; } Family, COCCIDÆ.

[Living upon various kinds of green-house plants and orange
trees ; a small scale-like insect, more or less covered with a
whitish, mealy powder.]

Fig. 247.—Female Mealy-bug, enlarged—
color, yellowish-white.

Fig. 247.

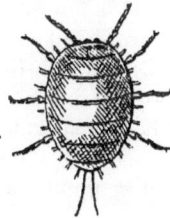

This species is commonly termed *the*
mealy bug, as if there was only one species. It is sometimes very troublesome in
hot-houses, conservatories, and also to some
kinds of garden plants. I have also received
specimens of oranges which were infested
by this, or a closely allied species. The
female (Fig. 247) measures from one line and a quarter to a
line and a half in length ; is of a whitish or yellowish color.
with a brown band on the middle of the back, and is covered
with a mealy powder which is excreted through pores situ
ated on various parts of the body. In addition to this there

is a woolly border around the edge of the body which is longest at the hind end of the body.

Fig. 248.

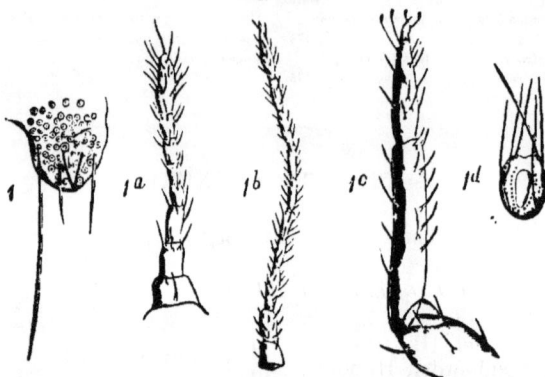

Fig. 248.—Common Mealy-bug; 1, a lobe of hind end of the body of the female; 1*a*, an antenna of the female; 1*b*, an antenna of the male; 1*c*, a leg of the female; 1*d*, anal ring of the female—all highly magnified.

The adult male is furnished with two wings, and is of a brownish color.

REMEDIES.—In conservatories use No. 88; on citrus trees seriously infested use No. 50, and pick off all fruit; when the foliage and fruit are off, spray thoroughly with No. 5 or 7; this should be done before the growing season, when the trees will send out a new foliage.

———

CHAPTER CLXXV.

The Destructive Mealy-bug.

(*Dactylopius destructor.*—Comstock.)

Order, HEMIPTERA; }
Sub-order, HOMOPTERA; } Family, COCCIDÆ.

[Infesting green-house plants and orange trees; a small yellowish bug, thinly covered with a mealy powder.]

Fig. 249.

Fig. 249.—Destructive Mealy-bug, female, greatly enlarged—color, brownish-yellow.

This species is sometimes very destructive to orange trees, and also to green-house plants. The female (Fig. 249) is of a dull brownish-yellow color, very slightly covered with a mealy powder; length, nearly two lines.

Fig. 250.

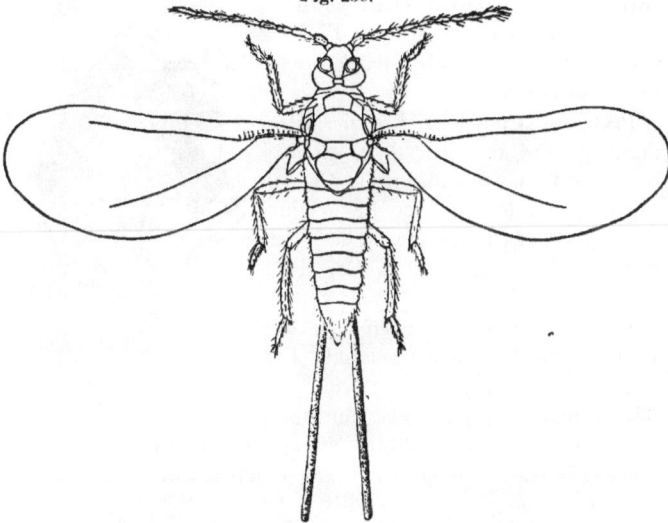

Fig. 250.—Destructive Mealy-bug, male, highly magnified—color, light olive-brown.

The winged male (Fig. 250) is of a light olive-brown color, and is marked with olive bands. The female deposits her eggs in a cottony mass which is excreted from the hind part of her body. The eggs and young bugs are of a light yellow color.

REMEDIES.—Same as in Chapter CLXXIV.

18

The Mealy-bug with Long Threads.

(Dactylopius longifilis.—Comstock.)

Order, HEMIPTERA ;
Sub-order, HOMOPTERA : } Family, COCCIDÆ.

The measurements of insects in this work are given in inches and lines. The above cut represents one inch divided into lines and fractions thereof.

[Infesting green-house plants ; a small yellowish bug thinly covered with a whitish mealy powder.]

Fig. 251.—Mealy-bug with long threads, female, greatly enlarged—color, yellowish.

Fig. 251.

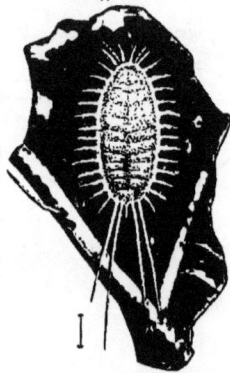

Professor Comstock has described another kind of mealy-bug which I have not found in California ; the following is condensed from his original description of this insect : "The female (Fig. 251) is from two to two and a half lines long, and is of a pale-yellowish color, sparsely covered with a mealy powder. The winged male (Fig. 252) is of a light olive-brown color. The female deposits eggs, but these are so far developed that the young bug issues shortly after the egg is laid. These insects are sometimes quite destructive to green-house plants, especially to ferns."

Fig. 252.

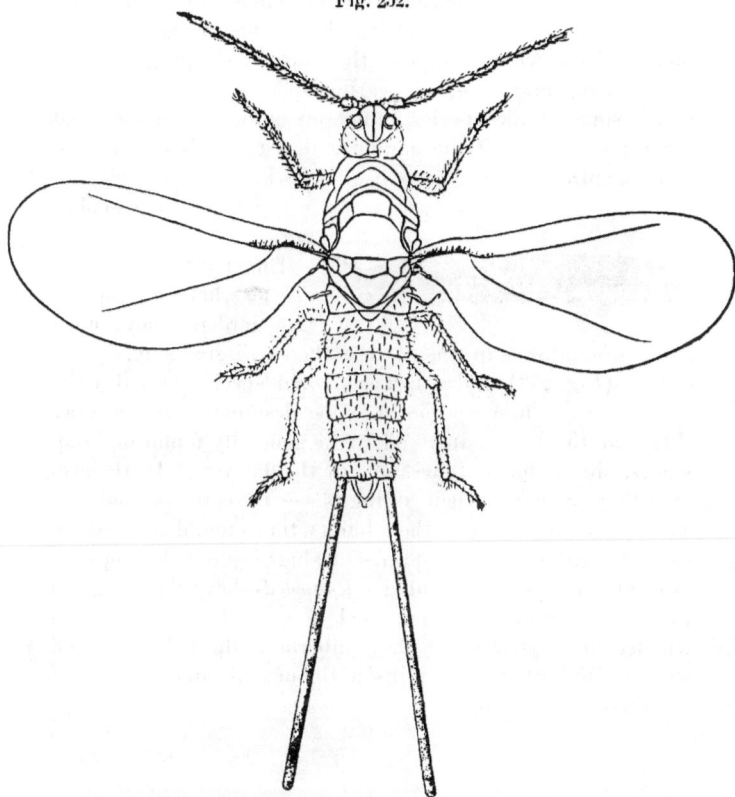

Fig. 252.—Mealy-bug with long threads; male, highly magnified—color, olive-brown.

REMEDIES.—Same as in Chapter CLXXVI.

CHAPTER CLXXVII.

Slugs or Snails. (Cal.)

Class, GASTEROPODA; } Family, HELICIDÆ.
Order, PULMONATA; }

[A small gray or black slug, feeding at night on conservatory and garden plants.]

Although the slugs or snails do not belong to the same sub-kingdom as insects do, *Articulata*, but belong to the sub-kingdom *Mollusca*, which comprise the soft-bodied animals of the shell-bearing and non-shell-bearing species; yet the damage which some of the species inflict on garden vegetables and other productions of the gardener, florist, etc., is sufficient to cause mention of these pests in this work.

Fig. 253.—Slug—color, gray.

Fig. 253.

The most destructive species that is found in the gardens, hot-houses and conservatories in this State is the small gray slug, *Lymax agrestis* (Fig. 253), and a darker colored species, probably the *Lymax ater*. These species may be designated as the gray slug and the black slug. They are generally found in damp places, hiding under stones, etc., in the daytime. In the evening they come from their hiding places in search of food, and as they are gregarious in their habits, they often do great damage. In Europe various species of slugs injure field crops.

The Gasteropoda—meaning *belly-footed*—bury their eggs in the ground; each egg is enclosed in a shell, and hatches in August or September. They hibernate through the cold weather, and attain their full size the next Spring.

REMEDY.—Use No. 87.

INSECTS INFESTING THE HOP PLANT.

CHAPTER CLXXVIII.

The Hopvine Plusia.

(*Plusia balluca.*—Geyer.)

Order, LEPIDOPTERA; Family, NOCTUIDÆ.

[The measurements of insects in this work are given in inches and lines. The above cut represents one inch divided into lines and fractions thereof.]

[Feeding upon the leaves of the hopvine; a green, fourteen-
legged caterpillar, marked with white streaks; the head green,
and destitute of black dots.]

This caterpillar arches up its back slightly when crawling.
When fully grown it measures about one inch and three lines
in length. It then crawls into some sheltered place and spins
a thin cocoon. The fore-wings of the perfect moth expand
about one inch and nine lines; they are almost entirely covered with metallic green scales, and are crossed by two oblique
dark lines. The hind wings are dusky gray.

I have not found this species in this State. The above
account is taken from the Canadian Entomological Report
for 1873.

REMEDIES.—Spray thoroughly with No. 5 or 7; or No. 64,
which is preferable. Read No. 106 carefully.

CHAPTER CLXXIX.

The Hopvine Snout Moth.

(*Hypena humuli.*—Harris.)

Order, LEPIDOPTERA; Family, PYRALIDÆ.

[Feeding upon the leaves of the hopvine; a naked, green, fourteen-legged worm, dotted with black and marked with from two to four white stripes, and having the head dotted with black.]

Fig. 254. — Hopvine Snout-moth; at the left, the moth—colors, dusky-brown, gray and black; *a*, the pupa—color, brown; above the pupa, the caterpillar—color, green, with white and dark lines and black dots.

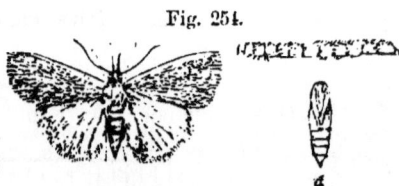

Fig. 254.

These caterpillars (Fig. 254) in walking arch up their backs like the span-worms, and when jarred from the leaves they usually hang suspended by a silken thread. They are very sprightly, frequently leaping sidewise to a distance of several inches when touched. When fully grown they are about an inch long. They then desert the plants and secrete themselves beneath the fallen leaves, etc., or enter the earth and form small cells in which to pass the pupa state (Fig. 254*a*).

The perfect moth (Fig. 254) expands about an inch and three lines. The fore-wings are of a dull brownish color, marked with darker spots and coal-black elevated points. The hind wings are pale dusky brown.

There are at least two broods of these worms produced in one season, and these are to be found from May to September. The manner in which these insects pass the Winter is not known; but the last brood of moths, which appear in September or October, probably hibernate, and deposit their eggs in the following Spring. This species is reported as occurring in this State, but I have never seen a specimen of either larva or perfect insect.

REMEDIES.—Spray thoroughly with No. 5 or 7; or No. 64, which is preferable. Read No. 106 carefully.

CHAPTER CLXXX.

The Semicolon Butterfly. (Cal.)

(*Grapta interrogationis.*—Fabricius.)

Order, LEPIDOPTERA ; Family, NYMPHALIDÆ.

[Feeding upon the leaves of the hopvine ; a browish cater-pillar, mottled with yellow and covered with red or light colored spines tipped with black, or the spines wholly black.]

The fully grown caterpillar is about one inch and six lines long ; the head is reddish-brown, thinly covered with small prickles, and on the top are two branching spines. When about to pupate, it suspends itself by the hind feet.

The chrysalis is ashy-brown, and the head is surmounted with two projections resembling ears ; on the back of the thorax is a nose-like prominence, and on the back are a number of silvery spots.

Fig. 255.

Fig. 255.—Semicolon Butterfly—colors, reddish-brown and black.

The butterfly (Fig. 255) which issues from this chrysalis has the outer margin of all the wings notched ; they are of a reddish-brown color, marked with black and dark brown spots, and with an outer brown border : in some the greater part of the hind wings are black. On the under side of each hind wing is a silvery character resembling a semicolon (;). The fore-wings expand from two inches and six lines to two inches and nine lines.

REMEDIES.—Should the caterpillars appear on the vines, spray thoroughly with No. 5 or 7; or No. 64, which is preferable. Read No. 106 carefully.

CHAPTER CLXXXI.

The Hop Aphis.   (Cal.)

(Phorodon humuli.—Schrank.)

Order, HEMIPTERA; } Family, APHIDIDÆ.
Sub-order, HOMOPTERA; }

[Living upon hopvines, usually near the terminal end, which they puncture with their beaks and imbibe the sap; small yellowish-white or green plant-lice, sometimes marked with black or brown.]

The wingless lice (Fig. 256, 3 and 4,) are yellowish-white or green.

Fig. 256.

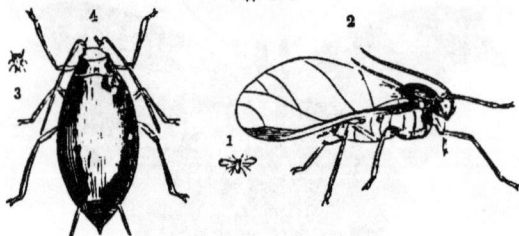

Fig. 256.—Hop Aphis; 4, a wingless aphis, enlarged; 3, the same, natural size—color, yellowish or green; 2, a winged aphis, enlarged; 1, the same, natural size—colors, green, brown and black.

The winged lice (Fig. 256, 1 and 2,) are green, and the females have the head brown or black, and the thorax and abdomen marked with dark brown or black.

REMEDIES.—Use No. 5 or 7; or No. 64, which is preferable. Read No. 106 carefully.

INSECTS INFESTING WHEAT.

CHAPTER CLXXXII.

Crane Flies.   (Cal.)

Order, DIPTERA ; Family, TIPULIDÆ.

[Feeding upon the roots of barley, corn, turnips, strawberries, etc.; a grayish footless grub, having the hind end of the body apparently cut squarely off and provided with tubercles.]

Fig. 257.—Crane Fly ; *1*, the larva or maggot—color, gray ; *2*, the pupa—color, brown ; *3*, the fly—color, brown ; *4*, the eggs.

In England the larvæ (Fig. 257, *1*,) of these flies are known by the name of "Leather Jackets," on account of their tough, leathery skins.   In that country they are sometimes very destructive to various crops by feeding upon the roots ; but I am not aware that they have ever been reported as being very injurious in this State, although quite a number of species are found here.

Fig. 257.

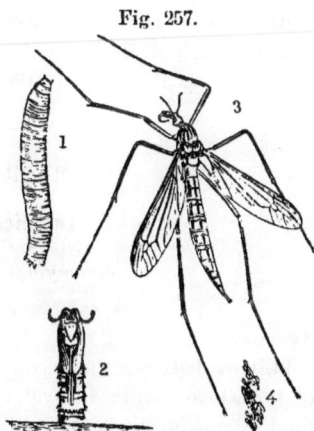

Fig. 258.

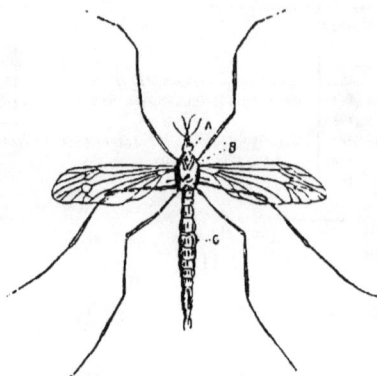

Fig. 258.—Crane Fly—color, brown.

These long-legged two-winged flies (Figs. 257, ♂, and 258), which somewhat resemble gigantic mosquitoes, are familiar to almost everybody, and known as the "Daddy Long-legs." The greater number of them live in decayed vegetable matter, and are hence not injurious to the agriculturist in this State, so far as at present known. The pupa (Fig. 257, ♀.) is somewhat cylindrical, and at the anterior end are two horn-like projections.

REMEDIES.—In gardens, use No. 55, A. In fields, deep plowing, thorough drainage, and cleaning the grounds. See No. 20, No. 32, A. B., and No. 106, A.

CHAPTER CLXXXIII.

The Hessian Fly.

(*Cecidomyia destructor.*—Say.)

Order, DIPTERA: Family, CECIDOMYIDÆ.

[Living between the leaves and the stalk on the lower part of the wheat plant; an oval, cylindrical, white, fleshy maggot, finally transforming into a brownish two-winged fly.]

Among the wheat pests dreaded in the Eastern States is the Hessian fly. I have not found this insect in California.

Fig. 259.—Female Hessian Fly, enlarged—color, black.

Fig. 259.

The Hessian fly (Fig. 259), named *Cecidomyia destructor* by Mr. Say, appeared in this country in 1776, and the general opinion was that it was imported in the stores by the Hessian soldiers in the employ of the British Government, from which it derived the name of "Hessian Fly." There is probably no other species of the insect kingdom that has occasioned so much discussion as to whether it is indigenous to this country. Sir John Banks, an English entomologist, reports to the British Government, in 1789, that no such insect could be found to exist in Germany. Its first appearance in this country was in 1776, on Staten Island, and at Flatbush, on the western extremity of Long Island. Some writers state that it travels about seven miles each Summer. However, Dr. Chapman discovered it in 1787 on the west side of the Alleghany Mountains, which would be about thirty miles each year, dating from Staten Island in 1776. Wheat, rye, barley, and even timothy grass, were attacked by them; and so great was their ravages in the larva state that the cultivation of wheat was abandoned in many places where they had established themselves. It has been a subject of general discussion as to where the female deposits her eggs. Some claim on the young leaves of the wheat, others on the grain before sowing; but the following opinion is evidently correct, as expressed by Dr. Chapman: "The Hessian fly lays her eggs in the small creases of the young leaves of the wheat." Mr. Havens states that the fly lays her eggs on the leaves. Mr. Herrick writes: "I have repeatedly, both in Autumn and in Spring, seen the Hessian fly in the act of depositing her eggs on wheat, and have also found that she selects for this purpose the leaves of the young plant. The eggs are laid in various numbers on the upper surface of the strip-shaped portion (or blade) of the leaf." Mr. Herrick also states that the number of eggs on a single leaf is

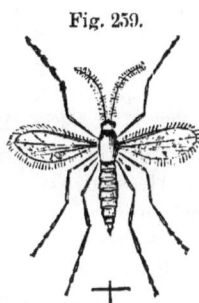

often twenty or thirty. The egg is about one fiftieth of an inch in length, and it hatches in less than fifteen days.

Fig. 260.

Fig. 260.—Larva of Hessian Fly, enlarged—color, yellowish-white.

Fig. 261.

Fig. 261.—The sheath removed, showing three hibernating larvæ or maggots of the Hessian Fly. They are now said to be in the *flax-seed* state, from the resemblance they bear to those seeds.

Harris says: "The maggot (Fig. 260) which proceeds from the egg is of a pale-red color. The maggots crawl down the leaf and work their way between it and the main stalk (Fig. 261), passing downward till they come to a point a little below the surface of the ground, with the head towards the root of the plant. Having thus fixed themselves upon the stalk they become stationary, and never move from the place till their transformations are completed. They do not eat the stalk, neither do they penetrate within it, as some persons have supposed; but they lie lengthwise upon its surface, covered by the lower part of the leaves, and are nourished wholly by the sap, which they appear to take by suction. As they increase in size they grow plump and firm; they become embedded in the side of the stem by pressure of their body upon the growing plant. (Fig. 262.)

Fig. 262.

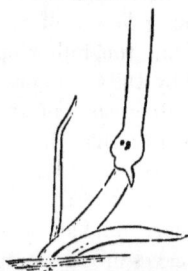

The maggot thus seldom destroys the plant; but where two or three are fixed in this manner around the stem they weaken and impoverish the plant and cause it to fall down or to wither and die." (Fig. 263, right.)

Fig. 262.—Lower part of an infested wheat plant, showing the swelling at the lower end of the sheath caused by the larvæ of the Hessian fly.

The maggot reaches maturity in about forty days, and measures about three twentieths of an inch in length.

Fig. 263.—At the left, a
healthy wheat plant; at the
right, a plant infested by the
maggots or larvæ of the Hessian fly.

The transformation of this
species is different from some
others in the pupa or chrysalis state, inasmuch as it first
passes through what is termed
by naturalists the "flax-seed"
state. It assumes the pupa
form only a few days before
the perfect insect emerges.

This pest produces two
broods each year—one in the
Spring and one in Autumn.

The perfect insect (Fig. 259)
is described as follows: "Col-
or, black, except that the abdomen is more or less tawny,

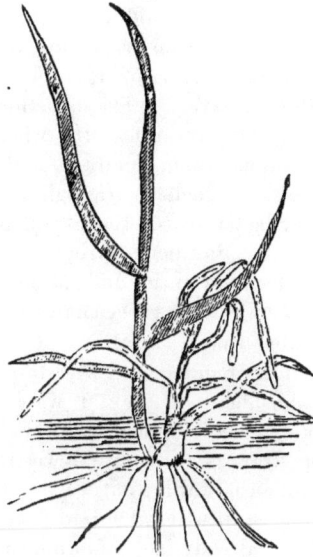

Fig. 263.

each ring being marked more or less with black; legs, pale-red
or brownish, with black feet; wings, three-veined, ciliate or
fringed. The length of the insect is about one tenth of an
inch, the expansion nearly one quarter of an inch." The
insect belongs to the order *diptera*, or two-winged flies.

Dr. Cyrus Thomas, State Entomologist of Illinois, writes as
follows:

"DOES THE HESSIAN FLY EMIGRATE?

" As regards the so-called emigration of this insect, we would
express our disbelief in any such movement from place to
place as is involved in the idea of the word *emigration*. The
history of the insect simply shows that it has steadily spread
from its original point of introduction to new sections of the
country as rapidly as they were settled and wheat became a
staple article of production.

" It is periodically abundant; most other noxious insects
are more abundant some years than others, becoming abun-

dant in some localities and scarce in others. It cannot, therefore, truly be said to *migrate* from one part of a State to another, or from one region to another."

Again, writing of the probable limits of the Hessian fly, Dr. Thomas says: "The question naturally arises whether this pest will ever infest the wheat regions of western Dakota, Montana, Utah, Colorado, and the Pacific States and Territories? We believe (though not aware that such a statement may be hazarded) that it was originally an inhabitant of Central and Southern Europe. It has become acclimated in the Eastern, Atlantic, and Middle States, in the Valley of the Upper St. Lawrence, and in the Valley of the Mississippi River; that it can thrive in the elevated dry Rocky Mountain plateau regions, and withstand the cool nights and dry hot atmosphere of the far west seems doubtful. At least, so slowly has it spread westward, so slight an amount of wheat or straw is transported westward (all produce of this kind going eastward), that we doubt whether, during this century at least, it will extend west of Kansas and Minnesota, where it has already had a foothold for several years."

From what has been said of the Hessian fly, it is obvious that the wheat-growers of California cannot be too careful in preventing the importation of this pest.

REMEDIES.—Should this pest appear in this State, a good preventative of its spreading would be the use of No. 56 or 20.

CHAPTER CLXXXIV.

The Joint Worm. (Cal.)

(*Eurytoma tritici.*—Fitch.)

Order, HYMENOPTERA : Family, CHALCIDIDÆ.

[Living in galls or swellings on the lower part of wheat plants; a footless pale yellow maggot.]

The parent flies appear in the latter part of April or beginning of May, and pierce the stalk in numerous places immediately above one of the joints, inserting an egg into each puncture. These eggs soon hatch into minute footless, yellowish grubs, whose presence in the plant near the joint soon causes the latter to swell, forming a sort of gall which is of a hard, woody texture. These grubs remain in their cells in the center of the stalk all Winter, and are changed to flies in the following Spring. The perfect insect (Fig. 264) is wholly black with the exception of the front shanks (tibiæ), which are pale yellow.

Fig. 264.—Joint Worm Flies, enlarged — color, black; *a*, the female; *c*, her abdomen, still more enlarged; *e*, her antenna, highly magnified; *b*, the male fly; *f*, his abdomen; *d*, his antenna.

Fig. 264.

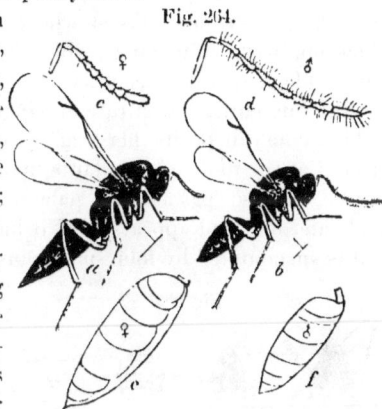

An insect very closely related to the preceding infests rye in the same manner as the latter infests wheat. This is known as the rye fly (*Eurytoma secalis*—Fitch). It is wholly black except the front and hind shanks (tibiæ), which are pale yellow.

Two other species infest barley in the same manner as the above insect infests rye. These are the black-legged barley fly (*Eurytoma hordei*—Harris)—of a black color, with only the knees and feet pale yellow; and the yellow-legged barley fly (*Eurytoma flavipes*—Fitch)—of a black color, with all the legs pale yellow. Some authors contend that the four species of flies described above are but varieties of one species, which was first described by Dr. Harris under the name of *Eurytoma hordei*.

REMEDIES.—Use Nos. 56 and 20.

CHAPTER CLXXXV.

The Chinch Bug.

(Micropus leucopterus.—Say.)

Order, HEMIPTERA ;   }
Sub-order, HOMOPTERA ;} Family, LYGÆIDÆ.

[Living upon the stalks of wheat, corn, oats, etc.; a small
black bug with white wings, which lie flat upon the back and
have a black spot on the outer edge of each; or a yellow, red
or gray bug entirely destitute of wings.]

The female deposits her eggs (Fig. 265, *a* and *b*,) in the
ground at the roots of the plants upon which her progeny are
to feed. These eggs are of a pale amber-white color, elongate
oval, and one end appears as if it had been cut squarely off
and is surmounted by four small rounded tubercles.

Fig. 265.

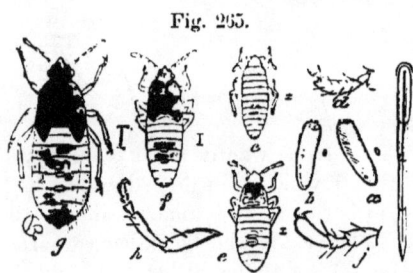

Fig. 265.—Chinch
Bug Larva, Pupa and
Eggs; *g*, the pupa
enlarged — colors,
brownish-black and
gray; *e*, *e* and *f*, the
larvæ in different
stages of their
growth—colors, yel-
low or red; *d*, one of
their feet, enlarged; *b* and *a*, two eggs, highly magnified—
color, whitish or pale brown; *h*, a leg of the adult, enlarged;
j, the foot of the same, still more enlarged; *i*, the beak of the
adult, highly magnified.

The young bugs (Fig. 265, *c* and *e*,) are variously marked;
some are yellow, with an orange stain upon each of the three
larger abdominal segments; others are reddish, with the two
basal abdominal segments pale or with a pale band across the
middle; and still others are of a dingy gray. These insects
obtain their nourishment by puncturing the plants with their
beaks and imbibing the sap; they hibernate in the perfect or
winged stage (Fig. 266).

Fig. 266.—Chinch Bug, enlarged—colors,
black and white.

Fig. 266.

I have had many specimens of insects
sent me, those forwarding them stating
that they thought them to be the chinch
bug, but so far I have not found a specimen
of this species in this State.

REMEDIES.—Should this pest appear in
this State, it can be prevented from spreading by trenching, as recommended in No.
86; use in the trenches No. 81 or 85. If
No. 85 is dusted on the perpendicular side of the trench, and
renewed every three or four hours, few of the insects entering
the trench will survive. See Nos. 20 and 56; also No. 106, A,
as recommended for cleaning hop fields.

CHAPTER CLXXXVI.

The Grain Aphis. (Cal.)

(*Aphis avenæ.*—Fabricius.)

SYNONYM.—*Siphonophora avenæ.*—Fabr.

Order, HEMIPTERA;
Sub-order, HOMOPTERA; } Family, APHIDIDÆ.

[Living upon the stems, leaves, etc., of wheat, barley, etc.;
small brown, green or yellow plant-lice, which puncture the
plant with their beaks and extract the sap.]

Fig. 267.—G r a i n
Aphis; *1*, the winged
louse, enlarged (natu-
ral size indicated at *2*)
—colors, green, yel-
lowish and black; *3*,
a wingless louse, en-
larged (natural size
indicated at *4*)—color,
green, yellow or brown.

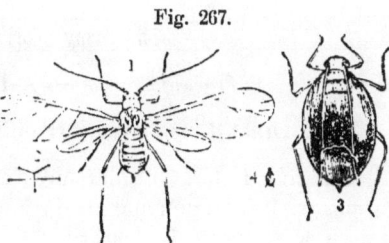

Fig. 267.

The wingless lice (Fig. 267, *3* and *4*,) are either dark brown
19

or dark green, with a reddish band across the abdomen at the base of the honey tubes; the antennæ, knees and feet are black.

The winged female (Fig. 267, *1*,) is either dark green or brown, with a row of black dots on each side of the body; the head and thorax are sometimes marked with black.

The present season, 1883, this species appeared in at least ten counties in this State, and in some instances did considerable damage to wheat, etc., but were prevented from spreading by the late rain, which also gave such impetus to the growth of the infested grain that in many fields the damage is scarcely perceptible. On the 29th and 30th of April, and on the 1st, 2d and 3d of May, unusual flights of the winged insects (Fig. 268) were noticed in this city (Sacramento) but they disappeared entirely with the advent of the rain of May 4, 1883.

Fig. 268.

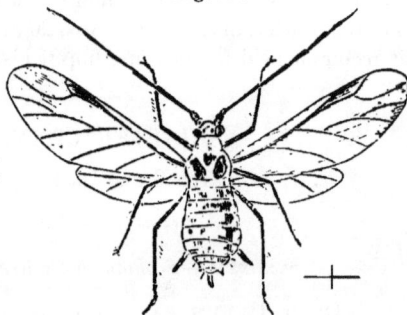

Fig. 268.—G r a i n Aphis (winged female, e n l a r g e d) — colors, green, yellowish and black.

REMEDIES.—Use Nos. 56 and 20.

CHAPTER CLXXXVII.

The Army Worm.

(*Leucania unipuncta.*—Haworth.)

Order, LEPIDOPTERA; Family, NOCTUIDÆ.

[Migrating in large armies and feeding upon the leaves of various kinds of grain and grass; a naked caterpillar having sixteen legs, the body marked with lines of dusky, black, white and yellow.]

This insect plays two roles—one as a cut worm, living concealed during the daytime and coming forth at night to feed, and the other as an army worm (Fig. 269) migrating from one field to another and devouring everything before it. It is in this latter capacity that this insect has figured the most conspicuously.

Fig. 269.—Army Worm— colors, dusky, black, white and yellow.

Fig. 269.

When fully grown it measures a little over one inch in length, is of a dingy velvety black color, with a white line on the back and four light lines on each side of the body, the two uppermost lines white, the two lower ones yellow; the head is yellowish, and marked in front with two curved blackish lines.

Fig. 270.

Fig. 270.—Pupa of Army Worm—color, brown. The caterpillars of this species which act as cut worms, attain a larger size than those which migrate from one place to another, and their colors are more obscure. Three broods or more are produced in one season, the last brood hibernating as partially grown worms. Before pupating they enter the earth and form small cells. (Pupa, Fig. 270.)

Fig. 271.

Fig. 271.—Army Worm Moth; color of fore-wings, light reddish-brown.

The fore-wings of the perfect moth (Fig. 271) expands about one inch and nine lines, are of a light reddish-brown color, dotted with black, and marked near the center with a black dot, from which they derive the specific *unipuncta* or one-dotted. The hind wings are smoky or blackish.

REMEDY.—Use No. 86.

CHAPTER CLXXXVIII.

The California Locust, or Grasshopper. (Cal.)

(*Œdipoda atrox.*—Scudder.)

Order, OTHOPTERA ; Family, ACRIDIDÆ.

[Feeding upon nearly every kind of grain and grass, and also the leaves of trees and nearly every species of plant : a six-legged locust, or grasshopper.]

This insect is closely related to the destructive Rocky Mountain locust—*Caloptenus spretus*—which is found from the Sierras to the middle of the State of Iowa, but has never been reported as occurring west of the Sierras. The California locust has at times appeared in immense swarms in various parts of this State, but has been especially destructive in the Sierra Valley. They sometimes fly in immense swarms, hiding the sun from view for hours at a time. Their usual time of appearance is during the months of June and July. Their flight is usually from the northeast to the southwest.

This insect measures from nine to fifteen lines to the tips of the wing-covers; general color, a dull brown, varied with darker brown : the cheeks and a spot on each side of the thorax are yellow; along the inner or upper edge of each wing-case is a yellow line : along the opposite edge of the wing are from one to three dark brown spots, the one nearest the base the largest, the others small, and sometimes obsolete : there are several other spots of the same color on the disk of each wing-cover; the hind wings are transparent : the hind femorae (thighs) are brown or yellow usually, but not always, marked on the outside with two oblique dark brown spots : the hind tibiæ (shins) are yellowish.

Fig. 272.—Eggs of a Locust, or Grasshopper—color, yellow.

Fig. 272.

The female locust deposits her eggs in masses in holes excavated in the earth (Fig. 272) by the aid of four horny plates which are situated at the posterior end of the abdomen. Several other kinds of locusts, or grasshoppers, are found in this State, but they have about the same habits as the above species, with this exception : that they have never been known to occur in such destructive numbers, or to migrate from one place to another, as the California locusts are known to do.

REMEDY.—Mr. R. B. Blowers has been successful in destroying locusts, or grasshoppers, in his clover and alfalfa fields that were seriously infested, by using an arrangement constructed as follows: He had a pan constructed of sheet iron, ten feet long and three feet wide, turned up a few inches on the sides and ends; this was strengthened by pieces of two by two inch Oregon pine. A board ten feet long, two feet wide and half an inch thick is used on the back of the pan, and fastened with braces. A light runner is placed under each end and in the middle, raising the pan about one inch from the ground. Ropes are attached to each of the front corners, and to these a horse is hitched. Coal tar to the depth of from half an inch to one inch is placed in the pan. A boy is then placed on the horse's back, and drives backward and forward over the infested grounds. If the tar is too thick, thin with petroleum. The dead bodies of the insects can be taken out of the liquid with a rake or some other implement.

CHAPTER CLXXXIX.

The Wheat Midge.

(*Diplosis tritici.*—Kirby.)

Order, DIPTERA; Family, CECIDOMYIDÆ.

[Living in the heads of wheat; an orange-yellow footless maggot.]

Fig. 273.—Wheat Midge and Larva; at the left the larva, or maggot, natural size; in the middle the same, highly magnified—color, orange; at the right the fly, or midge, with its wings closed—color, orange.

Fig. 273.

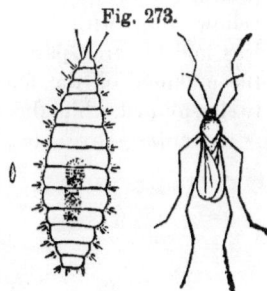

As we have never met with this insect in this State, we take the following condensed account from Packard's "Guide to the Study of Insects."

" When the wheat is in blossom the females lay their eggs
within the chaffy scales of the flowers, in clusters of from two
to fifteen or more.   In eight or ten days the eggs disclose the
transparent maggot, which, with age, becomes orange-colored,

Fig. 274.

and when fully grown (Fig. 273) measures about
one line and a half long.

[Fig. 274.—A green kernel of wheat infested
by the larvæ, or maggots of the wheat midge—
color of larva, orange.]

" They crowd around the germ of the wheat
(Fig. 274), which by pressure becomes shriveled
and aborted.   At the end of July, and in the
beginning of August, the maggots become full fed, and in a
few days cast their skins; shortly after this—and sometimes
before it—they descend to the ground, which they enter to the
depth of about an inch; here each one spins around its body
a minute silken cocoon.   They remain in their cocoons unchanged during the Winter, and
are changed to pupæ the following June.

[Fig. 275.—Wheat Midge, female,
enlarged and natural size—color,
bright orange.]

" When the perfect fly is about
to issue, the pupa works its way
to the surface; this usually occurs in June or July.   The fly
(Fig. 273 and 275) is of an orange-
yellow color, with long slender
legs, and two transparent wings;
the antennæ of the female are
twelve-jointed, while those of the male are twenty-four jointed."

Fig. 275.

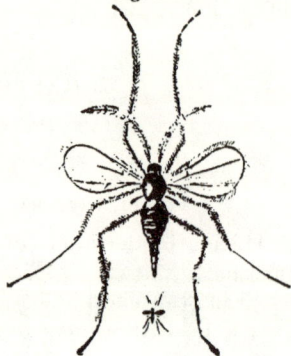

REMEDIES.—Use Nos. 20 and 56.

INSECTS INFESTING BARLEY.

The following insects infest barley, but are treated of elsewhere in this work:

The Crane Flies (*Tipula*).
The Black-legged Barley Fly (*Eurytoma hordei*).

The Yellow-legged Barley Fly (*Eurytoma flavipes*).
The Grain Aphis (*Aphis avenæ*).

INSECTS INFESTING RYE.

The Rye Fly (*Eurytoma secalis*) (See article on The Joint-worm, Chapter CLXXXIV.)

INSECTS INFESTING CLOVER AND ALFALFA.

CHAPTER CLC.
The Clover-root Borer.
(*Hylastes trifolii.*—Muller.)

Order, COLEOPTERA ; Family, SCOLYTIDÆ.

[Living in the roots of clover; a whitish six-legged grub, about one and a half lines long, with the head yellow, finally changing into a reddish-brown slightly hairy beetle.]

These insects usually reach maturity in October, and the

beetles hibernate in their burrows. In the following Spring they come forth from their Winter quarters, and the female, when about to deposit her eggs, first gnaws a large cavity in the crown of the roots, and then deposits therein from four to six pale whitish elliptical eggs, which hatch in about a week. The young larva begins to bore downward into the root, passing between the outer skin and the heart. When fully grown, it assumes the pupa form in the bottom of its burrow, and the perfect beetle issues in the course of a week or so.

I have found a grub in roots of alfalfa, but did not succeed in rearing the perfect insect.

CHAPTER CXCI.

The Clover-stem Borer.   (Cal.)

(*Languria mozardi.*—Latreille.)

Order. COLEOPTERA : Family, EROTYLIDÆ.

[Living within the stems of clover (Fig. 276, Plate 3.); a yellow six legged larva about three lines long, with two curved spines at the hind end of the body; finally changing into a blue-black beetle, having the thorax yellowish-red.]

The female beetle first gnaws a hole into the stem and then deposits therein a single egg (Fig. 277, Plate 3,) of a yellowish color, rounded at each end and somewhat curved. The young larva (Fig. 278, Plate 3,) burrows downward, consuming the central substance of the stem to a distance of six or eight inches. It assumes the pupa form (Fig. 279, Plate 3.) in the lower end of its burrow, and the beetle (Fig. 280, Plate 3.) which issues in the Autumn, deserts its burrow and, at the approach of cold weather, seeks some sheltered place in which to pass the Winter. I have found this, or a closely allied species, in alfalfa.

REMEDIES.—Clean cultivation : see Nos. 20 and 106, A.

CHAPTER CXCII.
The Clover-hay Worm.
(*Asopia costalis.*—Fabricius.)

Order, LEPIDOPTERA ; Family, PYRALIDÆ.

[Living in silken tubes in clover-hay ; an olive-brown or dull white sixteen legged-caterpillar.]

When fully grown these caterpillars (Fig. 281, *1* and *2*,) measure about nine lines in length, and shortly afterward assume the pupa form (Fig. 281, *4*,). There are probably two broods in one season, and the last brood passes the Winter in the larva state. In Europe the perfect moth (Fig. 281, *5* and *6*,) is known as the "golden fringe;" it is of a liliaceous or purplish color with a silky gloss; the front wings are tinged with gray and marked on the front edge with two large, bright, golden-yellow spots, which are narrowed behind into a slender line that extends across the wing; the hind wings are lighter than the front ones and are crossed by two light straw-colored lines; all of the wings are margined with deep orange; expands about ten lines.

Fig. 281. — Clover-hay Worm; *1*, back view of one of the worms; *2*, side view of the same (both suspended by silken threads) — color, brownish; *7*, a worm in its silken tube; *4*, the pupa —color, honey-yellow; *5*, the cocoon—color, whitish; *5*, the moth with its wings expanded; *6*, the

Fig. 281.

same with its wings closed—colors, grayish purple and yellow.

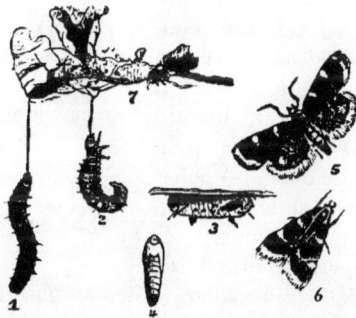

The larvæ or caterpillars of this moth are sometimes very destructive to clover-hay, especially to that which has stood in the stack for several years. They are most abundant in the bottom of the stack, where the infested hay will frequently be found to be webbed together by the silken tubes which they spin for a habitation in which to dwell.

CHAPTER CXCIII.

The Wire Worm.

Order, COLEOPTERA; Family, ELATERIDÆ.

[Feeding upon the roots of wheat, oats, corn, potatoes, and many other plants; a nearly cylindrical reddish or yellowish-brown six-legged larva or worm.]

Fig. 282. — Wire-worms and Snapping-beetles; 3, *Elater obscurus*, enlarged (natural size indicated at 4) — color, black or brown; 2, *E. lineatus*, enlarged (natural size indicated at 1) — color, grayish-brown, with darker lines; 5, *E. sputator*, natural size;

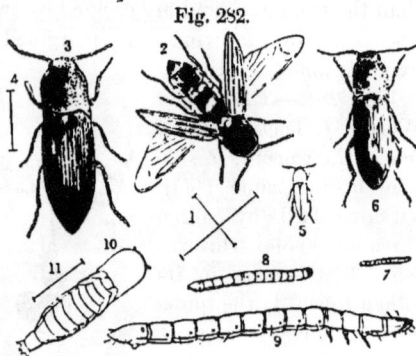

Fig. 282.

6, the same enlarged — color, blackish; 10, the pupa of a wire-worm, enlarged — color, yellowish; 8, the larva of *E. lineatus*, natural size; 9, the same enlarged — color, yellowish; 7, a wire-worm, natural size. All of these are natives of the Eastern continent.

This is the larva of a beetle known as the "snapping beetle," "skipjack," etc. (Fig. 282, 2, 3, 5, and 6). They derive this name from the method they adopt to attain an upright posture when they happen to fall upon their backs. Being unable to turn over, they make a sudden jerk or spring, accompanied by a clicking noise; hence the name.

The larva (Fig. 282, 7, 8, and 9,) has six legs, two beneath each of the three first segments, and a tubercle under the last segment of the abdomen, which it uses as a pro-leg.

It is an undecided question among naturalists as to the length of time these insects remain in the larva state; some say two years; others, as high as five; however, it is generally conceded to be about three years, which, like many other beetles of long larval lives, have an opportunity of doing immense injury to crops, etc.

Fig. 283. — Wire-worm—color, yellowish-brown.

Fig. 283.

Mr. Curtis writes: "Wherever grasses will grow, the wire-worm (Fig. 283) may be found." Dr. Fitch writes: "They abound alike on the roots of the coarsest sedges, and on other wild grasses, on the borders of marshes and on those of the most delicate pasture." They do great damage to crops, especially corn, but can be found feeding upon the roots of barley, cabbage, wheat, rye, potatoes, asparagus, carrots, oats, etc., of the field, and dahlias, pinks, carnations, etc., of the garden. A great many remedies have been published for destroying those pests, but are generally inapplicable to field crops.

Fig. 284. — Milli-pedes or Thousand-legged Worms—color, brownish or yellowish.

Note. — The milli-pedes, centipedes, or thousand-legged worms (Fig. 284), are sometimes mistaken

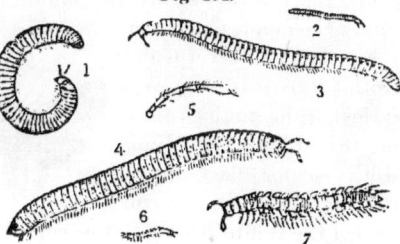

Fig. 284.

for wire-worms. They can easily be distinguished from the
latter by having one or more pairs of legs to each segment of
the body, whereas the true wire-worms have a pair of legs
attached to the first three segments only, the remaining segments being destitute of legs; compare Figs. 283 and 284. On
roots of trees, use Remedy No. 59; on roots of cereals, etc., No.
55, A.

REMEDY.—Use No. 55.

CHAPTER CXCIV.

Cut-worms.   (Cal.)

Order. LEPIDOPTERA: Family, NOCTUIDÆ.

[Feeding upon the roots of corn, melons, cabbages, etc.,
usually severing the plants; or ascending fruit trees, grapevines, rosebushes, etc., and devouring the buds; a naked,
greasy looking sixteen-legged caterpillar or worm.]

Fig. 285.   Fig. 286.

   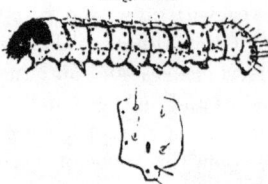

Fig. 285.—W-marked Cut-worm (*Agrotis clandestina*)—
color, ash-gray and black or brown.

Fig. 286.—Glassy Cut-worm (*Hadena devastator*); the lower
figure represents one of the middle segments of its body
enlarged—color, pale green.

Fig. 287.—Cut-worm and   Fig. 287.
Moth (*Mamestra renigera*):
at the left the moth (known
as the Figure 8 Minor
Moth)—color, dark gray;
at the right the worm

(called the White Bristly Cut-worm)—color, yellowish gray.

These worms (Figs. 285, 286 and 287) are very destructive to young plants of various kinds: their usual mode of procedure is to cut the young plants off and drag them into their burrows, there to devour them at their leisure. Others ascend various kinds of trees and shrubs and feed upon their leaves.

Fig. 288.

Fig. 288.—Pupa of Cut-worm in its earthen cell—color, brown.

They come forth to feed mostly in the night, remaining in their burrows during the daytime. When about to pupate they form smooth cells (Fig. 288) in which to undergo their transformations.

Fig. 289.—Gothic Dart Moth (*Agrotis subgothica*, the parent of the Dingy Cut-worm)—colors, gray, yellow and white.

Fig. 289.

The perfect moths (Figs. 289 and 287) are usually of an ashen-gray color, variously marked with dusky or black; their hind wings are mostly whitish or smoky.

REMEDY.—Use No. 73.

CHAPTER CXCV.

The White Grub. (Cal.)

(*Lachnosterna quercina.*—Knoch.)

Order, COLEOPTERA; Family, SCARABÆIDÆ.

[Feeding upon the roots of corn, potatoes, strawberry plants, nursery stock and various kinds of grasses; a white curved six-legged grub; or feeding upon the leaves of the apple and other trees: a robust reddish-brown beetle.]

The beetle (Fig. 290, 3 and 4,) which deposits the eggs from which these grubs (Fig. 290, 2,) hatch, is commonly known as the "May-beetle," or "June-bug." It is from eight lines to nearly an inch long, of a nearly uniform chestnut-brown color, and the breast is covered with yellowish hairs.

Fig. 290.

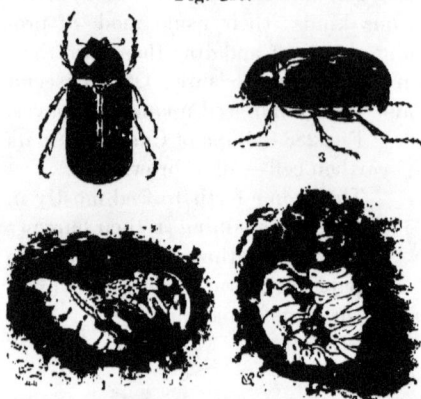

Fig. 290.—White Grub; 4, the beetle, back view; 3, the same, side view— color, chestnut- brown; 1, the pupa in its cell--color, yel- lowish-white; 2, the grub in its burrow in the earth—color, white, with the head brown.

The eggs are laid in the earth, and these soon hatch into small, whitish six-legged grubs, with brownish heads. When at rest they lie upon one side, the body being curved so as to bring the head and tail nearly in contact. They feed upon the roots of various kinds of plants, and are supposed to spend two or three years in this, their larval stage.

When about to pupate they form smooth oval cells in the earth (Fig. 290, 1,) in which to undergo their transformations. The perfect insects feed upon the leaves of the apple, cherry, and various other fruit and ornamental trees.

REMEDY.—Use No. 107.

CHAPTER CXCVI.

The Corn-worm, or Boll-worm.   (Cal.)

(Heliothis armigera.—Hubner.)

Order, LEPIDOPTERA : Family, NOCTUIDÆ.

[Living in the ears of corn and feeding upon the kernels, or burrowing into the bolls of cotton: a green or brownish sixteen-legged worm, marked with darker stripes.]

The body of this caterpillar (Fig. 291c) is sparsely covered with polished black elevated dots, and the head is brown.

When fully grown it measures about one inch and six lines in length. It sometimes assumes the pupa form in its burrow, but usually descends to the ground, which it enters and forms a smooth cell (Fig. 291*d*) in which to undergo its transformations.

Fig. 291.

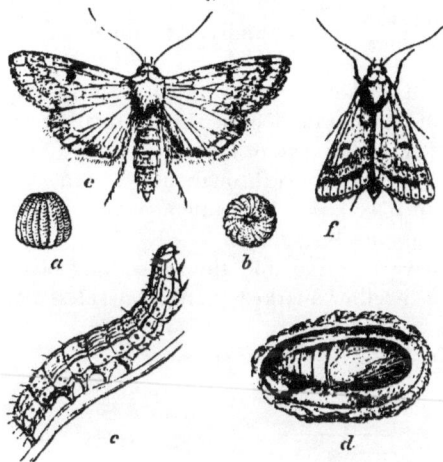

Fig. 291.—Corn-worm; *c*, the worm—colors, green and brown; *d*, the pupa in its cell—color, brown; *e*, the moth with wings expanded; *f*, the same, with wings closed—color, pale yellow tinged with olive-green; *a*, an egg, side view, highly magnified; *b*, the same, top view.

The fore-wings of the perfect moth (Fig. 291, *e* and *f*,) expand about one inch and six lines, are of a pale yellowish color, sometimes tinged with olive-green and reddish markings. Near the center of each wing is a blackish spot, and near the outer margin is a dusky spot. The hind wings are paler, with a blackish outer border, and containing a light colored spot in the center.

The caterpillar of this moth will bite human flesh. A case occurred in the city of Sacramento, where the arm of a baby was attacked by one of these worms until it bled freely; and several other cases have been reported.

REMEDIES.—Use No. 5 or 7; spray the ears of corn to prevent the moth from depositing eggs on them.

CHAPTER CXCVII.

The Corn Aphis.

(*Aphis maidis.*—Fitch.)

Order. HEMIPTERA ;　　}
Sub-order. HOMOPTERA ;　} Family, APHIDIDÆ.

[Living on the roots, stalks, or ears of corn, which they puncture with their beaks and imbibe the sap; a small greenish plant-louse, marked with dusky or black.]

The wingless lice are apple green throughout, excepting the front of the head, which is dark. The antennæ are brownish and the honey tubes are black.

The winged lice have the head and thorax shining black, the abdomen greenish-yellow marked along the sides with black dots.

REMEDY.—Use No. 108.

———

INSECTS INFESTING TOBACCO.

———

The Tobacco Worm (*Macrosila Carolina*) infests Tobacco, and is treated of in another part of this work.

INSECTS INFESTING COTTON.

CHAPTER CXCVIII.

The Cotton Worm.

(*Aletia argillacea.*—Hubner.)

Order, LEPIDOPTERA : Family, NOCTUIDÆ.

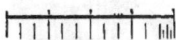

[The measurements of insects in this work are given in inches and lines. The above cut represents one inch divided into lines and fractions thereof.]

[Feeding upon the leaves of the cotton plant ; a pale greenish caterpillar, dotted with black and marked with yellow stripes.]

Fig. 292.

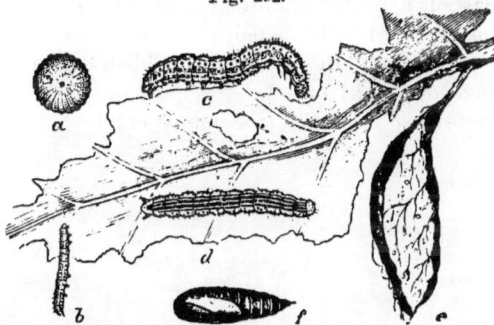

Fig. 292.—Cotton Worm : *a*, the egg, magnified—color, green ; *b*, *c*. and *d*, worms of different sizes—color, green, with

20

lighter lines; *f*, the pupa—color, brown; *c*, a rolled leaf, in which the worm assumes the pupa form.

Although this caterpillar (Fig. 292, *b*, *c*, and *d*,) is provided with a full complement of sixteen legs, yet in walking it is obliged to arch up its body, somewhat as the span-worms do. This is due to the fact that the front pair of legs beneath the middle of the body are so much shortened as to be of no use to the caterpillar. When about to pupate it rolls a leaf around its body (Fig. 292*c*), fastening the edges together with silken threads; it then lines the interior with a layer of silk, and soon afterward assumes the pupa form. (Fig. 292*f*.)

Fig. 293.

Fig. 293.—Cotton-worm Moths; *a*, with wings expanded; *b*, with wings closed — color, yellowish, with lilac lines.

The perfect moth (Fig. 293) has the fore-wings of a nearly uniform reddish-brown color, with a dark spot, centered with two pale ones, near the center of each wing. The head and thorax are also reddish-brown, while the hind wings are smoky brown.

REMEDY.—Use No. 103. (See note at end.)

INSECTS INFESTING THE SQUASH.

CHAPTER CXCIX.

The Squash Vine Borer. (Cal.)

(*Egeria cucurbitæ.*—Harris.)

Order, LEPIDOPTERA; Family, EGERIDÆ.

[Living in squash and pumpkin vines near the roots: a whitish sixteen-legged larva about one inch long.]

Fig. 294.

Fig. 294.—Squash-vine Borer—colors, white and brown.

The eggs from which these borers (Fig. 294) hatch are deposited on the vines near the roots, and as soon as hatched the young borers penetrate the stems and devour the pith, frequently occasioning the death of the vines. They reach their full growth in Autumn, and usually enter the earth and construct a sort of cocoon in which to pass their transformations; sometimes, however, they pass through their transformations within their burrows. They pass the Winter in the pupa state, and are changed to perfect moths in the following Spring.

Fig. 295.—Squash-vine Borer (the moth) —colors, black, olive and orange.

Fig. 295.

The perfect insect or moth (Fig. 295) has blue-black fore-wings, which expand from one inch to one inch and six lines; the hind wings are wholly transparent; the abdomen is deep orange, marked with black.

REMEDIES.—Spray with No. 5 or 7, or No. 68 or 104. (See note at end of No. 98.)

CHAPTER CC.

The Squash Bug. (Cal.)

(*Coreus tristis.*—DeGeer.)

Order, HEMIPTERA;   } Family. CORISIDÆ.
Sub-order. HETEROPTERA: {

[Living upon the leaves and fruit of the squash and pumpkin vines; a rusty-black elongated bug which punctures the plants with its beak and imbibes the sap.]

Fig. 296.—Squash Bug—colors blackish brown and dirty yellow.

Fig. 296.

The perfect or winged bugs (Fig. 296) pass the Winter in some sheltered situation; they are about seven lines long, of a dirty yellow color beneath and black above, the projecting edges of the abdomen spotted with pale yellow; the wing-covers are rusty black, with the thin overlapping ends black.

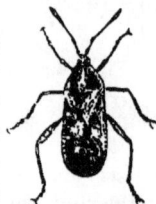

The females lay their eggs on the underside of the leaves, to which they fasten them with a gummy substance.

The young are of a gray color, and as they increase in size and by molting their skins, they change to a dull yellow color. The leaves on which the young feed soon wither and become dry and wrinkled: the bugs then change to fresh leaves, and in this manner the vine is eventually destroyed.

The squash bug can be found in all stages of its growth throughout the Summer season. It hibernates around fences, crevices of walls, among weeds, and in other sheltered places. In January, 1882, I found on pine trees, which were covered with ivy, immense numbers of squash bugs in all stages of their existence—larva, pupa and imago.

REMEDIES.—Use Nos. 20 and 19, or No. 64: thorough spraying with the latter (No. 64), one pound of the soap to each gallon of water, then adding the buhach, as described, will be effectual. See also No. 114.

INSECTS INFESTING THE PUMPKIN.

The following insects infest the pumpkin, and are treated of in another part of this work:

The Squash-vine Borer (*Egeria cucurbitæ*).
The Striped Cucumber Beetle (*Diabrotica vittata*).

The Squash-bug (*Coreus tristis*).
The Cucumber Aphis (*Siphonophora cucurbitæ*).

INSECTS INFESTING THE MELON.

The following insects also infest the Melon:
Cut Worms.

The Striped Cucumber Beetle (*Diabrotica vittata*).
The Cucumber Flea Beetle (*Haltica cucumeris*).
The Green Fruit Beetle (*Gymnetis nitida*).
The Pickle Worm (*Phacellura nitidalis*).

CHAPTER CCI.

The Melon Worm.   (Cal.)

(*Phacellura hyalinitalis.*—Linnæus.)

Order, LEPIDOPTERA: Family, PYRALIDÆ.

[Eating large holes in cucumbers, melons and pumpkins, and also feeding upon the leaves of these plants; a yellowish green sixteen-legged worm.]

The worms which occur upon the leaves (Fig. 297, Plate 4,) usually web up the latter with silken threads. When fully grown they measure about an inch and three lines in length; they then spin their cocoons among the leaves, and soon afterwards assume the pupa form. The last brood pass the Winter in the pupa state.

The perfect moth (Fig. 297, Plate I,) expands about an inch, is of a pearly white color with a black stripe along the front edge of the fore-wings, and a similar stripe along the outer edge of all the wings.

REMEDIES.—Use Nos. 20 and 14 ; spray thoroughly with No. 51 or 65.

INSECTS INFESTING THE CUCUMBER

CHAPTER CCII.

The Striped Cucumber-beetle.   (Cal.)

(*Diabrotica vittata.*—Olivier.)

SYNONYM—*Galeruca vittata.*

Order. COLEOPTERA ;   Family. CHRYSOMELIDÆ.

[Living in the stems of cucumber, melon and squash vines ; a slender whitish six-legged grub, about four lines long, with a brownish-black head, and a spot of the same color on the last segment ; finally changing into a yellowish leaf-eating beetle (Fig. 298) about three lines long, and having two black dots on the thorax and three black stripes on the wing-cases.]

Fig. 298.

Fig. 298.—Striped Cucumber-beetle—c o l o r s, yellow and black.

Fig. 299.—Larva of the Striped Cucumber-beetle, enlarged; 1, back view; 2, side view —color, yellowish-white.

In early Summer the grubs (Fig. 299) of this species are sometimes found in the stems of young cucumber vines, near the roots. They bore the stems in various directions, and after attaining their full size they desert the plants and form small cells in the earth, in which to pass the pupa state.

REMEDY.—Use No. 111.

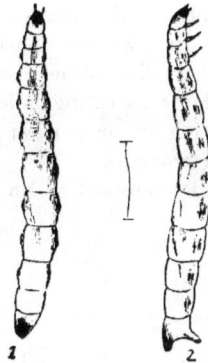

Fig. 299.

CHAPTER CCIII.

The Cucumber Flea-beetle. (Cal.)

(*Haltica cucumeris.*—Harris.)

SYNONYM.—*Epitrix cucumeris.*—Harris.

Order, COLEOPTERA; Family, CHRYSOMELIDÆ.

[Feeding upon the leaves of the cucumber, melon, potato, etc.; a small black beetle.]

Fig. 300.—Cucumber Flea-beetle, enlarged—color, black.

Fig. 300.

This beetle (Fig. 300) is less than a line long, and the antennæ and legs are yellowish, except the hind thighs, which are black and greatly swollen, enabling the insect to leap to a considerable distance. The larvæ, or young, are supposed to live in the leaves of various kinds of plants, and to enter the earth to pupate.

The perfect beetles pass the Winter in some sheltered place. These beetles attack the seed-leaves of cucumber plants as soon as they appear above the ground, thereby destroying them. They also attack tomatoes, potatoes, etc., and injure the plants by eating holes in the leaves—the injury to the plant being in proportion to the extent of surface of the leaves destroyed.

REMEDY.—Use No. 111.

CHAPTER CCIV.

The Cucumber Aphis.   (Cal.)

(*Siphonophora cucurbita.*—Middleton.)

Order, HEMIPTERA;
Sub-order, HOMOPTERA; } Family, APHIDIDÆ.

[Living on the under sides of the leaves of cucumber, squash and pumpkin vines, which they puncture with their beaks and imbibe the sap; small green plant-lice, sometimes marked with brown.]

The wingless lice are green, with a few darker markings. The winged lice are green, the head yellowish, the eyes brown and the thorax usually brownish.

REMEDY.—Use No. 111.

CHAPTER CCV.

The Pickle Worm.   (Cal.)

(*Phacellura nitidalis.*—Cramer.)

Order, LEPIDOPTERA; Family, PYRALIDÆ.

[Boring cylindrical holes into cucumbers, melons, and squashes; a pale greenish-yellow worm, with a reddish head.]

Fig. 301.—Pickle Worm; *j*, an infested cucumber; *a*, the worm—color, yellowish or green; *c*,one of the middle

Fig. 301.

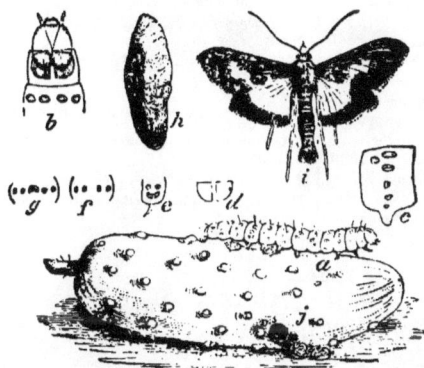

segments of its body, enlarged; *d*, the horny plate on the top of the first segment, called the *cervical shield; e*,arrangement of black spots on one side of the first segment; *f*, shows arrangement of black spots on top of the second and third segments; *g*, arrangement of black spots on top of the last segment; *b*, the head and fore part of the caterpillar's body, enlarged, back view; *h*, the cocoon—color, white; *i*, the male moth—color, yellowish-brown and dull golden yellow.

When fully grown this worm (Fig. 301*a*) is about an inch long; it then crawls beneath the leaves, etc., which lie upon the ground, and spins a slight whitish cocoon (Fig. 301*h*). The last brood passes the Winter in the pupa state.

The perfect moth (Fig. 301*i*) expands about one inch and three lines. The fore-wings are of a yellowish-brown color with a purplish reflection, and near the middle of the hind edge is an irregular semi-transparent dull golden-yellow spot. The hind wings are of the same brownish color, with their inner two thirds semi-transparent and dull golden-yellow.

REMEDIES.—No. 20; spray fruit and foliage with No. 5 or 7, to prevent moth from depositing its eggs.

INSECTS INFESTING THE POTATO.

CHAPTER CCVI.

The Potato-stalk Weevil.   (Cal.)

(*Baridius trinotatus.—*Say.)

Order, COLEOPTERA ; Family, CURCULIONIDÆ.

[Living within the stalks of potatoes ; a whitish footless larva, about three lines long ; finally transforming into a bluish-gray snout-beetle, which is marked at the base of the thorax with three black dots.]

Fig. 302.—Potato-stalk Weevil ; *a*, the larva enlarged—color, white ; *b*, the pupa in its burrow, enlarged—color, yellowish white.

Fig. 302.

The female weevil makes a slit in the stalk by means of her snout, and then deposits a single egg therein. The larva hatching from this egg (Fig. 302*a*) burrows downward into the stalk, sometimes extending its burrow even into the roots. It assumes the pupa form (Fig. 302*b*) in its burrow, and the perfect beetle issues in the latter part of the Summer or late in the Fall, and passes the Winter in some sheltered situation.

Fig. 303.— Potato-stalk Weevil—color, bluish-black.

Fig. 303.

This weevil (Fig. 303) is about two lines long, of an elongate-oval form, and is of a bluish-black color, with three black dots at the base of the thorax, the middle dot being situated upon the small wedge-shaped piece technically called the *scutel.*

In the southern part of this State these insects are quite frequently met with upon the Jamestown weed (*Datura stramonium*), in the stems of which they breed.

REMEDY.—Use No. 25.

CHAPTER CCVII.

The Stalk Borer. (Cal.)

(*Gortyna nitela.*—Guenee.)

Order, LEPIDOPTERA; Family, NOCTUIDÆ.

[Burrowing into the stalks of corn, potatoes, tomatoes, currant-bushes, etc.; a brownish sixteen-legged worm marked with white stripes.]

Fig. 304.—S t a l k Borer: *1.* the moth—color, gray; *2.* the c a t e r p i l l a r—colors, white and brown.

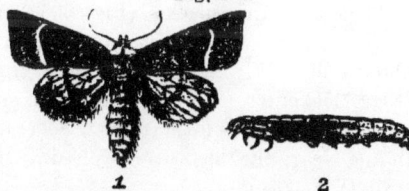

Fig. 304.

This b o r e r (Fig. 304, *2,*) when fully grown measures about one inch and three lines in length, is of a reddish-brown color, marked on the back with three white lines, the two lowest ones interrupted on the segments from the fourth to the seventh, inclusive; the underpart of these segments is reddish-brown.

while this part of the remaining segments is greenish-white ; the
head is yellowish-brown, usually with a dark dash upon each
side. It sometimes assumes the pupa form within its burrow,
but it usually enters the earth, where it forms a cell in which to
undergo its transformations. The fore-wings of the moth
(Fig. 304, *l*,) expand from an inch and one line to an inch and
six lines : they are of a mouse gray color, tinged with lilac,
and finally sprinkled over with bright yellow scales ; toward
the outer edge they are crossed by a yellow line. These borers
appear from April to August, there being but one annual
brood ; the moths are supposed either to hibernate, or to
deposit their eggs in the Fall, these not hatching until the following Spring ; the first supposition is probably the correct one.

REMEDY.—Use No. 25.

CHAPTER CCVIII.

The Colorado Potato Beetle.

(*Doryphora 10-lineata.*—Say.)

Order, COLEOPTERA : Family, CHRYSOMELIDÆ.

[Feeding upon the leaves of the potato and tomato : a six-
legged yellowish grub about six lines long, and marked with
two rows of black dots along each side of the body, with the
head and legs also black ; finally transforming into a robust
yellowish beetle, having a black spot on the head, the thorax
covered with dots and short streaks of black, and the wing-
cases marked with ten black lines.]

The egg of this species (Fig. 305a) are deposited in small
clusters upon the leaves, and hatch out in the course of
about one week. After attaining their full growth the larvæ
(Fig. 305b) enter the earth and form small cells in which to
undergo their transformations. Several broods are produced
in one year, the last brood passing the Winter in the pupa
state (Fig. 305c).

Fig. 305.—Colorado Potato Beetle : *a, a,* the eggs—color,
yellow ; *b, b, b,* the larva in different stages of its growth—
colors, yellow and black ; *c,* the pupa—color, yellow ; *d, d,* the
beetle—colors, yellow and black ; *e,* one of the wing-cases,
enlarged ; *f,* one of the hind legs, enlarged.

Fig. 305.

REMEDIES.—No. 103; spray thoroughly with liquid solution (see note at end of No. 103); good results have been obtained by hand picking, etc. (See No. 112). For further description of remedies, see Professor Riley's Seventh Missouri Report, pages 1 to 19.

CHAPTER CCIX.

The Three-lined Potato Beetle. (Cal.)

(*Lema trilineata.*—Olivier.)

Order, COLEOPTERA; Family, CHRYSOMELIDÆ.

[Feeding upon the leaves of the potato; a six-legged slug-like larva, which is finally changed into a yellowish beetle, having two black dots on the thorax and three black stripes on the wing-cases.]

Fig. 307.

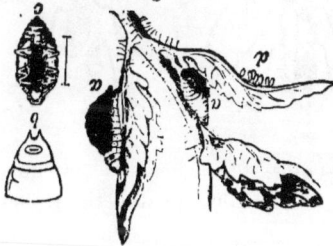

Fig. 306.

Fig. 307.—Larva. Pupa and Eggs of Three-lined Potato Beetle—; *a, a.* the larva—color, yellow; *b*, the tip of its body, enlarged; *c.* the pupa, enlarged —color. yellow; *d.* the eggs —color, yellow.

Fig. 306.—Three-lined Potato Beetle, enlarged—colors yellow and black.

The parent beetle (Fig. 306) deposits her eggs (Fig. 307d) in patches of from half a dozen to a dozen, usually placing them on the underside of the leaves; they are somewhat oval in shape and of a golden-yellow color. They hatch in about two weeks, and the larvæ (Fig. 307a) reach their full growth in a few weeks and then enter the earth and form small cells in which to pass the pupa state (Fig. 307e). Several broods are produced in one season, and the perfect beetles pass the Winter in some secluded place. The larva is of a dull yellowish color, with a black head; it has the habit of covering its back with its own excrements. This insect is very common in the southern part of the State.

REMEDIES.—Use same as in Chapter CCVIII.

CHAPTER CCX.

The Ash-colored Blister-beetle. (Cal.)

(*Cantharis cinerea.*—Fabricius.)

SYNONYMS.—*Lytta cinerea*—Fab; *Macrobasis unicolor*—Kirby.

Order, COLEOPTERA; Family, MELOIDÆ.

[An elongate ash-colored beetle, about six lines long, feeding upon the leaves of potatoes, etc.]

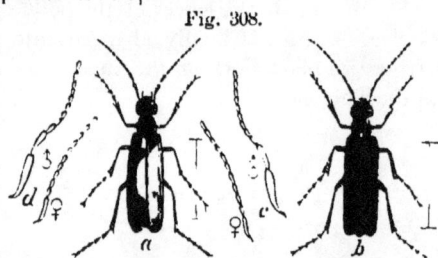

Fig. 308.

Fig. 308. — Ash-colored Blister-beetle; *a*, the beetle, enlarged — color, ash-gray; *d*, its antenna, enlarged — on the left, that of the male; on the right, that of the female; *b*, the black variety (*murina*), enlarged—color, black; *c*. its antenna. enlarged —on the left, that of the female; on the right, that of the male.

This species has done serious injury to the potato crop in one county in this State. It appears in the perfect state (Fig. 308) and attacks the potato by feeding upon the leaves. Parties sending specimens to me could not give any particulars concerning the natural history of this insect. As the specimens sent me were from portions of the State infested by grasshoppers and crickets, it may be that this species in the larva state feeds upon the eggs of these pests.

Professor Riley gives the natural history of this species, in substance as follows: The female lays her eggs in the nests of such locusts or grasshoppers as deposit their eggs in the ground. The larvæ produced from these eggs are of an elongate form and provided with six legs. They at once begin to feed upon the locust eggs, and at the approach of Winter they cast their skins and appear in an entirely different form, known as the semi-pupa. In the true pupa form the next change results in the exclusion of the perfect insect; but in the present case, as soon as the skin is cast, the insect appears again in the larval form. The semi-pupa differs from the true pupa in lacking the wing and leg-sheaths, but, like it, is incapable of moving about. It is of a pale yellow color, slightly curved, and beneath the fore part of the body are six short tubercles, which seem to represent the legs. In the following Spring it casts its skin and again appears in the larval form. Its body is now much curved, the head nearly coming in contact with the tail. After attaining its full size it assumes the pupa form, from which the perfect beetle issues in the course of a few weeks.

I have not found it in any locality not infested by grasshoppers, so it may be a friend as well as a foe.

REMEDY.—Use No. 103—liquid solution preferable—and No. 112.

CHAPTER CCXI.

The Striped Blister-beetle.   (Cal.)

(*Cantharis vittata.*—Fabricus.)

SYNONYM.—*Lytta,* or *Epicauta vittata.*

Order, COLEOPTERA ; Family, MELOIDÆ.

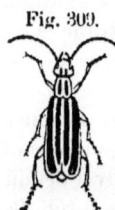

Fig. 309.

[Feeding upon the leaves of the potato ; an elongate yellowish beetle (Fig. 309), about six lines long, marked with two black spots on the head, two black stripes on the thorax, and two black stripes on each wing-case—the outer stripe the widest, and sometimes divided into two stripes by a yellow line.]

Fig. 309.—Striped Blister Beetle—colors, dull yellow and black.

The habits and natural history of this species are the same as those of the ash-colored blister-beetle—Chapter CCX.

REMEDIES.—Use No. 103—liquid solution preferable—and No. 112.

CHAPTER CCXII.

Small Potato Beetle.—No. 1.   (Cal.)

(*Epitrix subcrinita.*—Leconte.)

Order, COLEOPTERA ; Family, CHRYSOMELIDÆ.

[A small metallic colored beetle, feeding on the leaves of the potato.]

This species appeared early in August, 1882, on one side of a field of potatoes which contained three acres, and by the 20th had infested the whole field, entirely destroying the foliage.

The perfect insect (Fig. 310, Plate 4,) is oval in form, of a greenish-black color with a slight sub-metallic luster ; antennæ, ten-jointed : legs, pale brown ; posterior thighs, stout ; length, about one and one half lines. The leaves attacked by these pests were filled with holes similar in appearance to grapevine leaves attacked by the grapevine flea-beetle (*Haltica chalybea.*) Should they appear this season, I will endeavor to learn something of their natural history.

REMEDY.—Use No. 103—liquid solution preferable.

CHAPTER CCXIII.

Small Potato Beetle.—No. 2. (Cal.)

(*Epitrix hirtipennis.*—Mels.)

Order, COLEOPTERA; Family, CHRYSOMELIDÆ.

[A small reddish-colored beetle feeding on the leaves of potatoes.]

This species was found in company with the preceding—*Epitrix suberinita*—feeding on the leaves of potatoes, but was not so numerous as the latter. The perfect insect is reddish in color, with indistinct black markings; antennæ, ten-jointed; legs, pale brown; length of insect, about one and a half lines.

REMEDY.—Use No. 103—liquid solution preferable.

CHAPTER CCXIV.

The Potato Moth. (Cal.)

(*Gelechia Sp?*)

Order, LEPIDOPTERA; Family, TINEIDÆ.

[A small whitish caterpillar a little over six lines in length, the head and true legs black; feeding upon potatoes.]

In 1881 and '82 specimens of potatoes were received infested by the larvæ of a small moth. Length of larva, about six lines; color, yellowish-white; head and cervical shield black, with a whitish space between them; true legs, black. (Similar to larva, Fig. 111, Plate 1.)

21

Fig. 311.

Fig. 311.—Potato Moth—color, ash-gray.

The moth (Fig. 311) is of an ash-gray color: length of body, about four lines; spread of wings, seven to eight lines; fore-wings dark ash-gray, ciliated; hind wings lighter in color, and also ciliated. The moths appear about the first of July, and deposit their eggs in potatoes after the latter are gathered from the ground and placed in heaps or in sacks. The following letter was received from a gentleman who has suffered from this pest:

" Yours at hand, and in reply I will say that the larva does not attack the potatoes in the ground, but it is after they are taken from the ground (dug) and placed in bulk that the moth deposits her eggs on the potatoes, especially on those on the top of the heaps. Some are infested to such an extent as to appear like a honeycomb. Last year (1882) the moths appeared about the first of July, and were present in all stages of their existence until the first of February, 1883. When I noticed them first in 1881, the potatoes on top of the heaps were seriously infested. I covered the entire heaps with old sacks; as soon as the larvæ were full grown they left the potatoes and made their cocoons on the under side of the sacks, in which to pass their transformation. When the sacks were taken off there was not a space the size of a silver dollar but had one or more cocoons attached. In order to prevent its further spread I have burned the sacks, straw coverings and sheds, and removed the storing place to a distant part of my place."

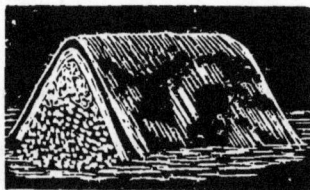

This pest is reported from three counties. There are evidently two or three broods each year. To prevent this pest from injuring the crop of potatoes after they are gathered, the potatoes should be covered with earth, or placed in what is called a pit, in England, which is made by piling the potatoes in a heap and putting a covering of from four to six inches of earth on them (See Fig. 311¼), or

Fig. 311½.

by covering the heap with old sacks, etc., and placing on the top a light covering of earth. In all cases the storing place should be as far as possible from localities already infested. Sprinkling slacked lime on the heaps will prevent the moths from depositing their eggs on the part covered by the lime.

INSECTS INFESTING THE SWEET POTATO.

CHAPTER CCXV.

Tortoise Beetles.

Order, COLEOPTERA ; Family, CHRYSOMELIDÆ.

[Feeding upon the leaves of the sweet potato; a flattened larva, having a row of spines along each side of the body; finally transforming into flattened tortoise-shaped beetles.]

Fig. 312.

Fig. 313.

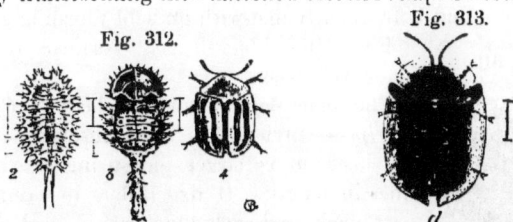

Fig. 312.—Two-striped Tortoise Beetle, larva and pupa ; *2,* the larva, enlarged—color, dirty white ; *3,* the pupa, enlarged —color, brownish ; *4,* the beetle, enlarged—colors, yellow and black.

In this country there are no less than four different kinds of tortoise beetles which are known to infest the sweet potato; they are more or less hemispherical in form, and the head is concealed beneath a thin, transparent extension of the front edge of the thorax.

Fig. 313.—Mottled Tortoise Beetle, enlarged—color, black.

The two-striped tortoise beetle (*Cassida bivittata*—Say: Fig. 312, 4.) is marked with two black stripes on each wing-case. In the three following species, the wing-cases are unmarked.

The mottled tortoise beetle (*Cassida guttata*—Olivier: Fig. 313.) has the shoulders blackish to the extreme outer edge of the wing-case. In the remaining two species the thin, transparent outer edge of the wing-cases and thorax is unmarked.

Fig. 314.

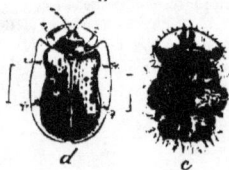

Fig. 314.—Golden Tortoise Beetle and Pupa: *d*, the beetle—color, yellow; *e*, the pupa—color, brown.

Fig. 315.—Egg of Golden Tortoise Beetle, enlarged—color, dirty white.

Fig. 315.

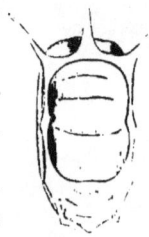

The golden tortoise beetle (*Cassida aurichalcea*—Fabricius; Fig. 314*d*,) is of a deep yellowish color, usually dotted with black, or it is of a transparent golden color, shining like a drop of liquid gold. It measures about two lines in length.

This is the only tortoise beetle that is known to occur in this State, and is frequently met with on wild morning glories.

Fig. 316.

Fig. 316.—Black-legged Tortoise Beetle—color, yellowish.

The black-legged tortoise beetle (*Cassida nigripes*—Olivier: Fig. 316) closely resembles the last, but is larger, measuring nearly four lines in length. It also differs in having the legs black, and each wing-case is marked with three black dots. The larvæ of these tortoise beetles are elongate-oval in outline, greatly flattened, and have a row of spines on each side of the body and two larger spines, which are situated at the hind end of the body; these are usu-

ally held over the back, and the excrements of the larva are
sometimes collected upon them.

The larva of the two-striped tortoise beetle (Fig. 312, 2,) is of
a dirty, or yellowish white color. It differs from the other three
mentioned below by not collecting its excrements upon its
anal spines.

Fig. 317.   Fig. 318.

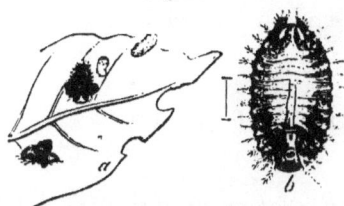 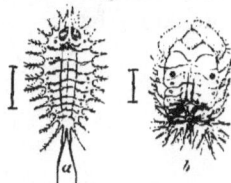

Fig. 317.—Larva of Golden Tortoise Beetle; *a*, several larvæ on a leaf; *b*, a larva, enlarged—color, dark brown.

Fig. 318.—Larva and pupa of Mottled Tortoise Beetle; *a*,
the larva, enlarged—color, green; *b*, the pupa, enlarged—color,
green.

The larva of the golden tortoise beetle (Fig. 317) is of a dark
brown color, with a pale shade upon the back.

That of the mottled tortoise beetle (Fig. 318*a*) is of a greenish color, while the larva of the black-legged tortoise beetle
(Fig. 319, *a* and *b*,) is of a pale straw color, marked with dusky
spots, the spines tipped with black.

Fig. 319.

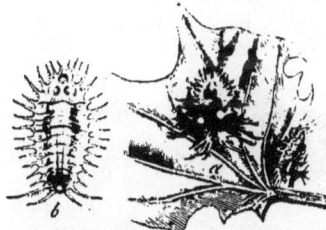

Fig. 320.

Fig. 319.—Larva of Black-legged Tortoise Beetle; *b*, the
larva, enlarged—color, yellow; *a*, two larvæ on a leaf.

Fig. 320.—Pupa of Black-legged Tortoise Beetle, enlarged—
color, brown.

When fully grown these larvæ attach themselves to some object by the hind part of the body and soon cast off their skins, which are worked backward and allowed to remain, enveloping the hind part of the pupa. (Figs. 320, 318b, 314c and 312, 3). The pupæ somewhat resemble the perfect beetles, but differ by being destitute of the hard wing-cases, and also in having a row of spines along each side of the body.

REMEDY.—Use No 113; if seriously infested, use No. 103.

INSECTS INFESTING THE TOMATO.

CHAPTER CCXVI.

The Tomato and Tobacco Worm.   (Cal.)

(*Macrosila Carolina.*—Linn; and *M. 5-maculata.*—Haw.)

Order, LEPIDOPTERA; Family, SPHINGIDÆ.

[Feeding upon the leaves of tomato, potato, and tobacco plants; a large green worm, having seven oblique white stripes on each side of the body, and a horn on the hind end.]

Fig. 321.

Fig. 321.—Potato or Tomato Worm—colors, green and white.

Fig. 322.

Fig. 322.—Pupa of Potato or Tomato Worm—color, brown.

The above two species very closely resemble each other in all of their stages. The full grown worm (Fig. 321) measures from three to five inches in length. When about to pupate, it enters the earth and forms a smooth cell in which to undergo its transformations.

Fig. 323.

Fig. 323.—Potato Worm Moth or Five-spotted Sphinx—colors, gray, black, and yellow.

The pupa (Fig. 322) is dark brown, and is furnished with a long tongue-case, which curves around from the forward end, its outer extremity resting upon the breast of the pupa, somewhat resembling the handle of a pitcher.

The perfect insects (Fig. 323) are commonly called "hawk-moths," from a habit they have of hovering over flowers in the evening while partaking of the nectar by means of their long proboscis. Their fore-wings expand from four to five inches, and are of a grayish color; the abdomen has a row of orange spots, surrounded by black, on each side.

REMEDIES.—Use Nos. 14, 100, and 101. All chrysalids (pupa) dug up or plowed up should be destroyed.

CHAPTER CCXVII.

The Tomato Aphis. (Cal.)

(*Megoura solani.*—Thomas.)

Order, HEMIPTERA;   }   Family, APHIDIDÆ.
Sub-order, HOMOPTERA; }

The measurements of insects in this work are given in inches and lines. The above cut represents one inch divided into lines and fractions thereof.

[Living upon the leaves and stems of tomato plants, which they puncture with their beaks and imbibe the sap; small greenish plant-lice, sometimes marked with yellow or brown.]

The wingless lice are pale green, with a dark green stripe along the back: head whitish, the eyes brown.

The winged lice are greenish, the thorax black or marked with black or brown.

REMEDIES.—Use No. 74.

INSECTS INFESTING THE CABBAGE.

CHAPTER CCXVIII.

The Cabbage Maggot.

(*Anthomyia brassicæ.*—Bouche.)

Order, DIPTERA ; Family, MUSCIDÆ.

[Boring into the roots of the cabbage, turnip and rutabaga ; a white maggot having a flattened hind end margined around with minute teeth.]

These maggots and the flies into which they are finally transformed so closely resemble the onion maggot and fly, and their habits are so similar, that the account given of the latter will apply equally well to the present species.

REMEDY.—Use No. 113.

CHAPTER CCXIX.

The Southern Cabbage Worm.   (Cal.)

(*Pieris protodice.* Bois Lee.)

Order, LEPIDOPTERA ; Family, PIERIDÆ.

[Feeding upon the cabbage and mustard ; a sixteen-legged greenish or bluish worm, dotted with black and marked with from four to six yellow stripes.]

Fig. 324.—Southern Cabbage Worm and Pupa—*a*, the worm—colors, greenish-blue and yellow; *b*, the pupa—color, gray.

Fig. 324.

The full grown worm (Fig. 324*a*) measures about one inch and three lines long. When about to pupate it suspends itself by the hind legs and a transverse loop of silken thread passed around the fore part of the body. The pupa (Fig. 324*b*) is grayish-brown, dotted with black; the head terminates in a conical prominence, and there is quite a large prominence on the back of the thorax; it is about eight lines in length.

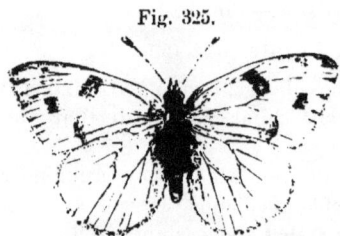

Fig. 325.

Fig. 325.—Southern Cabbage Butterfly, male—colors, white and black.

Fig. 326.—Southern Cabbage Butterfly, female—colors, white and black.

The wings of the butterfly (Figs. 325 and 326) expand about two inches, and are white, the fore ones marked with about eight blackish spots, and the hind wings sometimes have a border of triangular slate-colored spots with a zig-zag slate-colored line inside of them.

Fig. 326.

REMEDIES.—Spray once each week early in the season with No. 5 or 7; and should the caterpillars appear, use No. 83 or 85, or 64; the latter is preferable. (See also No. 114.)

CHAPTER CCXX.

The Imported Cabbage Worm.

(*Pieris rapæ.*—Linnæus.)

Order, LEPIDOPTERA: Family, PIERIDÆ.

[Feeding upon the cabbage, etc.; a green sixteen-legged worm dotted with black, and marked on the back with a yellow line, and with a row of yellow spots on each side of the body.]

Fig. 327.

Fig. 327.—Imported Cabbage Worm and Pupa; *a.* the worm—color, green with yellow lines; *b.* the pupa—color, greenish or gray.

When fully grown, this worm (Fig. 327*a*) is about one inch and three lines long; it then suspends itself by the hind feet and a transverse loop of silken threads passed around the fore part of the body. It soon sheds its skin and appears in the pupa state (Fig. 327*b*). Several broods are produced in one year, the last brood hibernating in the pupa state.

Fig. 328.—Imported Cabbage Butterfly, male—colors, white and black.

The perfect insect, or butterfly (Figs. 328 and 329) expands about one inch and nine lines, and is white or, yellowish-white, with the fore-wings tipped with black

Fig. 328.

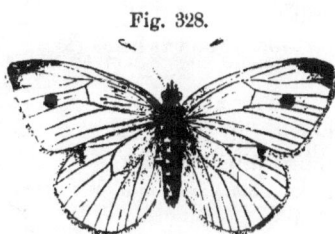

and marked with from one to three black spots; the hind wings have a blackish spot on the front margin.

Fig. 329.

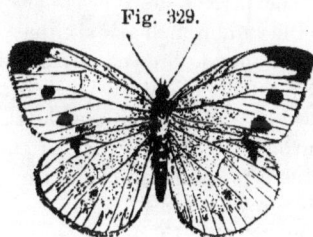

Fig. 329.—Imported Cabbage Butterfly, female—colors, white and black.

REMEDIES.—Same as in Chapter CCXIX.

CHAPTER CCXXI.

The Cabbage Plusia.   (Cal.)

(*Plusia brassicæ.*—Riley.)

Order, LEPIDOPTERA; Family, NOCTUIDÆ.

[Feeding upon cabbages and lettuce; a naked green twelve-legged worm, dotted with white and marked with white lines.]

This caterpillar (Fig. 330a) arches up its back slightly when walking; it is of a yellowish-green color, sparsely dotted with white, and marked on the back with a dark line, on each side of which are three whitish lines; the head is green, and is marked on each side with five black eyelets, which are scarcely noticeable with the naked eye.

Fig. 330.—Cabbage Plusia; a, the caterpillar—color, green with white lines: b, the pupa in its cocoon—color, brown: c, the moth —color, grayish-brown.

Fig. 330.

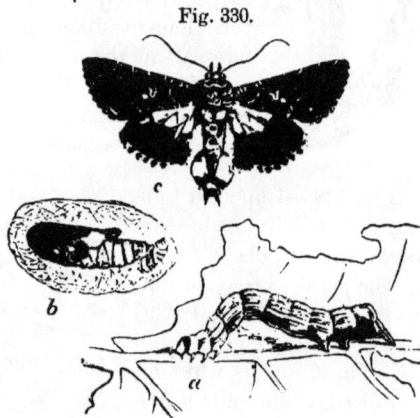

When about to pupate it spins a thin whitish cocoon (Fig. 330b) in some sheltered place, frequently among the leaves of the plant it infests. The perfect moth (Fig. 330c) expands from an inch and three lines to an inch and six lines: the fore-wings are dusky gray, inclining to brown, variegated with light grayish-brown, and near the middle of each is a small oval spot and a somewhat U shaped silvery mark. These insects may be found during the greater part of the year, there being at least three broods produced in one season; the last brood hibernating as half grown worms.

REMEDIES.—Use same as in Chapter CCXIX.

CHAPTER CCXXII.

The Yellow Bear Caterpillar.

(*Spilosoma Virginica.*—Fabricius.)

Order, LEPIDOPTERA : Family, BOMBYCIDÆ.

[Feeding upon the leaves of the pea, bean, beet, cabbage, grape, etc.; a hairy caterpillar of a yellowish-gray or greenish-white color, marked with from two to four lines, usually of a dark color, and covered with white, yellow or reddish hairs.]

Fig. 331.—Yellow Bear; *a*, the caterpillar—colors, white or gray, the hairs white or yellow; *b*, the pupa—color, brown; *c*, the moth —color, white with black dots.

Fig. 331.

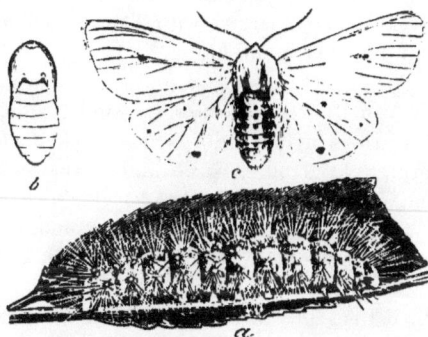

This caterpillar (Fig. 331*a*) varies greatly in its colors and markings; the young caterpillar is of a greenish-white color, with three white lines on the back : the more mature ones are pale yellow or dark gray, with two dark colored lines on the back, and sometimes there is a yellowish line low down on each side of the body. The hair is in spreading clusters, and is either white, yellow, reddish-brown, or the base is brown with the tips black. When fully grown it measures about one inch and six lines in length; it then creeps into some sheltered place and spins a thin cocoon, intermixed with the hairs with which the body was covered.

The perfect moth (Fig. 331*c*) is commonly known as the "white miller," and is of a pure white color, usually marked with a few black dots. The fore-wings expand from an inch and six lines to nearly two inches. The caterpillar of this moth can be found from the 20th of April to the 1st of October.

REMEDY.—Should these caterpillars appear in large numbers, see No. 14.

CHAPTER CCXXIII.

The Harlequin Cabbage-bug.   (Cal.)

*(Strachia histrionica.—*Hahn.*)*

Order, HEMIPTERA ;
Sub-order, HOMOPTERA ;   } Family. SCUTELLERIDÆ.

[Living upon cabbages, radishes, turnips, etc.; a wingless greenish bug, marked with black, or a winged bug, marked with blue-black and orange.]

The eggs (Fig. 332 *c, d*, and *e*.) from which these bugs hatch, are usually deposited on the under side of a leaf, in two or three rows of half a dozen each. They are of an elongate-oval form, light green or white, and marked with two black circles and a black dot, the latter placed above the lower circle (Fig. 332*d*.) They are placed upon one end, and on the upper end is a crescent-shaped black spot. (Fig. 332*e*.)

The perfect insects (Fig. 333) are about five lines long by three lines broad, and prettily marked with blue-black and polished orange. They pass the Winter in some sheltered place. Unlike most other plant-bugs, they do not give forth a disagreeable odor when crushed.

Fig. 332.   Fig. 333.

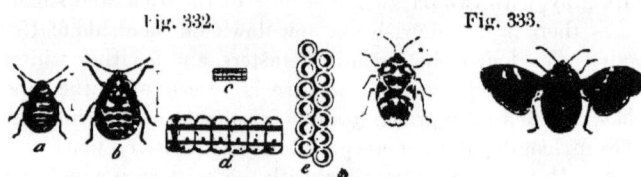

Fig. 332.—Larva, Pupa and Eggs of Harlequin Cabbage-bug : *a*, the larva, enlarged—colors, pale green and black ; *b*, the pupa, enlarged—colors, greenish or yellow and black ; *d*, side view of eggs, enlarged—colors, white and black ; *c*, the same, natural size ; *e*, the eggs as seen from above, enlarged.

Fig. 333.—Harlequin Cabbage-bug—colors, orange and black ;

at the left, the bug, with its wings closed; at the right, the same, with its wings expanded.

This species spread to such an extent in Sacramento County as to prevent the growing of cabbages.

REMEDIES.—Nos. 20, 106*a* and 36. If the cabbages, etc., are seriously infested by this cabbage-bug, use No. 64, or 83, or 85, of double the strength recommended.

CHAPTER CCXXIV.

The Cabbage Aphis. (Cal.)

(*Aphis brassicæ.*—Linnæus.)

Order, HEMIPTERA; Sub-order, HOMOPTERA; Family, APHIDIDÆ.

[Living upon the leaves of the cabbage, which they puncture with their beaks and extract the sap; small greenish-yellow plant lice, sometimes marked with black and covered with a bluish-white powder.]

Fig. 334.—Cabbage Aphis; *1*, the winged louse, natural size; *2*, the same, enlarged—colors, greenish-yellow and black; *3*, the wingless female, natural size; *4*, the same, enlarged—color, green or yellowish.

Fig. 334.

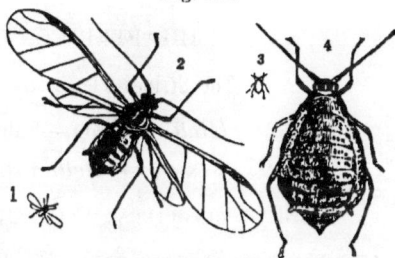

The wingless lice (Fig. 334, *4*,) are of a pale pea green or greenish-yellow color. The winged lice (Fig. 334, *2*,) are green or greenish-yellow, with the head black and a row of dots along each side of the abdomen of the same color.—Professor Thomas.

This species of plant-louse is the most injurious that the market gardener has to contend with.

REMEDY.—Use No. 74.

INSECTS INFESTING THE RADISH.

CHAPTER CCXXV.

The Radish Maggot.

(*Anthomyia raphani.*—Harris.)

Order, DIPTERA: Family, MUSCIDÆ.

[Boring into the roots of the radish; a white footless maggot, its blunt posterior end margined with minute teeth, of which the lower two are the largest and are notched at the tips.]

These maggots so closely resemble the onion maggots in all their stages, that the account given of the latter will apply equally well to the present species.

REMEDY.—Same as recommended in Chapter CCXVIII.

CHAPTER CCXXVI.

The Striped Flea-beetle. (Cal.)

(*Haltica vittata.*—Fabricius.)

SYNONYM.—*Haltica striolata.*

Order, COLEOPTERA: Family, CHRYSOMELIDÆ.

[Feeding upon the leaves of turnips, radishes, cabbages, etc.; a small black beetle, having a yellowish wavy stripe on each wing case.]

Fig. 335.—Striped Flea-beetle, enlarged—colors, black and yellow.

Fig. 335.

These beetles (Fig. 335) have the hind thighs greatly enlarged, which enables them to leap to some distance, like a flea, hence the name "flea-beetle." The larvæ or young live in the ground and feed upon the roots of various plants. When fully grown they form small cells in the earth, wherein to undergo their transformations. They are minute slender whitish grubs, provided with six legs and having the head light brown.

REMEDIES.—Should the beetles appear in large numbers, use No. 64, double strength; spray thoroughly, as often as the beetles appear; or Nos. 5 and 7 are effective. See also Nos. 20 and 106, A.

INSECTS INFESTING THE TURNIP.

The following insects infest the Turnip, and are treated of in another part of this work:

The Wire Worm.
Crane Flies (*Tipula*).
The Striped Flea-beetle (*Haltica vittata*).

The White-lined Sphinx (*Deilephila lineata*).
The Harlequin Cabbage Bug (*Strachia histrionica*).

22

INSECTS INFESTING THE ONION.

CHAPTER CCXXVII.

The Onion Maggot. (Cal.)

*(Anthomyia ceparum.—*Linnæus.)

Order, DIPTERA ; Family, MUSCIDÆ.

[Boring into the bulb of the onion, causing it to wilt and decay : a white footless maggot, which tapers at one end, the opposite end appearing as if obliquely cut off, and marked near the middle with two elevated eye-like brown dots.]

Fig. 336.

Fig. 337.

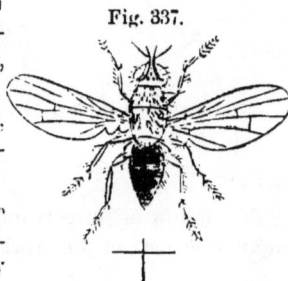

Fig. 336.—An Onion infested by Onion Maggots—color of maggots, white.

Fig. 337.—Onion Fly, enlarged—colors, ashy-gray and black.

When fully grown these maggots sometimes assume the pupa form in a cavity which they have gnawed in the onion (Fig. 336), but they usually burrow a short distance into the earth and form a small cell ; here the change to the pupa state takes place, and the perfect fly issues in the course of a week or so. This fly (Fig. 337) closely resembles the common house-fly, but is more slender : it is of an ashen-gray color ; the face is silvery, with a rusty black stripe between the eyes, and there is usually a black stripe or row of black spots along the back.

REMEDY.—Use same as recommended in Chap. CCXVIII.

INSECTS INFESTING LETTUCE.

CHAPTER CCXXVIII.

The Lettuce Aphis.

(Siphonophora lactucæ ?)

Order, HEMIPTERA ;   } Family, APHIDIDÆ.
Sub-Order, HOMOPTERA ;

[The measurements of insects in this work are given in inches and lines. The above cut represents one inch divided into lines and fractions thereof.]

[Living upon the leaves of the garden lettuce, which they puncture with their beaks and imbibe the sap; small green plant-lice, usually marked with black or brown.]

The wingless lice are green or brown, usually marked with black dots.

The winged lice are green, the head and thorax sometimes black or brown.

REMEDIES.—Spray with No. 3 or No. 4. or use No. 83 or No. 85.

INSECTS INFESTING THE PEA.

CHAPTER CCXXIX.

The Pea Aphis.   (Cal.)

(*Siphonophora pisi*—Kaltenbach.)

Order, HEMIPTERA ; }
Sub-order, HOMOPTERA ;}   Family, APHIDIDÆ.

[Living upon the stems, pods and leaves of the pea, which they puncture with their beaks and imbibe the sap: small green plant-lice, sometimes marked with yellow.]

The wingless lice are green, with the eyes black.

The winged lice are green, the thorax brown or yellow, the eyes black or brown; the antennæ are longer than the body.

REMEDIES.—Use No. 3 or No. 4, or No. 83 or No. 85.

CHAPTER CCXXX.

The Pea Weevil.   (Cal.)

(*Bruchus pisi.*—Linnæus.)

Order, COLEOPTERA ; Family, BRUCHIDÆ.

[Living in peas; a small footless, deep yellow grub, with a black head, finally transforming into a small gray snout-beetle, which is striped with black and white.]

The female weevil deposits her eggs upon the young pods;

the larva (Fig. 338c), as soon as hatched, burrows through the pod and into a pea. Here it remains until it becomes fully grown, feeding upon the interior of the pea. When about to pupate, it gnaws a round hole under the shell of the pea, leaving the latter untouched. It assumes the pupa form (Fig. 338d) in the Autumn, and is changed to a beetle (Fig. 338b) during the Winter or early in the following Spring. The beetle gnaws a hole in the shell of the pea, through which it makes its escape, although it not infrequently remains in the pea until the latter is planted. During all this time it seldom injures the germ of the pea, so that peas infested by it will grow with almost as much certainty as those not infested.

Fig. 338.

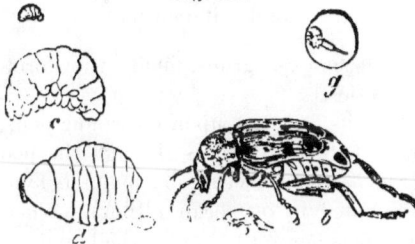

Fig. 338.—Pea Weevil; c, the larva, side view, enlarged (natural size indicated above)—color, yellowish; d, the pupa, back view, enlarged (natural size indicated below and to the right—color, yellowish; b, the weevil, enlarged (natural size indicated below) —color, rusty-black and white; g, an infested pea.

The beetle, or weevil, is about two lines long, and is of a grayish color, the wing-cases being marked with alternate stripes of gray and black, and behind the middle of each is an oblique white stripe. The exposed tip of the abdomen is covered with a whitish down, and is marked with two black spots placed near the center.

Since writing the above, I have met a gentleman from a section of this State where the pea weevil is doing considerable damage. He expressed his belief that the pea weevil was produced by "spontaneous generation," as he had never seen the beetle in his field when the crop was growing. The existence of the weevil in the perfect state is of short duration, and as it probably works only at night, this would account for its not being seen.

REMEDY.—Use No. 115.

CHAPTER CCXXXI.

The Bean Weevil. (Cal.)

(*Bruchus fabæ.*—Riley.)

SYNONYM—*Bruchus obsoletus.*—Say.

Order, COLEOPTERA; Family, BRUCHIDÆ.

[Living in beans; small footless grubs, finally transforming into grayish-brown beetles.]

The female weevil deposits her eggs upon the young pods, and the larva, as soon as hatched, burrows through the pod and enters the bean (Fig. 339*b*); sometimes as many as fourteen larvæ have been counted in one bean (Riley). When fully grown they gnaw a passage to the hull or shell, and are soon changed to pupæ; the transformation to the perfect state sometimes takes place in the Fall, but usually not until the following Spring.

Fig. 339.

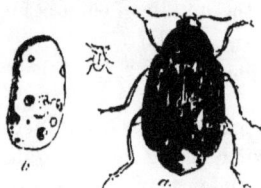

Fig. 339.—Bean Weevil: *a*, the weevil enlarged, the small figure at the left being the natural size—color, brownish-gray; *b*, a bean infested by the larva of this weevil.

The perfect weevil (Fig. 339*a*) is of a brownish-gray color, and measures about one line in length.

REMEDY.—Use No. 115.

CHAPTER CCXXXII.

The Bean Aphis. (Cal.) .

(*Aphis rumicis.*—Linnæus.)

Order, HEMIPTERA ; } Family, APHIDIDÆ.
Sub-order, HOMOPTERA ;

[Living upon the leaves and stalks of the bean, which they puncture with their beaks and imbibe the sap; small blackish plant lice.]

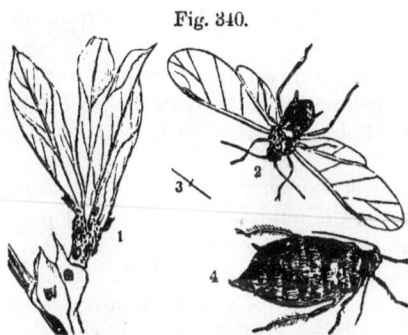

Fig. 340.

Fig. 340.—Bean Aphis; *1*, a stalk infested by the aphides ; *4*, a wingless aphis, enlarged—color, blackish ; *2*, a winged aphis, enlarged—color, black ; *3*, natural size of No. *2*.

The wingless lice (Fig. 340, *4*,) are black, the head and thorax sometimes greenish.

The winged females (Fig. 340, *2*,) are wholly black. In England this species is called the "Collier," "Black Dolphin," "Black Fly," etc.

"The bean aphis sometimes appears in such vast numbers as to smother the beans, making them look as if they were coated with soot. The attacks are begun by a few wingless females establishing themselves near the top of the bean-shoots, where they produce living young. These in their turn are soon able to produce another living generation and so on and on, till the increase is numerous and from the numbers of the 'black fly' and the sticky juices flowing from the punctures which they have made with their suckers, the plant becomes a mere dirty infested mass, with a few infested leaves sticking out from amongst the plant-lice."—Miss Ormerod.

REMEDY.—Use No. 116.

INSECTS INFESTING ASPARAGUS.

CHAPTER CCXXXIII.
The Asparagus Beetle.

(*Crioceris asparagi.*—Linnæus.)

(Order, COLEOPTERA; Family, CHRYSOMELIDÆ.)

[Feeding upon the leaves of asparagus; a robust, ash-gray six-legged larva, about three lines long, with a row of black dots along each side of the body, the head and two spots on the first segment also black.]

This beetle (Fig. 341, Plate 4,) measures about three lines in length; its thorax is reddish and usually marked with black; the wing-cases are pale-yellowish, marked with blue-black as in the figure.

The perfect beetles pass the Winter in some sheltered situation and come forth early in the following Summer to deposit their eggs. These are placed upon the stalks of the asparagus (Fig. 342, Plate 4,) and somewhat resemble in form a grain of wheat, but are much smaller and are of a blackish-brown color. After the larvæ (Fig. 343, Plate 4,) attain their full size, they desert the plants and enter the earth, where each one forms a small cell in which to undergo its transformations.

REMEDIES.—Use No. 19, or No. 83, or No. 85, or No. 3, or No. 4.

INSECTS INFESTING RHUBARB, OR PIE PLANT.

The following insects infest the Rhubarb plant, and are treated of in another part of this work:

Cut Worms *(Agrotis).*   The Frosted Leaf-hopper *(Pœcilopterа pruinosa).*

CHAPTER CCXXXIV.

The Parsley Worm.

*(Papilio asterias.—*Fabricius.)

Order, LEPIDOPTERA ; Family, PAPILIONIDÆ.

[Feeding upon the leaves and blossoms of parsley, carrots, parsnips, etc.: a whitish or greenish-yellow sixteen-legged worm, banded with black and marked with black and yellow spots.]

Fig. 344.—Parsley Worm—colors, greenish-yellow and black.

Fig. 344.

When fully grown, this worm (Fig. 344) is about one inch and six lines long. Before assuming the pupa form it suspends itself by the hind feet and a transverse loop of silken threads passed around the fore part of the body.

Fig. 345.

Fig. 345.—Asterias Butterfly—colors. black, blue and yellow.

The pupa varies in color from pale green to yellowish, or ash-gray; at the anterior end are two ear-like projections, and there is a smaller projection on the back of the thorax.

The butterfly (Fig. 345) expands from three inches and six lines to four inches, is of a black color, with two rows of yellow spots near the outer edge of the wings; the hind wings are tailed, and are marked with several blue spots, while above the angle near each tail is an orange spot centered with black.

REMEDIES.—Same as recommended in Chapter CCXIX.

HOUSEHOLD AND STOREHOUSE PESTS.

CHAPTER CCXXXV.

The Clothes Moth.   (Cal.)

(*Tinea flavifrontella*.—Linnæus.)

Order, LEPIDOPTERA; Family, TINEIDÆ.

[Living in silken tubes on carpets, woolen goods, etc.; a small, pale, sixteen-legged worm.]

Fig. 346.—Clothes Moth—color, light buff.

The perfect insect (Fig. 346) is a small moth or miller, of a uniform light buff color: the wings are long and narrow, with the most delicate fringe of silken hairs.

Fig. 346.

Fig. 347.

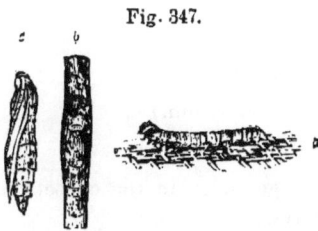

Fig. 347.—Caterpillar, Pupa, and Case of the Clothes Moth: *a*, the caterpillar—color, whitish; *b*, its case; *c*, the pupa—color, brown; all enlarged.

This moth deposits her eggs in carpets, woolen goods, furs, etc. As soon as hatched the young larva (Fig 347*a*) immediately begins to construct for itself a nearly cylindrical tube (Fig. 347*b*), formed by fastening the gnawed pieces of the cloth together with silken threads. In this tube the larva lives, and instead of dragging its habitation over the hairs, etc., it first cuts these off, thus doing more injury than if it merely fed upon the cloth, fur, etc. When fully grown, the caterpillar closes both ends of the tube and soon assumes the pupa form (Fig. 347*c*), from which the perfect moth issues in the course of a few weeks.

Fig. 348.—Larva and Cases of the Carpet Moth.

Fig. 348.

Closely related to the above species is the carpet moth, or woolen moth (*Tinea tapetzella*—Linn), whose larva also lives in a silken tube (Fig. 348) and is sometimes very destructive to carpets, etc.

Fig. 349.—Carpet Moth—colors, black, yellowish-white and gray.

Fig. 349.

The perfect moth (Fig. 349) is blackish at the base of the fore-wings, the remainder being yellowish-white; the hind wings are dark-gray, and the head is white.

REMEDIES.—No. 82, 90 and 117.

CHAPTER CCXXXVI.

The Carpet Beetle.   (Cal.)

(Anthrenus scrophulariæ.—Linn.)

Order, COLEOPTERA : Family, DERMESTIDÆ.

[Living beneath, and eating large holes in the carpet; a small hairy brownish larva or worm.]

This larva (Fig. 350a), when fully grown, measures a little over three lines in length, and is of a brownish color, the sutures of the segments whitish.   On various parts of the body are tufts of hair—that at the hind end the longest, and frequently as long as the body itself.   When about to pupate, it crawls into some sheltered place; here it remains perfectly quiet, and is changed to a pupa (Fig. 350c) within the old larval-skin.   A short time before the perfect beetle emerges the larval-skin is rent on the back, disclosing the included pupa; soon after this takes place, the skin of the pupa is also rent on the back, giving us a glimpse of the partially inclosed beetle.

Fig. 350.

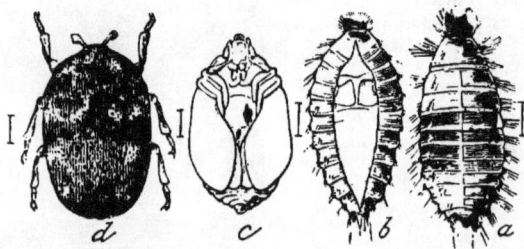

Fig. 350.—Carpet Beetle, enlarged : *d*, the beetle — colors, black, white, and scarlet; *c*, the pupa—color, yellowish : *b*, the same, in the old larva skin : *a*, the larva—colors, dark and light brown.

The beetle (Fig. 350*d*) crawls out of its environment in the course of a day or so.   It is only about one and a half lines long by one line broad.   Its colors are black, white and scarlet, the latter forming a line along the middle of the back.

REMEDIES.—Use Nos. 82, 90, and 117.

NOTE.—This beetle frequents flowers, and can then be destroyed by using No. 81.

CHAPTER CCXXXVII.

The Cockroach. (Cal.)

*(Blatta germanica.—*Linnæus.*)*

Order. ORTHOPTERA ; Family, BLATTARIDÆ.

[Infesting houses, and feeding upon cloth, etc. ; a flattened reddish-brown six-legged insect.]

The female cockroach lays her eggs in a reddish-brown elongated capsule or pod, each capsule containing about thirty eggs. The young cockroach closely resembles the adult, but is entirely destitute of wings, although in the adult female the wings are greatly aborted, and are sometimes reduced to short wing-pads.

These insects are nocturnal in their habits, remaining hidden during the daytime and coming forth at night to feed. Although they are sometimes very troublesome, yet they partially atone for their ill-doings by ridding the house of bedbugs and similar vermin, which they devour.

REMEDY.—Use No. 121.

CHAPTER CCXXXVIII.

The Mosquito. (Cal.)

(Culex Sp?)

Order. DIPTERA ; Family, CULICIDÆ.

[A small two-winged blood-thirsty insect, sometimes very troublesome to both man and beast.]

It is only the females of this species which manifest the blood-thirsty propensity; the males are perfectly harmless. The eggs are laid in masses upon the water—usually in some stagnant pool. The young larva (Fig. 351), as soon as hatched, makes its way to the bottom of the pool, where it acts as a scavenger, by feeding upon the dead and putrefying vegetation. It rises occasionally to the surface for air, which it inhales through a tube situated near the tail. In the course of a fortnight it attains its full growth, and soon afterwards assumes the pupa form.

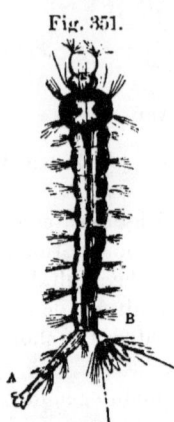

Fig. 351.

Fig. 351—Larva of Mosquito, greatly enlarged —color, whitish.

Fig. 352.—Male Mosquito, highly magnified —color, brown.

Fig. 352.

The pupa remains near the surface of the water, and has the power of wiggling about. In a few days it is changed into the perfect insect, or mosquito (Fig. 352). The latter is too well known in California to require further description.

REMEDIES.—Use Nos. 82, 117 and 118.

NOTE.—Care should be taken to prevent stagnant water being kept or allowed to remain near the family residence, as it is in such places the mosquitoes deposit their eggs, and the young pass their lives as larvæ and pupæ.

CHAPTER CCXXXIX.

The Dried Fruit Moth.   (Cal.)

Order, LEPIDOPTERA; Family, TINEIDÆ.

[A small larva, feeding upon dried fruit, etc.]

The small larvæ commonly known as "dried fruit worms," are a great annoyance to those who dry fruits, and to raisin-makers; and also to the merchants and dealers handling dried fruits and raisins which are not properly protected against these insect pests.

The natural history of these insects is not known to me further than as follows:

They hibernate in the larva state, the larva spinning a light cocoon in the cavity of the fruit, etc., upon which it has fed. Early in Spring they change to pupæ, and in about fifteen days the perfect insect emerges. The eggs are laid on the fruit while in the course of drying, especially such as is dried by sun heat; but often in fruit while it is getting ready to be packed, or after packing, if the package is not moth-proof.

Fruit in sacks has not any protection, as the moth can deposit the egg through the cloth. If the fruit is packed in boxes, and the boxes have not close joints. the moth deposits the eggs in the seams, and the larva as soon as hatched makes its way to the fruit in the package.

In January last (1883), I received a small box filled with pits and decayed parings of apricots and peaches, which were infested by the larvæ of these moths—the person sending them making the inquiry: "Are they the larvæ of the codlin moth?"

In September, 1881, a choice lot of dried plums, pears and peaches were exhibited at the State Fair, neatly packed in thirty pound boxes with glass covers (the fruit was dried by artificial heat). It was placed on exhibition in a fruit store until the following Spring. In one of these packages of plums under the glass I found seventy-three moths which had emerged from chysalids in the fruit.

At present I have specimens of what appear to be two different species, but they may be merely varieties of one species. The larva measures from five to eight lines in length, is of a bright yellow color, and tapers slightly toward each end; stomata, faintly bordered with brown; head deeply notched above, yellowish-brown; cervical shield yellowish-brown. The larva is slender, and from its tapering form can be readily distinguished from the larva of the codlin moth.

Perfect insect. No. 1 (Fig. 353, Plate 4); length of body, nearly four lines; spread of wings, nine lines; color, head and thorax dark reddish-brown; fore-wings, inner third yellowish-white (forming a bar across base of wings when at rest), balance of wing to apex dark brown, with two oblique blackish stripes and blackish dots, the darker parts sprinkled with whitish scales; cilia, ash-brown; hind wings, silvery white; cilia, darker than the wings.

No. 2 (Fig. 354, Plate 4); length, four lines; spread of wings, ten lines; color, head and thorax dark ash; fore-wings, mottled dark ash with a wavy blackish line across about one third the length of the wing from the base; a second blackish line reaching half way across, nearly eqidistant between the first blackish line and the apex: hind wings, silvery-white; cilia, a purer white than the wing.

REMEDIES.—Use Nos. 15, 16, 17 and 18.

CHAPTER CCXL.

The Grain Weevil. (Cal.)

(Calandra granaria.—Linn.)

(Order, COLEOPTERA; Family. CURCULIONID.E.)

[Living in the kernels of wheat; a reddish-brown weevil.]

This species of weevil is very destructive to grain, especially in store-houses and mills in California, and although small, is a very formidable pest. The female makes a hole in the kernel, or grain of wheat, with her beak and deposits therein a single egg. The larva which is produced from this egg lives upon the farinaceous part of the grain, leaving only the hull. When full grown it is nearly one line in length, and changes into a pupa within the hull in which it has lived. It remains in the pupa state about nine days, when the perfect beetle (Fig. 355, Plate 4,) appears and gnaws its way out of the grain. This pest devours the inside parts of the grain, not only in the larva but also in the perfect state. It is said that a single pair will produce six thousand in five months; also, that from the time the egg is deposited until the perfect insect appears is about forty-five days. It is claimed that at a temperature of less than forty-five degrees (Fahr.) these insects are incapable of multiplying their species, and in order to escape the cold they hide themselves in the cracks of floors, walls, roofs, etc., and remain there until warmer weather. It is probable that as the temperature is over sixty-five degrees for at least eight months of the year in central California, that this may account in part for their rapid increase in this locality. I have before me now (January 2d, 1883,) a sample of Tuscany wheat (crop of 1882) received in this city about the 15th of November last, in a close-fitting paper box and apparently in perfect order. It was kept in an office where the temperature is generally above sixty-five degrees, for about three weeks, or until about the 10th of December. On examination, it appears as if every kernel produced a weevil. I have kept them since that time and they still appear healthy. It is not by examining the surface of the grain heaps that the presence of this pest can be detected, but by examining at a depth of say four or more inches below the surface. By putting some of the grain in a vessel and then covering it with water, those kernels infested, or which the larva or perfect insect have eaten out, will float.

Since writing the above, I have made some investigations, which are not yet complete, (August 27th, 1883), but I have reason to think that the grain weevil (*C. granaria*) infests the wheat when growing in the field, and that the germ or egg is laid in the wheat (or, at least, in some of it) before it reaches the granary.

On the 14th of August, 1883, I found the male insects of the rice weevil (*Calandra* [*Sitophilus*] *oryzæ*) in considerable numbers among the wheat weevils; and also found them pairing with the females of the latter.

Description of Rice Weevil: Male not so large as wheat weevil, and has two red or yellowish-red spots on each wing-cover.

REMEDIES.—No. 72, No. 75 and No. 76, and careful selection of seed; general remedy for weevil, No. 119.

23

CHAPTER CCXLI.

The Angoumois Grain Moth.

(*Gelechia cerealella.*—Linnæus.)

Order, LEPIDOPTERA : Family, TINEIDÆ.

[Living in grains of wheat and corn: a minute white larva.]

This pest is very destructive to stored wheat and corn, especially to the former. It eats out the interior of the grain, leaving nothing but the empty hull. It is said that a single grain furnishes sufficient food for a larva from the time it issues from the egg until it becomes fully grown. It assumes the pupa form within the grain, or hull. The perfect moth (Fig. 356) has the head of a dull ochre color, the fore-wings pale shining ochre, with a grayish or brownish streak at the base of each wing; the hind wings are also grayish-ochre.

Fig. 356.   Fig. 357.

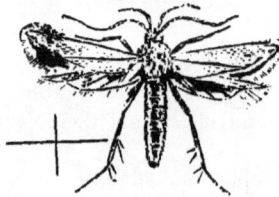
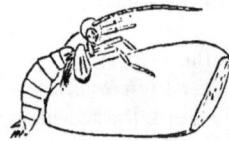

Fig. 356.—Angoumois Grain Moth, enlarged—color, yellowish-brown.

Fig. 357.—The Angoumois Grain Moth just from the pupa.

Closely related to the above is the common grain moth (*Tinea granella*—Linn), sometimes called the "grain wolf." Its larva differs from that of the Angoumois moth by fastening several grains together with silken threads and afterwards eating out the interior of each grain. The perfect moth is of a creamy-white color, with six brown spots on each fore-wing.

REMEDIES.—No. 82, No. 90, No. 117, No. 119. As these remedies apply to destroying moths and beetles, and their larvæ or grubs, any of them may be applied in this case that is practicable.

CHAPTER CCXLII.

Bran and Flour Bugs. (Cal.)

(*Silvanus quadricollis* and *S. surinamensis.*)

Order, COLEOPTERA: Family, CUCUJIDÆ.

[Living in stored grain, bran, flour, sugar, etc.; small reddish slender beetle.]

These beetles are very small, none of them measuring over two lines in length. They are of a chestnut-brown color; *quadricollis* (Fig. 358, Plate 4,) as its name implies, has a nearly square thorax, while in *surinamensis* the thorax is rounded, and has several teeth on the outer edge.

In their larval state these insects usually live in the grains of wheat or corn, which they frequently hollow out until nothing but the hull remains. They assume the pupa state within the grains. Besides wheat and corn they are also found in bran, or middlings, in flour, sugar, and in various other situations.

REMEDY--Use No. 120. When in grains, same remedy as recommended in Chapter CCXL.

CHAPTER CCXLIII.

The Rawhide Beetle. (Cal.)

(*Dermestes lardarius.*—Linn.)

Order, COLEOPTERA; Family, DERMESTIDÆ.

[A brown hairy larva, feeding upon rawhides, finally changing into a blackish-brown beetle, with a whitish bar across the base of the wing-cases.]

The larva (Fig. 359*a*) of this species is found to be very troublesome at certain seasons of the year, in stores where hides are stored. It measures about nine lines in length, and is covered with stiff hairs. When full grown it assumes the pupa form, and the perfect insect (Fig. 359c) emerges in about three weeks.

Fig. 359. — Rawhide Beetle and Larva; *a*, the larva, enlarged—color, brown; *b*, one of its hairs, enlarged; *c*, the beetle, enlarged—colors, black and gray.

Fig. 359.

Description of Beetle: Shape, oblong-oval; color, dark-brown or blackish; on the base of the wing-cases is a broad whitish or buff-colored band, on which are some brown or black spots; the under side of the body is blackish, and covered with a whitish-colored powder or scales; length, four lines; width, about two lines. This species feeds also on bacon, hams, etc.; and several other species which belong to the same family are very destructive to furs.

REMEDY.—Use No. 91.

CHAPTER CCXLIV.

The Black Horse-fly. (Cal.)

(*Tabanus atratus.*—Fabricius.)

Order, DIPTERA ; Family, TABANIDÆ.

[Biting and annoying horses, cattle and other animals; a large black two-winged fly, having the back of the thorax covered with a bluish-white powder, and the wings smoky dark brown or black.]

Fig. 360.—Black Horse-fly, L a r v a and Pupa; *a*, the larva—color, greenish-white; *b*, the pupa--color, brown; *c*, the fly—colors, black and bluishwhite.

Fig. 360.

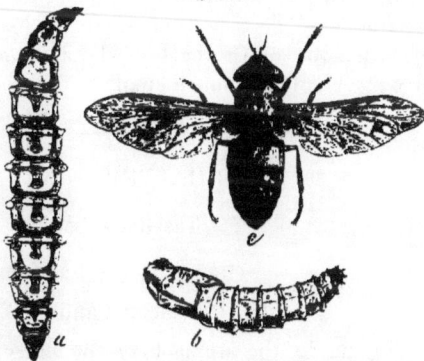

The body of this fly (Fig. 360*c*) is about t e n l i n e s long, and the wings expand about two inches.   Like the mosquito, it is only the females which attack animals; the males are destitute of mandibles, and live upon the sweets of flowers. The larva of this fly (Fig. 360a)lives in the vicinity of fresh water streams; they have been found beneath submerged stones in a small stream of running water, among floating pieces of wood, and on dry land less than a rod from a small permanent stream of water. It appears that a certain degree of moisture is necessary for their existence, although they are not strictly aquatic. They feed upon snails, and probably earth worms. These larvæ measures from one inch and nine lines to two inches and three lines in length, and are nearly cylindrical, but taper at each end; they are of a transparent greenish or yellowish color, and furnished above and below with large rounded sponge-like tubercles which are extended or retracted at the will of the insect. They reach their full size in mid-Summer, and then transform into pupæ (Fig. 360b), within their cells in the earth. In the course of a week or so they are changed into perfect flies.

Several other horse-flies occur in this State; the one oftenest met with is known as the "green-headed horse-fly"—Tabanus lincola—Fabr.; and may easily be distinguished by having a whitish line on the back of the abdomen; the head is usually green. It is a much smaller species than the above.

CHAPTER CCXLV.

The Horse Bot-fly.

(*Œstrus equi.*—Fabricius.)

Order, DIPTERA; Family, ŒSTRIDÆ.

[Living in the stomach of the horse; a yellowish-red or whitish grub, thinly covered with small bristles or spines.]

Fig. 361.

Fig. 361.— Horse Bot-fly, male —colors, grayish-yellow and black.

The eggs from which these grubs hatch are deposited in patches by the female fly, and each egg is attached to the hair by a sticky fluid which is deposited with it. In a few weeks the grub hatches, and is conveyed to the mouth by the horse licking or biting the place where the egg has been deposited; it then passes down the horse's throat with the food. After reaching the stomach, it attaches itself to the inner lining by means of two curved hooks with which the head is provided. Here it remains until fully grown, when it lossens its hold and is carried onward and expelled with the excrements. Upon reaching the earth it at once buries itself, and soon contracts to a reddish-brown pupa, from which the perfect fly issues in the course of a few weeks.

Fig. 362.—Horse Bot-fly, female—colors, gray, black and yellow.

Fig. 362.

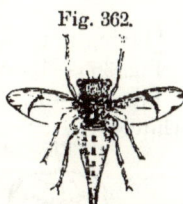

The perfect fly (Figs. 361 and 362) is of a pale yellowish color, spotted with red, the thorax banded with black or red, and the wings, which are only two in number, are of a whitish color reflecting a golden tint, and are crossed by a dark band with two reddish spots at the tips.

Figuier writes: "In fact, it is not in the egg state, but really in that of the larva, that the horse, as we shall explain, takes into his stomach these parasitical guests to which nature has allotted so singular an abode. When licking itself the horse carries them into his mouth, and afterward swallows them with his food, by which means they enter the stomach. It is a remarkable fact that is sometimes seen, other insects, as the *Tabani.* for instance, that by their repeated stinging cause the horse to lick himself and thus to receive his most cruel enemy. In the perilous journey they have to perform from the skin of the horse to his stomach, many of the larvæ of the *Œstrus,* as may be supposed, are destroyed—ground by the teeth of the animal, or crushed by the alimentary substance."

Symptoms shown by a horse seriously infested by bots: He does not eat heartily, and therefore loses flesh, and has a stiff and staggering walk, and to use a common phrase, appears "consumptive."

REMEDIES.—To prevent the eggs from reaching the horse's mouth, clean off daily by scraping. There exists a vast difference of opinion in regard to remedies; some persons recommend drenching with oils and bleeding; others claim that the animal is benefited by the presence of the bot. Consequently, I can only advise consulting a veterinary surgeon.

CHAPTER CCXLVI.

The Ox Bot-fly.

(*Œstrus boris.*—De Geer.)

Order, DIPTERA; Family, ŒSTRIDÆ.

[Living in tumors on cattle; a whitish footless maggot.]
Fig. 363.—Larva, or maggot of Ox Bot-fly—color, white.
Fig. 364.—Ox Bot-fly, enlarged—colors, black, yellow and white.

Fig. 363.   Fig. 364.

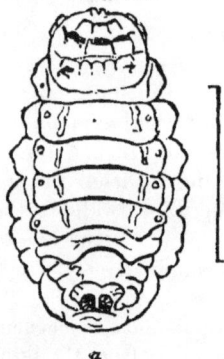 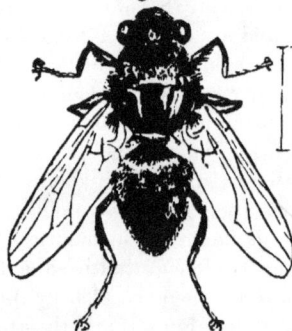

a

The parent fly deposits her eggs upon the backs of the cattle, or, according to some authors, she first punctures the skin, and then deposits therein a single egg; the first hypothesis is the more reasonable one, and we may suppose, with very good reason, that the grub, or maggot (Fig. 363), as soon as hatched, burrows through the skin and takes up its abode just beneath it, where its presence results in the formation of a tumor, in which the maggot lives. After attaining its full size it deserts its former abiding place and falls to the ground, which it enters and soon contracts to a dark-brown pupa, from which the perfect fly issues in the course of a few weeks.

The perfect fly (Fig. 364) is of a black color, thickly covered with hairs except on the thorax, which is twice broadly banded with yellow and white; at the base of the abdomen is a white or yellow band, and at the tip is a band of reddish hairs. This species causes great annoyance to cattle in pastures. It is said that the buzzing noise made by the fly terrifies work oxen to such an extent that they at times become unmanageable.

HARE, or JACK RABBIT.

A species of *Œstrus* attacks the hare, or jack rabbit; the habits and natural history of this œstrus are probably similar to that of the *œstrus boris*.

REMEDY.—The opening of the tumor may be enlarged with the point of a sharp knife, and the maggot extracted, or forced out; this can be done without much pain to the animal.

CHAPTER CCXLVII.

The Sheep Bot-fly.

(*Œstrus ovis.*—Linnæus.)

Order, DIPTERA; Family, ŒSTRIDÆ.

[Living in the heads of sheep; a whitish footless maggot.]

It is thought that the parent insects (Fig. 365, *1* and *2*,) are viviparous, and that the maggots (Fig. 365, *4, 5* and *6*,) are brought forth alive, and are deposited in or near the nostrils of the sheep, up which they crawl until reaching the frontal sinuses, where they attach themselves by means of the hooks with which their heads are provided. Here they remain until attainning their full size, when they loose their hold and make their way to the opening of the nostrils and then fall to the ground, which they enter, and soon contract to dark brown pupa (Fig. 365, *3*,) from which the perfect fly issues in a few weeks.

Fig. 365.—Sheep Bot-fly; *1*, the fly with its wings closed; *2*, the same with its wings expanded —color, ashen-brown; *3*, the pupa—color, brown; *4* and *5*, the maggot or grub—color, whitish; *a*, the head; *b*, the anal plate; *6*, the young maggot; *c*, its spiracles or breathing pores.

Fig. 365.

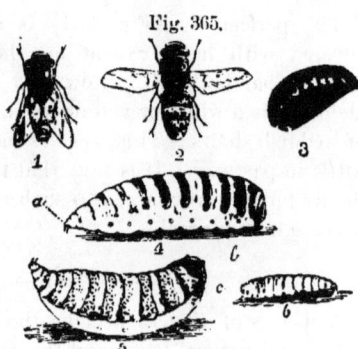

The fly (Fig. 365, *1* and *2*) is of a dirty ash color, with four black lines on the thorax, and the abdomen is spotted with black. The grubs or maggots sometimes make their way even into the brain, and to their presence is due the disease known as "grub in the head."

Concerning this species, Figuier writes as follows: " Even at the sight of this insect the sheep feel the greatest terror. As soon as one of them appears the flock becomes disturbed; the sheep that is attacked shakes its head when it feels the fly on its nostril, and at the same time strikes the ground violently with its forefeet. It then commences to run here and there, holding its nose near the ground, smelling the grass, and looking about anxiously to see if it is still pursued. It is to avoid the attacks of the *Œstrus* that, during the hot days of Summer, sheep lie down with their nostrils buried in dusty ruts, or stand up with their heads lowered between their fore-legs and their noses nearly in contact with the ground. When these poor beasts are in the open country they are observed assembled with their nostrils against each other and very near the ground, so that those which occupy the outside are alone exposed."

REMEDIES.—Use Nos. 93, 94 and 95.

NOTE.—The above remedies have been used, with excellent resuls by Messrs Green and Trainer, of Sacramento, at their farms in Placer County.

CHAPTER CCXLVIII.

The Scab Mite.

(*Psoroptes equi.*)

Class, ARACHNIDA: }
Sub-class, ACARINA; } Family, ACARIDÆ.

The following is taken from the Seventeenth Illinois Report by Professor Cyrus Thomas:

"The scab mite of the sheep (Figs. 366 and 367), which is now believed to be the same species as that infesting the horse and ox, belongs to the family Acaridæ, and sub-family Sarcoptinæ, which also contains the human itch mite, or mite that produce the disease in man known as the *itch*."

Fig. 366.   Fig. 367.

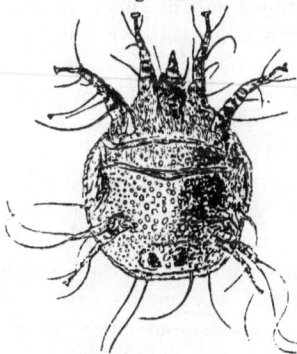
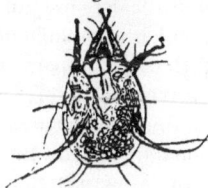

Fig. 366.—Scab Mite (adult), highly magnified, ventral view—color, whitish.

Fig. 367.—Scab Mite (young), enlarged, ventral view—color, whitish.

"The scab, as all are aware, is a skin disease analagous to the mange in the horse and itch in man, and, like these, is produced by a very small mite. It was, for a very long time, supposed that this mite was produced spontaneously by an unhealthy and unclean condition of the flocks, or from some insufficient or improper food, etc., and not from a preceding parent. Our parasite has received various scientific names, according to the fancy or opinion of authors, but, as the further history of its classification is unnecessary now, I will mention but two of these.

" First—*Dermatodectes ovis;* sheep itch-mite of Gerlach, by
whom it was considered as peculiar to sheep; but the most
recent authority restores the name *Psoroptes equi,* horse itch-
mite of Gervais, considering the species infesting the horse,
cattle and sheep as identical. It is distinguished from the
itch-mite of man (*Sarcoptes scabei.*—Latr.) by two or three
important characters. The four anterior legs and two of the
hind ones, at least in the males of each species, are furnished
with sucking disks placed on comparatively long pedicels or
stems. In the itch-mite of man, as far as I can ascertain,
these pedicels are only one-jointed. But the most important
difference is that the mouth of the human itch-mite is furnished with scissor-like jaws or nippers with which they can
readily cut into the skin and form their subcutaneous burrows.
The sheep mite is furnished with comparatively slender lancet-like mandibles, to the sides of which the little palpi or
mouth feelers are glued, thus forming a sort of tube, one part
of which is capable of piercing. But while this peculiar form
of the oral apparatus enables them to pierce and suck, it
deprives them of the power of cutting, and hence, notwithstanding the general opinion to the contrary, they are not
subcutaneous in their habits and do not form true burrows,
as the human itch-mite. It lays its egg on the surface of the
skin, to which they adhere by a gluey matter. The length of
time these require to hatch in such situations is not positively
known, but some placed in a bottle and kept to the warmth
of the body hatched in fourteen days. The young, which are
produced from these have only six legs, but after several
changes of skin or moultings they acquire eight, which is the
normal number of this class. With the little sucking disks of
their feet they are enabled to cling firmly to the skin of the
sheep. By piercing the skin with their lancet-like mandibles,
irritation and a species of inflammation of the skin follows
and an exudation takes place which ultimately forms
the scab. As stated by a writer on this subject: ' Examination will disclose spots on the skin white and hard, the
center marked with yellow points of exudation which adheres to the wool, matting the fibres together. The wool
may be firm on these spots and no scales are seen in this

stage. Then the yellow moisture evaporates, giving place to a yellow scab which adheres firmly to the skin and wool.' Raw places appear at points which the animal can reach with its teeth and hind feet. The disease is aggravated in Summer by the presence of the larvæ of the blow-fly, the maggot burrowing into the scab."

REMEDY.—Use No. 92.

CHAPTER CCXLIX.

The Liver Fluke. (Cal.)

(*Distoma hepaticum.*)

The following is taken from the Seventeenth Illinois Report, by Professor Cyrus Thomas :

"Of the intestinal worms that attack sheep, we notice as the most important the Liver Fluke (*Distoma hepaticum*). This species belongs to the class *Scolecida*, order *Tremoloda* or ' *Suctorial* worms,' as given above, inhabits the gall-bladder or ducts of the liver in sheep and, it is believed, causes the disease known as the 'rot.' It derives its common name from its resemblance in form to the flounder, of which ' fluke ' is a Scotch and old English name. It is somewhat broad and flattened, of an elongate-ovate form, somewhat pointed at each end and is usually nearly an inch long, often much less, but occasionally more ; its breadth at the widest part, which is toward the front, is about half its length. Its color is usually that of the organ in which it resides. It belongs to a very low type of beings, having neither eyes, true respiratory organs, heart, or any other organ of special sense. The sexes are not even distinct and the alimentary canal does not even pass through the body, but dividing and sub-dividing, permeates all parts of it, distributing the imbibed nourishment, which needs little or no assimilation to adapt it to use in forming the materials of the body. It is proper to state, however, that Youatt and other writers on sheep distinctly affirm that flukes have eyes and even figure them. But what possible use they have for these organs, in the situation they occupy, it is impossible to say ;

and moreover, without nerves, of which there are but mere traces, these organs would be entirely useless, even if they were in the light. That they do possess eye-like spots at a certain stage of their life is true; but there is nothing to show that these are organs of sight or eyes in any true sense. What this author considers the heart and circulatory system is probably the water vascular system, found in these and all other animals belonging to this class, which is supposed by many to represent the respiratory system in the higher grades. He also supposes that the eggs or spores, after being cast off, remain undeveloped until taken into the stomach by the sheep with its food, which, as will be seen, is an error.

" The species belonging to this order vary considerably in their transformations and habits, some passing through a cycle of six forms, while others present only three or four. Some infest the liver or hepatic ducts of vertebrates; others infest the intestines of birds and batrachians, the gills of fishes or paunch of ruminants, while others are found imbedded in the vitreous humour and lens of the eyes of certain fresh water fishes, such as the perch.

" The cycle of changes through which the liver fluke of the sheep passes has not been fully traced, but the life-history of *Distoma militare*, another species of the same genus, which inhabits the intestines of water-birds, has been nearly completely traced and from it we may, with what we know of the life of the liver fluke, form a somewhat correct idea of the history of its transformations.

" This species, as stated, in its perfect or mature state, resides in the intestines of certain water-birds. The ova or spore-like eggs which it produces are few, some eight or ten in number. From each egg issues a ciliated larva, which still retains something of the character of an egg, although active, as there is an outer envelope in which is the real animal, or in which it developes, its history at this point of its life being yet imperfectly known. From this egg-like larva proceeds the second larva form, which is known as a *Redia*. Its mode of development in this form is not fully known. It is now found attached usually to the body of some water-snail (*Paludina*), the cilia of the first larva having now disappeared. When the

Redia, or second lava form, has acquired its complete growth, it is somewhat of an arrow-head shape, consisting of a sac, within which is suspended a tubular bag containing colored masses, which Huxley supposes are alimentary. The head is represented by a kind of a crown, and near the other extremity are two lateral projections. In the body cavity, external to the tabular sac, vesicles now appear, which rapidly increase and assume the form of *Cercaria*, the name given to the third larval stage. The *Redia* now bursts and these new zooids escape. The multiplication at an intermediate and incomplete stage (before sexual characters have appeared) is very remarkable and introduces to our view a strange feature in animal life.

" The *Cercaria* resembles a peanut, with a slender tail attached to one end; it also has lateral membraneous attachments, by means of which it swims after the manner of a tadpole. After swimming free for a certain length of time, it finally fixes itself upon and usually bores its way into the body of a water-snail or some other mollusk. The tail then drops off and the body incloses itself in a cyst. The coronal hooklets of the perfect form now appear. It now remains quiescent, unable to develop further in its present situation, awaiting for some water-bird to swallow the mollusk in which it is imbedded. As soon as this is done and the cyst is set free in the alimentary canal of the bird, further development begins and the complete or *Distoma* form is assumed. The body elongates and narrows anteriorly, the suckers move nearer the head and the circle of hooklets being complete, it attaches itself by these to the walls of the intestine. Such is the strange life history of this intestinal worm; and although that of the liver fluke may vary in some respects, yet it is doubtless similar in a general sense.

" The following outline, given in my address before the Illinois Wool-growers Association, September 20, 1877, is probably substantially correct :

' They produce a kind of spore, or egg, but its subsequent progress, so far as it is at present known, presents one of those singular life histories occasionally met with in the lower order of animals. In some way, not well understood, this egg or germ spore makes its way to the external world; its history from this time until it is hatched is unknown, but moisture in some form is probably necessary to its development. It is next found in the body of some mollusk, as the snail, or some aquatic insect, where its form is so different from that of its after-life that it was long considered as appertaining to an entirely different group of animals. From these, in some way not yet ascertained, it passes into the sheep. It probably escapes from the mollusk or insect to herbage in moist places, or water, and is taken into the stomach of the sheep with its food or drink, and passes through the lacteals, and makes its way into the ducts of the liver.

'As sheep do not feed on mollusks as the water-birds do, it is difficult to imagine how the cercaria, if it becomes encysted in the body of the mollusk, makes its way into the stomach of the sheep. It is possible this may be explained in one of two ways. First, as has been stated above, the cercaria is for a time a free swimmer, and hence may be taken into the stomach when drinking, or attached to herbage in damp places. Second, it has been ascertained by Van Beneden that some species of this group pass to the mature state directly from the redia stage without undergoing the intermediate or cercaria stage; as these are free they may be taken into the stomach in water or on damp herbage.

'As before stated, these internal parasites are supposed to be the cause of rot in sheep, though many persons are inclined to believe they are a consequence rather than a cause of disease. But all appear to agree that this disease is connected with the condition of the soil, or the state of weather, moisture being the element most likely to produce it. This corresponds exactly with the theory of its life history which I have presented, and indicates the best means of preventing it, to wit: Give them well-drained, open, airy pastures, and proper protection in damp and rainy seasons. It is more probable the condition of sheep is often attributed to this disease, when it is due to other causes.'"

INSECTS INFESTING THE APIARY.

CHAPTER CCL.

The Bee Moth. (Cal.)

(*Galleria cereana.*—Fabricius.)

Order, LEPIDOPTERA ; Family, PYRALIDÆ.

[Living in silken tubes in bee-hives, and feeding upon the wax and young bees; a small grayish worm.]

Fig. 368.

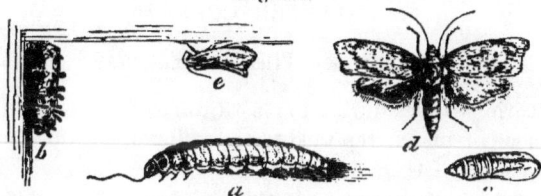

Fig. 368.—Bee Moth ; *a*, the worm or caterpillar—color, ash gray ; *c*, the pupa—color, brown ; *d*, the moth, with its wings expanded ; *e*, the same, with its wings closed—colors, gray and brown ; *b*, the cocoon—color, whitish.

This is by far the worst enemy with which the bee-keeper has to contend. The female moth, if prevented from entering the hive, will deposit her eggs in cracks or any opening in the hives. As soon as hatched, the young worm (Fig. 368*a*) enters the hive, and at once protects itself by spinning around its body a silken tube ; as it increases in size it enlarges the tube, feeding the meanwhile upon the wax and young bees. When

24

fully grown, it creeps into some corner of the hive, or into some other sheltered place, and spins a tough white cocoon (Fig. 368b), intermingled with its own black excrements. The perfect moth (Fig. 368, d and e,) has dusky gray forewings. which are scalloped at the outer end, and are sprinkled and dotted with brown.

REMEDY.—Use No. 122.

MISCELLANEOUS INSECTS.

CHAPTER CCLI.

What are they, Friends or Enemies?

At various times during the last two years, I have received specimens of insects reported to be feeding upon fruits, and I refer to some of them as follows:

Fig. 369.—Larva of Lace-winged Fly—color, gray.

Fig. 369.

1st. Mr. Scott, residing near this city (Sacramento), brought me some ripe peaches in which were small insects, eating in holes made through the skin of the fruit. The holes were less than one line in diameter, and from two to four lines in depth. By removing the insects from the holes in the fruit, I found that the supposed new pest was the larva of a *Chrysopa* (Fig. 369) or lace-winged fly. That they were feeding upon the fruit is beyond question; but probably the opening had been made by a species of plant-bug, or by the striped cucumber beetle (*D. vittata*), which infested the trees. I could not find any plant-lice on the trees. therefore the larva of the *Chrysopa* may have entered the punctures made by some other insect to feed upon the fruit.

2d. I have also received specimens of a species of ladybird, said to attack cherries when ripe. It is the *Coccinella 5-notata* var. *Californica*. (Fig. 370, Plate 4.)

Description: Form, ovate; length, three lines; color, head black, with two white dots; thorax black, with a white mark on each anterior angle; wing-cases, light orange, with a minute white mark at each side of the scutellum.

In answer to inquiries made by me, Mr. A. P. Crane, a prominent fruit grower at San Lorenzo, Alameda County, writes as follows, under date of June 12, 1883:

"DEAR SIR: Yours of the ninth instant came to hand last evening. I send you specimens of lady-bird by this mail. They troubled our Black Tartarian Cherries some years ago. We allowed the fruit to become too ripe before picking, and the species I send you destroyed considerable. I have not noticed them to be so plentiful in three or four years past as they are this year (1883), but they are not doing any harm as yet that I can observe. Last Wednesday, while looking through our apple orchard. I found a large number of these lady-birds devouring the woolly aphis. This morning, in looking for these specimens which I send you, I could not find any on the trees near the house, neither any aphis, but found the lady-birds on boxes containing cherries. The lady-birds are often observed feeding upon ripe apricots, but some growers think that the striped or spotted *Diabrotica* first punctures the fruit before the lady-bird will touch it. I believe the lady-birds will puncture and eat the fruit, if it is soft and ripe.

"A. P. CRANE."

3d. Mr. Welty, whose orchard is located on the Sacramento River, a few miles below this city (Sacramento), brought me a specimen of a sand wasp, belonging to the genus *Priocnemis*. (Fig. 371, Plate 4.)

Description: Length, nearly one inch; color, body black, with bluish or greenish shades; wings, brick-red, with black tips.

Mr. Welty states that these insects have destroyed the crop of twelve peach trees (variety, "Alexander,") for two years in succession—1881 and 1882. As soon as the fruit begins to get ripe, these pests attack it in such numbers as would destroy the entire crop if not picked before it is ripe. He also states that the other varieties of peaches are not attacked by this species.

4th.   Specimens of peach branches were brought me by Mr.
Dye, of Walnut Grove, in this county (Sacramento).   On each
branch were three or four mud nests (Fig. 372, Plate 4,) each
one containing a larva feeding upon a species of small spider
placed in the nest by the parent insect.   They were placed in
boxes.   The wasps did not mature entirely, but nearly completed their changes.   One of the nests produced a golden-
green fly, belonging to the *Chrysididæ*—a beautiful specimen.

Description: Length of body, nearly four lines; color,
ultramarine-blue; ovipositor, blackish; wings, hyaline (transparent).

I do not think that this insect is injurious to fruit.

5th.   Bees.—A great difference of opinion exists among fruit-
growers as to whether bees attack fruit and grapes.   Some
growers make positive assertion that they do attack fruit and
grapes, while others are equally positive that they do not until
the fruit or grapes are punctured by other insects, such as
plant-bugs, wasps, etc.

In my investigations, I found an orchard and vineyard
located at least two miles from any other orchard or vineyard.
In this place are kept a large number of bee-hives.   The owner
informs me that his grapes are used for making wines, and
that the bees do not touch them until the picking of the grapes
commences; then they feed upon the broken berries and such
as are attacked by other insects; also, that they do not attack
the fruit until it is punctured by other insects.   My informant
gives this as the result of many years experience.

Fig. 373.   Fig. 373.—*Blapstinus lecontei;* ventral and
dorsal views—color, black.

6th.   In 1882, Mr. R. B. Blowers, of Woodland, sent me specimens of a small dark-colored
beetle (Fig. 373), about one third of an inch in length, found
in some vineyards near Woodland, feeding upon the leaves
and young growth of grapevines.   I ascertained them to be
Blapstinus lecontei.   This species had not been reported as being
injurious heretofore.   It belongs to the same family as the
following species—*Tenebrionidæ.*

Fig. 374.—*Eleodes quadricollis*—color, black.

Fig. 374.

7th. Early in July, 1883, I received from J. W. Minturn, of Madera. Fresno County, specimens of a black beetle (Fig. 374)—the *Eleodes quadricollis* of Leconte—about three fourths of an inch in length, reported to be feeding upon the foliage of grapevines. As this species is considered harmless, I doubted that the damage to grapevines was done by it, and wrote, asking that full investigation be made, to learn if the damage was not done by some other species of insect, and received the following answer, dated July 18th: "I have received your letter, and herewith return answer. As to those beetles eating the foliage of the vines, it is simply solved beyond any doubt, as far as I am concerned; and it is more than probable that a majority of men would be quite satisfied on this point, had they been up the greater part of ten nights and seen on each vine that they came to from two to a dozen of these beetles, besides thousands of them on the ground over an area of one hundred acres. If further proof is required, I think the destruction of thirty-five acres within the above named time (there being no other insects, or rabbits, or squirrels to be found) would have the tendency of convincing any person. They have also appeared in some alfalfa fields, and I am informed by a gentleman lately returned from the Tejon Pass, that there they are in such numbers that they have completely devoured every green thing. Those that passed over my vineyard were traveling from east to west.

"P. S. The gentleman above referred to tells me that the Mexicans over there say these beetles precede a very heavy winter, and that they have been seen in the same numbers many times before."

In reply to further inquiries, the following letter was received, dated July 27th:

"The beetles, as nearly as I can remember, first came about the 5th of July; within three days they had damaged some ten acres, and were evidently rapidly on the increase at this time. About the 8th or 9th inst. I discovered them at night. Until that time I had been at a loss to know what was eating

my vines, as I had a rabbit-proof fence around my field of 160
acres, which is looked after each day. As far as I know, they
had not paired during the period that they were here, though
during the day they seek the shade of clods of earth, holes in
the checks, under some water gates that connect the checks,
and in these places large numbers seem to congregate. For
instance, this afternoon my brother turned up a couple of
clods, each say four or five inches in diameter, and under
them was at least sixty or seventy beetles. The greater part
of the beetles have disappeared; whether they have died or
migrated I cannot be sure, but think the latter, for if they had
died in such numbers I think I should have discovered them.
In a slough that is used as a water channel there are pools of
water somewhat stagnant, and at some of these I found thousands of dead beetles—still not anything like enough to account for the disappearance of the army that was here."

I sent a specimen of this beetle to Prof. Riley for determi-
nation, but he being absent the specimen was referred to Prof.
E. A. Schwarz, who replied as follows, under date of August
7th:

"The tenebrionid you send, and which has destroyed thirty-
five acres of grapevines, is *Eleodes quadricollis*—Lec.; a very
common species in the more northern part of your State. The
species of that genus so numerous and abundant in the region
west of the Rocky Mountains are all known to feed upon
decaying vegetable matter, and none have hitherto been reported as doing damage to cultivated plants. In fact your
communication, if correct, would indicate a change of habit
hitherto unprecedented in the history of economic entomology; and unless further proof be brought forth, I can hardly
believe that the species referred to is the real author of the
damage to grapevines."

8th. Ants.—Since the Spring of 1881, occasionally some
fruit-growers would report that the ants were among the insect
friends of the fruit-growers. In 1881, the late James B. Saul,
of Oak Shade Orchard, Yolo County, wrote me that in his
investigation he discovered that he was losing some larvæ of
the codlin moth, and could not account for the loss. The
larvæ were taken from the apples and pears before they had

attacked the seed-bag, or core of the fruit. I visited the orchard, at his request, but could not fathom the mystery. A few weeks afterward Mr. Saul wrote me that he had detected the thief. That the larvæ were taken from the fruit by ants. I wrote him to write a letter on the discovery, and I would have it published; but the letter was not written before his sudden illness and death, which took place a short time afterward. In 1882, in company with Mr. C. W. Reed, of Washington, Yolo County, I witnessed the ants taking the larvæ of the codlin moth from pears; also carrying away larvæ placed on leaves. It is generally understood that the larvæ of the codlin moth generally leaves the apple by eating a burrow in a different direction from the one by which it had entered the fruit. In noticing the absence of the larvæ without finding the burrow for its escape, was what caused the investigation by Mr. Saul. Since that time I have detected the presence of the apple curculio, which may account for some such holes in apples, since the curculio does not attack the seed.

REMEDIES.

REMEDY NO. 1.

To mix sulphur with whale oil soap or home made soft soap :

Boil the sulphur in water for ten or fifteen minutes (if concentrated lye or other formulas are to be added they should be dissolved in water and added when the sulphur and water are boiled the required time, and allowed to boil for five minutes) ; then add the soap.

Example.—To mix 10 lbs. of whale oil soap, 3 lbs. of sulphur and 1 lb. of No. 11, or 12, or 13 in 18 gallons of water. Boil say four gallons of water ; then add the sulphur and boil ten or fifteen minutes (the lye compound, Nos. 11, 12 or 13, should be dissolved in one gallon of water), then add the lye and boil for five minutes. The soap is then added and allowed to boil five minutes. When the mixture is boiled together as directed, it may be placed in a barrel, thirteen gallons of water added, and it is then ready for use. It should be applied to the trees at a temperature of 130° Fahrenheit.

All mixtures with sulphur should be made at least one week before using; the longer it is kept the more fetid it becomes—therefore, the more efficient.

REMEDY NO. 2.

Concentrated lye (American Lye Co. brand). The following analysis is given by Prof. Hilgard, of the State University of California : Caustic potash, 8.3 ; caustic soda and some carbonate of soda, 91.7—100.0. Taking the concentrated lye as a basis, one pound of the above brand dissolved in one gallon of water (tested by an alkalimeter for heavy liquids) gives: density, 1074; alkalimeter, 10. Cost per case of 48 lbs., $4.50, or about 9⅓ cents per pound.

Cost of caustic soda (English), in drums, 5½ cents per pound; commercial potash, in drums, 9 cents per pound.

Formula No. 1.

One pound of caustic soda (Eng.)

Two ounces of commercial potash (⅛ pound).

Dissolved in one gallon of water (tested with the alkalimeter, as above): density, 1090; alkalimeter, 12.

This formula produces a solution of a higher density than the concentrated, lye at a cost of

42⅔ pounds of caustic soda, at 5½ cts., $2.35.

5⅓ pounds of commercial potash, at 9 cts., .48—Total, $2.83. (Less than six cents per pound.)

Formula No. 2.

One pound of caustic soda (Eng.)

Four ounces commercial potash (¼ pound).

Dissolved in one gallon of water (tested with alkalimeter as above): density, 1094; alkalimeter, 12½.

38½ pounds caustic soda (Eng.) at 5 cts., $2.12.

9½ pounds commercial potash, at 9 cts., .85½—$2.97½. (Less than 6¼ cts. per pound.)

This formula produces a solution of a higher density than either of the above, and is not only a superior insecticide, but a richer source of fertilization for the tree.

NOTE.—The data from the alkalimeter is merely given to show the difference in density as indicated by one instrument.

To dissolve the above material thoroughly the time required is about forty-eight hours; use boiling water. To make a certain quantity of No. 12, one ninth of the total weight of material required is potash and eight ninths caustic soda.

Example.—To make forty-five gallons of solution No. 12 (one pound to each gallon of water) it requires forty-five pounds of material (i. e., caustic soda and potash); ⅑ of 45 pounds equals 5 pounds of potash; ⅛ of 45 pounds equals 40 pounds, giving 5 pounds potash and 40 pounds caustic soda in 45 pounds of the mixture.

To make 45 pounds of No. 13, one fifth of the total weight should be potash and four fifths caustic soda.

Example.—⅕ of 45 pounds equals 9 pounds; ⅘ of 45 pounds equals 36 pounds, giving 9 pounds of potash and 36 pounds of caustic soda in 45 pounds of the mixture.

As many fruit-growers are not acquainted with the use of the alkalimeter, and mistakes are liable to occur in the use of the alkaline solutions, I recommend the following simple test which will at least be a partial protection against the use of solutions of sufficient density to injure the tree, or, on the contrary, deficient in density to give the result required:

Fig. 375.

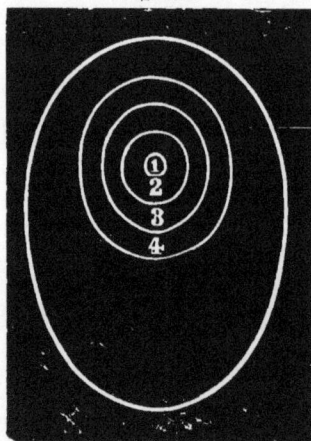

Fig. 375 represents an egg (hen's egg). The circular lines marked *1, 2, 3* and *4* are intended to show gauges of density of the solutions.

In all cases the solution must be thoroughly stirred from the bottom of the vessel in which it is dissolved, before testing or taking away any portion of it to use on trees, etc.

When the mixture is thoroughly dissolved, stir well, and then test the density by placing in the solution the hen's egg. (In all tests of this kind the egg used should not be more than twenty-four hours laid, and as near to a globular form as possible.) If the surface of the egg above the solution is more than the size of No. 4, the solution is too strong for use. This can be reduced to 3½ by adding water. If the surface of the egg above the solution is equal to No. 3½ it is just the right strength, and equal to Remedy No. 12. If the surface of the egg above the solution is the size of No. 3, it is equal to Remedy No. 11. If the egg just floats or shows a speck above the solution the size of No. 1½, it is equal to one pound of Remedy No. 12 to 1¼ gallons of water. In lye made from wood ashes the egg will just float in a solution equal to one pound of lye to one and one quarter gallons of water. Space above water equal to $3\frac{1}{2}$ represents one pound of lye to each gallon of water.

The above tests should be made with the solution at a temperature of 60° Fahrenheit. Should any fruit grower wish to get an alkalimeter, I will have one that is properly adjusted forwarded to him on receipt of one dollar.

REMEDY NO. 3.

Home made soft soap, made with lye leached from wood ashes. One pound to each gallon of water used. Apply at a temperature of 130° Fahr.

NOTE 1.—To mix with sulphur see No. 1.

NOTE 2.—In making soft soap, if one ounce of glycerine is added to each gallon of soap, it will make an excellent insecticide.

REMEDY NO. 4.

Whale oil soap (residue from bleaching whale oil and sold by Allyne & White, Nos. 112 and 114 Front street, San Francisco, in barrels, about $3\frac{1}{2}$ cents or 4 cents per pound), one pound to each gallon of water used. Apply at a temperature of 130° Fahr.

NOTE.—See Chapter IX.

REMEDY NO. 5.

One pound of whale oil soap and $\frac{1}{4}$ of a pound of sulphur; mix as directed in Remedy No. 1. One pound of this mixture to each gallon of water used. Apply at a temperature of 130° Fahr.

NOTE 1.—The weight of the soap and sulphur used in making this solution must be added together in computing the quantity of water required.

NOTE 2.—To procure genuine whale oil soap, see Remedy No. 4.

REMEDY NO. 6.

One pound of whale oil soap; one third of a pound of sulphur; one and one half ounces of Nos. 11 and 12; mix as described in Remedy No. 1. One pound of this mixture to each gallon of water used, and apply at a temperature of 130° Fahr. See notes in No. 5.

REMEDY NO. 7.

One pound of the whale oil soap and sulphur mixture to each gallon of water used. (This mixture is manufactured and for sale by Allyne & White, Nos. 112 and 114 Front street, San Francis co.) Apply at a temperature of 130° Fahr.

REMEDY NO. 8.

One pound of whale oil and paraffine soap to two gallons of water. Spray thoroughly. (Manufactured and sold by Allyne & White, Nos. 112 and 114 Front street, San Francisco.)

NOTE.—Paraffine can be substituted for coal oil in Remedy No. 41; see also No. 77.

REMEDY NO. 9.

Boil thirty pounds of tobacco leaves in thirty gallons of water, and apply at a temperature of 130° Fahr.

NOTE.—Every fruit-grower should grow a small patch of tobacco upon his premises. Mr. Ellwood Cooper, of Santa Barbara, grows all the tobacco he requires for insecticides, at a cost of about two cents per pound. Refuse tobacco can be bought at cigar manufactories, or nicotine of tobacco at Leibes Bros., Nos. 14 and 16 Fremont street, near Market, San Francisco.

See Remedy No. 123.

REMEDY NO. 10.

Boil six pounds of arsenic in sixty gallons of water (or one pound of arsenic to ten gallons of water) until the arsenic is dissolved; when ready to use, add ninety gallons of water, making one hundred and fifty gallons in all, or one pound of arsenic to twenty-five gallons of water. (Spray the foliage.) One or two pounds of potash dissolved and boiled in water before putting in the arsenic, will make the arsenic dissolve quicker. For every one pound of potash used, five gallons of water can be added.

NOTE.—In cases where trees or vines are seriously infested by canker-worms, beetles, etc., and the prospects of a crop ruined, use No. 10, or No. 89. Thorough spraying will effectually destroy them. Care should be taken that this application should not be put on trees that the fruit is to be used for food in any way, as it is dangerous to those eating the fruit. Only use this remedy when all other means fail. See Remedies No. 22 and No. 123. Great care should be taken to keep children away from where this solution is prepared or used.

See Remedy No. 123.

REMEDY NO. 11.

One pound of American concentrated lye to each gallon of water used.

See Remedy No. 2.

REMEDY NO. 12.

One pound of caustic soda (Eng. brand) and two ounces of common potash.

NOTE.—The weight of caustic soda and potash used in making this solution must be added together in computing the quantity of water required.

See Remedy No. 2, directions for preparing solutions, and No. 123; or lye made from wood ashes; should be boiled until it will float a hen's egg when cool. If the surface of the shell above the liquid is of a larger size than a silver three cent piece, or ring No. 2, Fig. 375, reduce with water to that gauge; then add one quart of water for every three quarts of lye; apply to trees when dormant. (This equals No. 12).

REMEDY NO. 13.

One pound of caustic soda (Eng. brand) and four ounces of common potash. (See note No. 12.)

See Remedy No. 2, directions for preparing solutions, and No. 123.

REMEDY NO. 14.

[Remedy for cottony grape scale, No. 14 should be No. 11.]

Where the caterpillars infesting trees or vines are large (two and a half inches in length and upwards) and do not live in colonies, they can readily be destroyed by cutting in two with a pair of scissors. This is preferable to hand-picking.

See Remedy No. 100 and No. 101.

REMEDY NO. 15.

All parings and other debris made by preparing apples, pears, etc., for drying, should be scalded or burned so as to destroy the larvæ they contain.

See Remedy No. 16.

REMEDY NO. 16.

Early in Winter, destroy by burning or scalding, all peach pits, apricot pits, etc., also the pearings, etc., left from drying fruit that has accumulated around the orchard and store-houses, as the larvæ hibernate in such material.

See Remedy No. 15.

REMEDY NO. 17.

In store-rooms, packing-rooms, etc., infested by the dried-fruit moths, put shelves across the windows and dust them over with buhach; renew the powder at least every second day, but better results will be obtained by renewing it daily.

REMEDY NO. 18.

For cleaning dried fruit infested by the larvæ of the dried-fruit moth, various methods are used; some place the fruit in ovens, others scald it. A practical plan used by extensive dealers, especially when the infested fruit is in sacks or boxes that are not moth-proof, is as follows: A cylinder about twelve feet in length, constructed of two circular ends eighteen inches in diameter, on which are nailed strips of wood twelve feet long, two inches wide and nearly one inch thick; between the strips an opening of about three eights of an inch is left (the openings at the lower end are large enough to let the fruit pass through). The cylinder is placed on a frame, one end of which is a few inches lower than the other end; it rests on a central shaft, to one end of which is attached a crank, by which the cylinder is turned around. The fruit is placed in the cylinder through an opening at the higher end; by turning the cylinder the worms (larvæ) are shaken out of the fruit and fall through the openings between the slats. The excrements and webs of the worms is also cleaned off of the fruit. All larvæ taken out of fruit in this way should be destroyed by scalding, or otherwise.

REMEDY NO. 19.

Ten pounds of fine sulphur and one pound of buhach; mix thoroughly and apply with a sulphur duster after sunset. Or if the vines are not in bearing, and the attack is not of a serious nature, use No. 103, or No. 10, or No. 89. Read No. 123 carefully.

Note.—The sulphur, if applied with buhach, will prevent mildew.

REMEDY NO. 20.

In orchards, vineyards, vegetable gardens, etc., grasses, weeds, and rubbish of every kind, including fallen leaves, should be carefully gathered off of the ground from around fences and buildings and burned early in the Fall, thus depriving the insects of a shelter wherein to hibernate; and the ground should be kept entirely clean of weeds, such as purslane, etc., in the Spring and Summer seasons. All grasses and weeds growing on the banks of sloughs, water ditches, etc., in or around orchards, vineyards, etc., should be burned, or otherwise destroyed.

See Remedy No. 32, A and B.

REMEDY NO. 21.

A smooth bark, free from moss, etc., on fruit trees, grapevines, etc., is an imperative necessity in a warfare against insect pests. This can be effected by scraping and using alkaline washes, such as Nos. 11, 12 or 13, when the tree is dormant; or Nos. 3, 4, 5, 6 or 7 in Summer. In all cases the scrapings should be burned.

Read Remedies Nos. 35, 123 and 124 carefully.

REMEDY NO. 22.

Early in the month of October wrap a band of thick paper (the thickness of medium building paper), from six to eight inches wide, around the base of the trunk of the tree close to the ground, and fasten by a piece of baling rope passed around about the center of the band: gather or mound the earth around the band below the rope. This is to prevent the wingless female from depositing her eggs on the tree, under or below the band. Take some coal tar and spread on the upper part of the band, above the rope—two inches wide is sufficient. A little castor oil mixed in the tar will prevent it from hardening as rapidily as if only tar is used. The tar should be renewed every two or three days, as it is necessary to keep it moist to prevent the females from passing over it. Such of the females as will not go near the tar, and are prevented from getting to the tree under the band, will deposit their eggs on the band. The bands should be kept on the trees until the apple trees begin to put forth their leaves, and should then be taken off and burned, ropes included.

NOTE.—There are at least three species of canker worms in orchards at the present time, and as the natural history of all the species as regards dates of emerging from the ground is not known, it is the safest plan to put on the bands early in the season. Should the caterpillars appear on the trees, spray thoroughly with No. 64 ; at the same time it would be well to experiment with Nos. 83 and 85.

See Nos. 10, 89 and 103. Read No. 123 carefully.

REMEDY NO. 23.

Put around the tree, about two feet from the ground, a band of cloth, such as muslin butter cloth ; cover with tallow after it is placed on the tree, and put some lard over the tallow ; the lard will keep soft and prevent caterpillars from passing over it. When the caterpillars are swept off the tree they will creep up the trunk again until they reach the greased band ; they gather below the bands in large numbers, and can then be destroyed by the use of clubs, etc.

See Chapter XXXIII. Read Remedies No. 35 and 124 carefully.

NOTE.—The caterpillars are swept off the branches with brooms.

REMEDY NO. 24.

A large number of varieties of deciduous fruit trees, vines and plants are liable to be infested by the caterpillars of various species of moths, commonly known as "leaf-rollers." An excellent remedy for preventing the spread of these pests is to pick off the rolled leaves as soon as noticed, and destroy them

25

by burning or otherwise. This can be profitably done on young
trees and nursery stock. In some cases the insect fastens the
leaf to the branches, and hibernates in them throughout the
Winter. All such nests should be gathered off the trees in the
Winter season, and burned.

Read Remedies Nos. 35 and 124 carefully.

REMEDY NO. 25.

Cut off and burn all branches showing punctures, swellings,
etc., on vines, branches of trees, or stalks of any plant infested
by larvæ, grubs, or beetles. This pruning, combined with
Remedy No. 20, will prevent the spread of various species of
insects.

See Remedies Nos. 26, 27, 60, 66 and 67.

REMEDY NO. 26.

Twig or branch pruning. Twigs and branches of peach,
apricot, blackberry, raspberry, currant, etc., infested by the
larvæ of moths or grubs of beetles boring into the new growth,
can be readily detected by their withered appearance, and
should be cut off and burned or otherwise disposed of; in this
manner the larvæ or grubs are destroyed.

See Remedies Nos. 27, 28, 60, 66 and 67.

REMEDY NO. 27.

Branches and twigs of apple, apricot, peach, pear and olive
trees, grapevines, etc., infested by the burrowing beetles, or
twig-borers, *P. confertus* and *B. bicaudatus*, should be cut off
and burned. Great care should be taken to capture the beetles.
Remember, in these species it is the perfect insect that attacks
the trees.

See Remedies Nos. 25, 26, 28, 60, 66 and 67.

REMEDY NO. 28.

Branches and twigs of apple, almond, apricot, plum, etc. infested by the eggs of the buffalo tree-hopper, or tree-crickets, or *Cicadas*, should be well cut out in pruning and the prunings burned, and while the tree is dormant, thoroughly spray with No. 12 or 13—one pound to each gallon of water. In this manner many of the eggs will be destroyed. Early in July the branches should be sprayed with No. 5 or 7, to prevent the female from depositing her eggs therein. No. 4 is also very effective.

See Remedies Nos. 25, 26, 27, 60, 66 and 67.

REMEDY NO. 29.

The eggs of certain species of moths are laid in rings encircling the new growth or branches; in some species (as Fig. 54) they partly encircle the branch. These rings of eggs are generally found on the branches of trees infested by the caterpillars of the moths the previous Spring.

To prevent the spread of such pests, immediately after the foliage has fallen off the trees the egg-rings should be collected and destroyed by pouring boiling water on them or by burning them. This work, if thoroughly done, will well repay the fruit-grower. (See Chapter XXXIII.) This also includes the eggs of the katy-did. (Chapter CXVIII.)

REMEDY NO. 30.

The egg-clusters of the tussock moth are generally found on the top of cocoons from which the wingless females emerged, and can readily be gathered and destroyed by burning, etc. The cocoons are found in the crotches and indents of trees infested by the caterpillars the previous Summer.

REMEDY NO. 31.

The nests of the tent caterpillars may be destroyed by holding a torch under them at a certain time of day when the caterpillars are not feeding, and are gathered in their tent or nest; in this way the whole colony of caterpillars can be readily destroyed. A torch such as is used in torchlight processions is excellent for this work.

See Remedy No. 29, and read No. 35 and No. 124.

REMEDY NO. 32.

For effective work the warfare against the grapevine-hoppers should be commenced early in the Fall season:

A.—By taking or scraping off all of the loose bark and thoroughly spraying the trunk of the vine with Remedy No. 51, or 64, or 65; this will dislodge and destroy the insects, and at the same time destroy the fungus spores on the vine. As soon as this work is done, a flock of sheep should be placed in the vineyard to eat the fallen leaves and keep down the growth of *alfilaria* and grasses upon which the insects, such as may escape, will feed on warm days. Grasses and weeds around fences, etc., should be destroyed by burning or otherwise.

B.—The vines should be cleaned as directed in section A. If sheep cannot be procured, the leaves should be raked in rows between the vines, that the vine-hoppers may lodge in them. When the leaves are perfectly dry, apply fire to the rows in places not more than eight or ten feet apart. Before lighting the fires the vines should be jarred, and the vine-hoppers around them drove into the dry leaves. This method is very effective.

See Remedies Nos. 20, 33 and 106, A.

REMEDY NO. 33.

Mr. R. B. Blowers, of Woodland, Yolo County, California, has succeeded in conquering the vine-hoppers this season, and kindly furnishes me with the following information:

"The mixture is as follows: In a fifty gallon barrel put thirty pounds of whale oil soap and fifteen pounds sulphur, and mix it up thoroughly. Then add water until the barrel is nearly full, and allow it to remain one day without anything else being done to it, except stirring it two or three times. The object of this is to allow the ingredients to be completely blended together, so as to form one homogeneous mixture. The next day take three pailfuls of this mixture and six of water, or in that proportion, until a barrel is filled containing forty or forty-five gallons. Then just previous to using add three quarters of a pound of buhach, and mix it thoroughly. After the addition of the buhach it should to be applied immediately. The best manner of conveying it to the vineyard is to place a barrel of it on a sled, so that it will not slop over. Apply this mixture to the vines with a fountain spray pump. Two men are necessary to do this successfully; one man with a spray pump on each side of the vine. The men should stand about six feet from the vine when making the application, and they should both apply the spray at the same moment. If they do not both make the application to the vine at the same moment, the insects may some of them escape on the opposite side of the vine from where the spray is applied. This is certain to destroy ninety eight per cent of the insects of any and every variety that may trouble vines or fruits or flowers."

A few days later Mr. Blowers wrote me as follows:

"WOODLAND. June 3d, 1883.

"DEAR SIR: The vine-hoppers were so plentiful this Spring that they destroyed many of the leaves on my vines before they were half-grown. Had I depended on my destroying the eggs and young hoppers, my entire crop would have been ruined. The cold weather being favorable, I commenced the work of spraying as I wrote you in my last. The buhach retained its power longer as the solution did not dry so rapidly, the vines being small. I did not get over my entire vineyard before the warm weather came, and I found it necessary to stop spraying, as the sun would burn the leaves.

"The balance I have given a heavy dose of powdered

sulphur and buhach to kill the old hoppers, but sulphur effectually destroys the eggs; the buhach is death to the young hoppers. All insects yielded to the effect of the spray, the sphinx moth (*P. achemon*) included. Yours, etc.,

"R. B. BLOWERS.

"P. S.—The vineyard never looked better."

I prefer mixing the whale-oil soap and sulphur, as in Remedy No. 1.

REMEDY NO. 34.

An effective way of destroying red and yellow spiders (or mites) is to spray the foliage of the tree or vine thoroughly with No. 5 or No. 7, one pound to each five quarts, or one gallon and one quarter of water used; add one gallon of No. 9 to every eight gallons as soon as the mites appear on the leaves. The spraying should be repeated as often as they appear on the foliage.

In cases where the tree or vine is bearing tender fruits, these pests can be prevented from spreading by thoroughly drenching the foliage with water.

Either of the above can be applied to nursery stock infested by red or yellow spiders (mites). When the tree is dormant use No. 8 or 44 on the ova, and follow in twenty-four hours with No. 13.

See Remedies No. 9 (in relation to tobacco), and also No. 20.

REMEDY NO. 35.

a.—100 pounds of bone dust.

b.—50 pounds of commercial potash.

c.—75 pounds of new lime (unslacked).

A.—Add to the potash just enough of water to liquify it (at first one half gallon to ten pounds of potash, then add as required for forty-eight hours).

B.—Use just enough water to slack the lime.

C.—Prepare enough of barrels or bins to hold the whole 525 pounds.

D.—Spread the bone dust on the floor.

E.—Dampen the bone dust thoroughly with the liquid potash, but be careful not to wet any part of it so that it will cake (which spoils it); when thoroughly mixed, fill in barrels, etc., leaving about six inches on the top of each package to fill with the lime.

F.—Let it remain in the barrels for ten days, then spread it on the ground around the vines or trees, say four hundred pounds per acre.

The above material, prepared as above described, makes an excellent fertilizer. Where a large quantity is required, it can be prepared in large bins.

G.—Use from 350 pounds to 500 pounds per acre, and even 1,000 pounds if the vines or trees are seriously infested.

H.—The combination of the potash and bone-dust produces ammonia, which is absorbed by the lime placed on top.

I.—Fruit trees that have been infested by scale insects, woolly-aphis, etc., can be greatly improved by the use of this fertilizer.

Wood Ashes.

J.—Fruit-growers residing in the vicinity of any town or city, or any place where wood ashes can be secured at a nominal cost, should collect every bushel available. The ashes should be thoroughly dampened and mixed with an equal quantity of bone-dust (bushel for bushel when dry), then placed about three feet deep in a bin, or enclosure of some kind, having the sides and bottom tight; spread on top from four to six inches of fresh slacked lime.

K.—If necessary, another layer of bone-dust and ashes, and one of lime, may be placed on top of the first layers.

L.—Let it remain for three weeks, and an excellent fertilizer will be obtained.

M.—Apply on orchard or vineyard grounds when thoroughly mixed.

NOTE.—In Sacramento several hundred bushels of ashes are wasted daily, which could be collected at a very trifling cost. Remember, the ashes should not be allowed to get wet before being gathered, as it then becomes partly leached, thus impairing its value.

Stable Manure.

As to the effect of stable manure, it may be approximated as
regards its utility as a fertilizer for orchard or vineyard purposes. In some of the best authorities I find that one ton of
stable manure contains about four pounds of phosphoric acid,
eight to nine pounds of potash, and about eight pounds of
nitrogen (ammonia and nitric acid). Stable manure for agricultural purpose is excellent, but for orchard or vineyard
purposes it is deficient in restoring the ash constituents so
much required by the trees. (See Remedy No. 124.)

I have briefly mentioned the available fertilizers, and the
necessity of their application to many of the orchards and
vineyards, especially those planted in California in what is
termed the "early days."

To restore the producing power of exhausted lands, to
strengthen and increase the growth of young orchards and
vineyards that are backward in producing a good growth, to
produce the highest yield from each and every acre planted.

Read No. 123 and No. 124.

REMEDY NO. 36.

Early in the Spring, or as soon as the insects appear on the
plants or vines, place loose straw, hay, or other like
material, around or under plants or vines so that the insects
can take shelter at night: in the morning before sunrise
remove the material laid down and burn it.

I recommended this for the destruction of the false chinch
bug on grapevines. The vine owner reported success, but he
sprinkled some coal oil on some straw which he placed on the
ground under the straw taken from around the vines, and in
this way prevented the insects from escaping into the ground
while the straw was burning, making the application a complete success.

Clean cultivation.—See No. 20 and No. 106, A.

Read No. 35, No. 123, and No. 124.

REMEDY NO. 37.

In all cases for preventing the spread of borers in fruit, ornamental or forest trees, the coating of the trunk and larger limbs with soft soap, common bar soap, or whale oil soap and sulphur, will prevent the female from depositing her eggs on the parts thus treated. Two pounds of soap to each gallon of water used; apply with a brush.

Trees infested with the larvæ or grubs of borers, the burrows should be searched for, and when found a wire should be forced into the burrow, and in this way destroy the larva or grub.

Wounds or sunburned patches on the bark of trees should be dressed with some of the above soaps as soon as noticed.

NOTE.—Some persons cut out the grubs with a knife; such a method is not safe for the tree.

See No. 60, for borers in branches.

REMEDY NO. 38.

A.—Spread a sheet or cloth under the tree at night, when the beetles are feeding; by shaking the tree the beetles will fall off, and can then be gathered off of the sheet or cloth and destroyed. Some assert that early morning is the best time, as the beetles are not inclined so much to fly. See Remedy No. 102.

B.—For such beetles as feed upon the foliage in daytime, the spreading of the sheet under the tree and shaking should either be done after sunrise, or about noon. If the beetles in either instance are numerous, dipping the sheet in coal oil will destroy such beetles as may fall upon it.

REMEDY NO. 39.

Early in the Fall season dig a trench around the tree, uncovering the roots; in the trench put two or three shovelfuls of new lime if the tree is large, and less in proportion to smaller

trees; add enough water to slack the lime, and cover up with
earth. The Winter rains will convey the lime around the
roots, and destroy the woolly aphis living upon them. Wood
ashes and lime mixed together make an excellent application
applied as above.

See Remedies Nos. 40, 41, 42, 43, 45 and 58.

REMEDY NO. 40.

Early in the Fall season dig a trench around the tree and
place in it dried tobacco leaves; wet the leaves with a solution consisting of one pound of saltpetre, dissolved in three
gallons of water, and then cover the leaves with earth; the
Winter rains will carry the tobacco water around the roots and
destroy the lice. Chimney soot makes an excellent mixture
with the tobacco and saltpetre.

See Remedies No. 9, in relation to tobacco: Nos. 39, 41, 42,
43, 45 and 58. Read No. 47.

REMEDY NO. 41.

In a trench around the tree pour water heated to a temperature of 130 degrees, in which is dissolved one pound of No. 12
or 13 to each gallon of water used. (Nos. 39 and 40 are preferable to this.

It is claimed that hot water (say 130°) poured around the
roots, produce good results, but my experiments in using it
did not warrant any recommendation.

See Remedy No. 98.

REMEDY NO. 42.

Two pounds of home made soft soap to one and one half
gallons of water poured around the roots of nursery stock
(young apple trees), destroyed the woolly aphis, the earth being first cleaned away from the trees.

The roots of young apple trees should be dipped in one of
the above solutions before planting.

See No. 47 and No. 58.

REMEDY NO. 43.

Early in February the tree should be thoroughly scraped, and the crotches and crevices cleaned out, and the limbs and trunk thoroughly washed with No. 44, and if the trees have been infested by the apple-leaf aphis (*Aphis malifolia*), wash the branches also; then, in twenty-four hours, wash or spray with No. 11 or 12, using one gallon of water to each one pound in weight of the mixture, or No. 13, one pound to each five quarts of water used.

See also Remedy No. 77.

REMEDY NO. 44.

Dissolve two ounces of borax in four gallons of water; the borax may be dissolved in two quarts of water, and then add enough water to make four gallons. To this add one quart of any kind of animal oil (i. e., lard oil, neatsfoot oil, etc., but whale oil is preferable), and one quart of coal oil; stir up properly, and it is ready for use. See heading No. 77. Paraffine may be used instead of coal oil, but only the best grades of either should be used. See Remedy No. 44. Read Remedy No. 47.

REMEDY NO. 45.

Young apple trees with roots infested by woolly aphis, should be dug up and burned, as they will not mature healthy trees. If you find any swellings or knotty excrescences on the roots of nursery trees, do not buy them. See Remedy No. 47.

REMEDY NO. 46.

Young peach trees having swellings on the roots, will not mature healthy trees. If such are offered at any price, no matter how low, do not buy them. See Remedy No. 47.

REMEDY NO. 47.

The greatest care should be taken in procuring nursery stock that is not infested by woolly aphis, leaf aphis, scale insects, etc. In all cases, before planting, deciduous fruit trees should be dipped in No. 11 or 12—one pound of the mixture to each one gallon of water used; or No. 13—one pound to each 1¼ gallons of water used, excepting the roots, which should be dipped in No. 5 or 7—one pound to each one gallon of water used. See Remedy No. 45.

REMEDY NO. 48.

To prevent the spread of the black scale, soft orange scale, red scale, cottony cushion scale, etc., on citrus trees, the branches should be thinned out, to give free access to light and air, and also to give an opportunity for thoroughly spraying. All prunings taken from the infested trees should be immediately burned.

REMEDY NO. 49.

The best season for destroying scale insects on citrus and evergreen trees is when the young larvæ are just hatched, and are creeping over the fruit, foliage, etc. Then apply No. 5 or 7, four pounds of the mixture to every five gallons of water used, excepting for red scale, when one pound of the mixture to each gallon of water used will be more effective. The sulphur is necessary for destroying the black smut or fungus. The application should be repeated when the second brood of larvæ appears. Effective results have been obtained by using No. 4. Care should be taken not to spray the trees in very warm weather, or in the heat of the day, as it may cause some of the foliage and fruit to fall off. But to get rid of the pests some inconvenience must be expected. Apply the above solutions at a temperature of 130° Fahr.

See Remedy No. 77.

REMEDY NO. 50.

In January or February, make a strong brine of salt and water, and spray the trees thoroughly; repeat in two weeks. This will denude the trees of foliage, but will effectually destroy the red scale, etc. I have tried this, and found it an excellent remedy, and the tree threw out a new growth of leaves. To every six gallons of brine add one gallon of No. 13. This is the only effectual remedy for mealy-bugs on citrus trees. As soon as the leaves have fallen, spray thoroughly with No. 4, 5 or 7.

Read Remedy No. 52.

REMEDY NO. 51.

To thirty gallons of No. 9, add twenty pounds of No. 5 or 7, dissolved in twenty gallons of water. If convenient, add one half pound of buhach by stirring it in the solution immediately before application. If whale oil soap is not at hand, or cannot be had, strong home made soft soap will do. See Remedy No. 3. Spray thoroughly.

The application should be made and repeated when the young larvæ are moving. This remedy can be used to good effect in Winter, on the vine-hopper. See Remedy No. 33.

NOTE.—In any case where this remedy should fail. use No. 44, or 77, or 8.

REMEDY NO. 52.

After spraying as directed in No. 51, a small trench should be dug around the tree and filled with fresh slacked lime, mixed with strong wood ashes, and allowed to remain uncovered to prevent the insects falling on the ground, and those under the surface of the ground from ascending the tree (the lime and ashes will also be useful as a fertilizer).

REMEDY NO. 53.

When the trees are dormant, spray thoroughly with No. 11 or 12, one pound of the mixture to each gallon of water used; or No. 13, one pound of the mixture to each five quarts of water used.

Read Remedy No. 123 carefully.

REMEDY NO. 54.

When the tree is in leaf use No. 6, four pounds of the mixture to each five gallons of water used: spray thoroughly. This solution will not injure the foliage or fruit and will effectually destroy mildew and young scale insects, and prevent the females of the codlin moth, curculios, etc., from depositing their eggs on the fruit. The spraying should be repeated in about two weeks. No. 5 or 7 may be used, one pound to each gallon of water.

NOTE.—Pears on trees sprayed twice matured ten days earlier than the pears that were not sprayed in the same orchard. See Remedy No. 69, E.

REMEDY NO. 55.

Various remedies have been recommended for destroying wire worms. The following I consider the most practical, at least on farm lands in California:

A.—In cases where garden flowers and plants, vegetable or grass-plots are infested, cut potatoes in two or more pieces, according to size, and cut out the eyes to prevent them from growing. In each piece of potato stick a piece of rod or pointed stick; bury one or more pieces of potato near the roots of the infested plants, one or two inches below the surface of the ground. Examine every second day, and destroy the larvæ or grubs eating the potato. This has been tried in this vicinity (Sacramento), and proved an excellent remedy. A patch of ground that was so badly infested by wire-worms that lettuce could not be grown upon it, was cleaned in ten days so that it yielded an excellent crop this season (1883).

For the protection of field crops, summer-fallow and clean out the weeds. Applications of salt, soot, etc., have been recommended, but probably the most practicable for large tracts is as follows:

Dr. Fitch quoted an article from an English paper reporting the success of white mustard, which concluded as follows:

"I am therefore under strong persuasion that the wireworms may be successfully repelled and eradicated by carefully destroying all weeds and roots, and drilling white mustard seed, and keeping the ground clean by hoeing."

The writer also adds that after the mustard crop he raised the best crop of wheat he had in twenty-one years.

Some of our California farmers express a doubt in regard to sowing mustard, and stating their fears that the remedy would be as troublesome to get rid of as the disease. I am credibly informed that such is not the fact.

The following is taken from the Special Report on Wire-Worms, 1883, by Miss Eleanor A. Ormerod, F. M. S., Consulting Entomologist of the Royal Agricultural Society of England:

1. "It has been found by practical experience that the growing and plowing of white mustard will get rid of the wireworm."

The following recommendations are reported by Miss Ormerod:

"The use of mustard as a growing manure crop is not sufficiently resorted to.

"CHARLES CASWELL, Peterborough."

2. "Mustard sown thickly and allowed to grow to a considerable height, and then plowed in, has been found to be a good preventive, and at the same time adding considerably to the fertility of the land. RIGHT HON. EARL OF PORVIS.

"Per ADAM LEE."

3. "Found mustard a good preventive, and sometimes the only safe crop to sow where wire-worms prevailed.

"RIGHT HON. VISCOUNT PORTMAN,

"Per J. FORESTER."

4. "For fallow after clearing, sow rape or mustard seed, about the end of July, and plow it under when about a foot or eighteen inches high.   Jo Craig."

5. "The wire-worm has been known to disappear after a crop of white mustard, of which one half was eaten on the ground by sheep.   Joseph Paget."

6. "I have no doubt of mustard being a good remedy where it can be applied so as to plowed under for the crop.

"F. R. Hulbert."

7. "White mustard sown and allowed to stand until it comes into flower, is a very useful preparation—where plowed in and pressed—for all corn, and I think prevents the wire-worm doing so much mischief.   Joseph Addison."

8. "I have never known wire-worms troublesome after mustard or vetches.   M. Lock Blake."

The following remedies are also discussed in this valuable report of Miss Omerod's:

Rolling and treading the land with sheep.

Salt and kainite.

Kainite.

Gas lime and alkali waste.

Gas tarwater.

Rape cake meal etc.

Hand-picking.

Application of sea-weed.

Rooks, etc.

The weight of testimony is in favor of mustard as a remedy, and it is probably the most practical for California.

REMEDY NO. 56.

Burn the stubble and loose straw, and all the weeds and grasses in and around fields that were infested, and also the surrounding fields. The burning should be done early in Autumn. This will apply to the joint-worm, wheat aphis, etc.

See Remedies Nos. 20 and 106, A.

REMEDY NO. 57.

Fifty pounds of scrap iron thrown into a barrel containing thirty gallons of water, twenty-five pounds of No. 11, 12 or 13, and two pounds of sulphur, makes an excellent solution for application to trees that are not healthy from the presence of the scale insects, gum disease, etc. This mixture should be allowed to stand fourteen days, and be applied when the tree is dormant. When the barrel is emptied of the liquid, fill again with water, and add one half the amount of No. 11, 12 or 13; also of the sulphur. In fourteen days it will be ready for use. Lye leached from wood ashes is excellent, but care should be taken that it is not applied too strong; test with the egg, carrying one and a half. See Remedy No. 2.

REMEDY NO. 58.

Dig the earth from around the roots of the vine or bush, and apply No. 4 or 3, four pounds of the mixture to each five gallons of water used; then fill the earth around the roots. This should be applied in May, when the eggs of the scale insects are hatching. If for woolly aphis, early in April.

See Remedy No. 59.

REMEDY NO. 59.

In Summer dig the earth from around the collar of the tree, and apply No. 5 or 7, one pound of the mixture to each gallon of water used; repeat every four days, until the bark begins to heal. If the tree is dormant, use No. 13, one pound of the mixture to each five quarts of water used. This will destroy any insect attacking the tree under the surface of the ground, and also heal the bark destroyed by insects, alkali, or from any other cause.

See Remedies Nos. 35, 123 and 124..

26

REMEDY NO. 60.

As soon as the stems are noticed beginning to wither, cut them out and destroy by burning. The currant bushes should be sprayed thoroughly in the Spring, using No. 5 or 7, one pound to each gallon of water used. The top of the roots should be thoroughly saturated. The spraying is to prevent the females from depositing their eggs on the wood.

See Remedies Nos. 25, 26, 27, 28, 60, 66 and 67.

REMEDY NO. 61.

As soon as the larvæ appear on the leaves spray or thoroughly drench the foliage with No. 5 or 7, four pounds of the mixture to each five gallons of water used; add one pound of No. 11 or 12, dissolved in one gallon of water, to each eight gallons of the former. The object is to destroy the caterpillars and the unhatched eggs. Or use No. 6, one pound to each five quarts of water used. See Nos. 63, 62 and 65. Repeat the spraying as often as the larvæ appear.

Read Nos. 35, 123 and 124.

REMEDY NO. 62.

If solutions cannot be readily procured when the larvæ or caterpillars appear on the leaves, build mounds of sand or fine earth around the tree. Early in the morning shake the branches and the caterpillars will fall off on the ground and cannot get to the tree over the mound. If material to make proper mounds is not convenient, put on bands as recommended in Remedy No. 23. (This will not apply to the pear-slug as regards the shaking.)

See Remedy No. 63. Read Remedy No. 98.

REMEDY NO. 63.

In case solutions cannot be had when the slugs appear, fine dust of any kind thrown on them will prevent their maturing.

Sulphur and lime, mixed in the proportion of two pounds of slacked lime to one pound of sulphur, dusted on the foliage, is an excellent remedy, but disagreeable for application.

REMEDY NO. 64.

Use No. 5 or 7, one pound to each two gallons of water used; or one pound of No. 3 or 4 to each one and one half gallons of water used. Mix in solution, when ready to use. one quarter of a pound of buhach to each ten gallons; application should be made in the cool of the evening; drench thoroughly. This is especially applicable to plant-lice (*Aphis*) on flowers and plants.

REMEDY NO. 65.

Use mixture No. 5 or 7, one pound to each gallon of water used, and then add an equal quantity of No. 9; mix together and spray thoroughly. Apply at a temperature of 130° Fahr.
See Remedy No. 61.
NOTE.—No. 6 may be used instead of No. 5 or 7.

REMEDY NO. 66.

Cut out the infested canes early in the Spring and destroy by burning before the beetles escape. Thorough pruning out of infested canes will be necessary to prevent the spread of the beetles, etc.
See Remedies Nos. 25, 26, 27, 28, 60 and 67.

REMEDY NO. 67.

Early in the Spring all branches of peach and other trees on which the buds are infested by larvæ or beetles, should be pruned out and immediately burned, and also all new growth showing signs of withering at the outer end, should be cut off

and burned.   This should be done thoroughly, as the safety of
the crop depends on the destruction of the early broods.

See Remedies Nos. 25, 26, 27, 28, 60 and 66.

REMEDY NO. 68.

Where the plants are seriously infested, dig out all infested
plants and burn them, and replace with plants not infested.
(In this case "an ounce of prevention is worth a pound of
cure.")

See Remedy No. 104.

REMEDY NO. 69.

A.   From observations it is evident that the destruction of
this pest must be consummated while it is in the caterpillar
state.

At any time between the first day of November and the first
day of March of each season, all the apple, pear and quince
trees, in any orchard infested by codlin moth, should be carefully scraped and all loose bark removed, as follows:

Fig. 376.

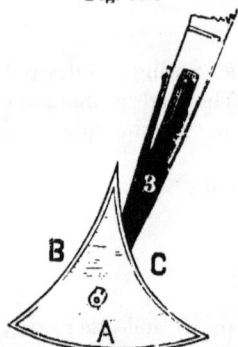

Provide some small ship scrapers
and grind two of the edges to a con-
cave curve (Fig. 376) so that they will
fit the trunk of the tree better than a
straight edge can (B and C, Fig. 376).
Scrapers having a length of side of
four inches will be large enough: use
handles to suit.   Procure a cloth made
of old sacks, or any material convenient: spread on the ground around
the tree as far as the scrapings are
likely to fall; then commence on the
tree as far up as there is any rough
loose bark, and scrape it carefully off.

Also examine and scrape all crevices in the bark or those
formed in the crotches of the tree.   Continue scraping until

you reach the ground. This done, gather the scrapings carefully off the cloth, so that they can be burned or otherwise destroyed immediately.

Be careful that you do not neglect gathering carefully the scrapings and destroying them, as on this point depends a great deal of your success. By thus burning the debris taken from the trees the larvæ hibernating in the debris are destroyed.

B. After having completed scraping off the loose bark, the trunk and limbs should be thoroughly washed or sprayed (providing the tree is only treated against the codlin moth). If the woolly aphis or scale insects are present, the whole tree should be sprayed with No. 11 or No. 12, one pound of either mixture to each gallon of water used. If properly applied, this will destroy any larvæ on the tree, and also produce a new smooth bark. See No. 21.

C. Not later than the 10th of May, bands should be placed on the trees as follows:

Cut old grain sacks or cloth in strips from six to eight inches wide, and place a band on each tree near the ground. (It is expected that the rough bark has been scraped off between the band and the ground). The fastening cord or wire should be as near the upper edge as possible, allowing the lower edge to spread out from the tree, say a quarter of an inch or so. Paper will do for bands, but cloth is preferable. Pieces of old sack, or rags, should be placed in the crotches. The larvæ, after leaving the fruit, when looking for a place in which to pass their transformations, will hide under the bands on the trees, or the rags in the crotches, and make their cocoons or nests. The bands and material in the crotches should be examined every seventh day, without fail, and all the larvæ found on them picked off and destroyed.

D. NOTE.—Notwithstanding the fact that recommendations have been made that it is not necessary to examine the bands, etc., every seventh day, the following fact would seem to indicate otherwise; a larva which I caught changed to a pupa on the 28th of June, 1883, and the moth emerged from the pupa on July 6th; length of time spent in the pupa state, less than eight days.

E.—As soon as the apples or pears are formed (say the size of marbles), spray the fruit and foliage with No. 6, four pounds of the mixture to each five gallons of water used; repeat the spraying within twenty days. This will prevent the moth from depositing her eggs on the young fruit, and also destroy mildew and invigorate the tree.

F. Note.—Infested fruit should be picked off of the tree and destroyed by boiling, or otherwise. Fruit falling off the trees should be gathered daily, and all that is infested destroyed.

G.—The spraying should be repeated at regular intervals to protect late varieties of apples, pears, etc.

H.—By following the above directions in a thorough manner, the codlin moth's ravages need not be feared. Half done work will have but little effect.

See Remedy No. 54.

REMEDY NO. 70.

All empty fruit packages returned from market or used in shipping fruit in any manner, should be thoroughly disinfected before being taken to the orchard, by dipping in boiling water containing one pound of mixture No. 13, or one pound of commercial potash to each twenty gallons of water used— the packages to be left remaining in such solution at least two minutes. If only boiling water is used, the package should be kept in it at least three minutes.

Note.—The necessity for the disinfection of return packages is beyond question. (See Chapter IV.)

REMEDY NO. 71.

Empty fruit packages kept in store from the previous year should be thoroughly disinfected before the 1st day of April, as recommended for return packages.

See Remedy No. 70.

REMEDY NO. 72.

Fruit-houses, store-rooms, etc., in which fruit is stored, packed or sold, should be thoroughly cleaned in the month of March of each year. In cases where they cannot be closed up, as described in Chapter XII. page 39, a solution should be made by saturing chloride of lime with coal oil, all that it will take up, then thinning with water until it can be used with a brush. All cracks, crevices, or seams in walls, skirtings, floors, etc., should be thoroughly saturated with this solution. This will penetrate the cocoons or nests, and effectually destroy the larvæ and pupæ of the codlin moth.

See Remedy No. 75.

REMEDY NO. 73.

A.—A most effectual remedy to prevent cut-worms from ascending a tree is to fasten a piece of tin or zinc around the tree, just above the ground. It should be cut out in a circular form, so that when placed on a tree it will be in the shape of an inverted funnel. The caterpillar cannot creep over the smooth surface of the tin. If the tin is not perfectly smooth, it should be made wider.

B.—It is also an excellent trap to dig holes with perpendicular sides in the ground around the trees; the caterpillars falling into the holes cannot get out until captured. Holes made in the ground with a stick are sometimes used. They may also be captured in the manner recommended in Remedy No. 22.

C.—Cabbage leaves spread around the roots of plants infested by cut-worms are excellent traps, as the pests take shelter under the leaves and do not enter the ground. By examining the leaves in the morning the pest can be found and destroyed.

D.—I have been very successful in destroying cut-worms by dusting, or applying in solution, Paris green, or London purple, or arsenic, on the lower side of cabbage leaves placed on grounds infested by cut-worms, invariably finding all that had

fed on the leaves dead.   Where land is thoroughly cultivated,
the cabbage leaves spread over the ground, prepared as above,
is an excellent remedy; but great care should be taken that
poultry, etc., be kept off of the grounds treated in this manner.
Poisons such as Paris green, London purple and arsenic should
be used only when all other remedies fail, and then the greatest care should be taken to prevent any accident.

E.—Where plants are infested the cut-worm can be found
close to the plant, three or four inches below the surface of the
ground.   Early in the Spring, when the buds are opening, the
cut-worms can be jarred off of the trees upon a sheet; about
midnight is the best time for doing this.

See Remedy No. 38.

REMEDY NO. 74.

For destroying aphis (plant-lice) on cabbage plants, etc.,
dust the plants with snuff—Scotch snuff is preferable—or dust
with No. 80, or spray with No. 83 or No. 85; or with No. 5 or
7, one pound to each one and one half gallons of water used;
or with No. 5 or 7, as above directed, adding an equal quantity
of No. 9; mix well and apply.   See No. 65.

The spraying should be done in the evening, when the sun
is near setting.

REMEDY NO. 75.

For cleaning granaries of weevil before storing grain in them,
use No. 72 freely, and then whitewash.

See Remedy No. 72.

REMEDY NO. 76.

A.—Grain in storehouses or granaries should be piled so as
to allow a free circulation of air around the rows of sacks.

B.—It is generally conceded that the grain weevil cannot
breed if the temperature is kept in the building or storehouse
lower than 65° Fahrenheit.

6.—Thorough ventilation is important, as I have reason to believe that at least part of the wheat is infested before it reaches the granary; therefore the necessity of preventing the eggs from hatching.

D.—Clean thoroughly, as directed in Remedy No. 75, and provide against a high temperature. and good results will follow.

E.—Grain kept in bulk should be turned over occasionally by shoveling, to prevent becoming heated.

F.—Stored grain should be kept free from damp, and kept perfectly covered, to prevent becoming wet from rains, etc.

REMEDY NO. 77.

Coal oil or kerosene emulsions. Personally, I am opposed to the use of mineral oils on trees or foliage, but deem it proper to give Professor C. V. Riley's remedy for scale insects, known as the " Kerosene Emulsion," which he reports as giving excellent results. It is as follows :

" The process of forming a perfectly stable emulsion of kerosene and milk is comparable to that of ordinary butter-making, and is as follows : The oil and milk in any desired proportions are poured together and violently dashed or churned for a period of time varying with the temperature from fifteen to forty-five minutes. The churning. however, requires to be more violent than can be effected with an ordinary butter churn.

" The Aquapult force-pump (the Gregory pump will answer this purpose.—M. C.) may be used satisfactorily for this purpose where moderate quantities are only required. The pump should be inserted in a tub or pail containing the liquid, which are then forced into union by continuous pumping back into the same receptacle through the flexible hose or spray-nozzle. * * * On continual churning through the pumps, the liquid finally curdles and suddenly thickens, to form a white and glistening butter, perfectly homogeneous in texture, and stable. This kerosene butter mixes readily in water, care being taken to thin it first with a small quantity of the liquid. * *

At a temperature of 60° the butter will be made in from thirty
to forty-five minutes.   *   *   At 75°, in about fifteen minutes.

" The following proportions are recommended : 2 quarts of
refined kerosene : 1 quart of fresh cow's milk (but sour).   This
will make an emulsion of kerosene 66⅔ per cent, cow's milk
33⅓ per cent.   Where cow's milk is not readily obtained, condensed milk can be used.   A can of milk as sold in stores
contains about twelve fluid ounces (three fourths of a pint).
Kerosene, 2 quarts equals 4 pints—64 per cent ; condensed
milk, 1 can equals ¾ pints, water (double the quantity of condensed milk), 2 cans equals 1½ pints—36 per cent.   In applications for scale insects on citrus trees it should be used at the
rate of one part of butter to from twelve to sixteen parts of
water, or, in other words, one part should be diluted with
water from twelve to sixteen times."

See United States Agricultural Report for 1881 and 1882,
pages 112 to 127.

See Remedies Nos. 8 and 44.

REMEDY NO. 78.

By dusting London purple around the stems of trees and
plants so that cut-worms or caterpillars will have to pass over
it ; by licking their feet they are poisoned.   I have also found
that it has the same effect on beetles, where it gets upon their
feet, antennæ, etc.; by cleaning it off their feet, mandibles,
etc., it poisons them.   London purple should not be used, in
the Summer season on fruits or vegetables used for food, at
least in California where there are no rains to wash it off.

See Remedies Nos. 79 and 103.

REMEDY NO. 79.

Paris green, dusted around the stems of trees and plants,
produce the same effects upon insects, and the same objections
are offered against its use in the Summer season as Nos. 78
and 103.

REMEDY NO. 80.

Buhach is a powder made from the flower of *Pyrethrum cinerariæfolium* (Fig. 377).

Fig. 377.

The experiments I have made with this powder, and the results obtained, have warranted me in recommending its use in many cases, for destroying insect pests of the household, orchard, vineyards, etc., without fear of injury to any person or animal. The plants from which this powder is made are grown on the farm of the Buhach Manufacturing Company, of Stockton, California.

Letters addressed as above will be immediately answered, giving such information as may be required as to prices, etc.

See Remedies Nos. 81, 82, 83, 84, 85, 118 and 123.

REMEDY NO. 81.

Dry Buhach powder can be dusted around a room to destroy mosquitos, house flies, etc., by a small bellows. (Price, 25 cts.)

It can also be applied to plants infested by aphides in the same manner. Should it be required to be used extensively on the grounds, it should be thoroughly mixed, one pound of the buhach to five pounds of flour, and dusted as sulphur is applied on grapes.

See Remedies Nos. 80, 82, 83, 84, 85, 118 and 123.

REMEDY NO. 82.

For fumigating clothes, etc., infested by moths, the buhach powder, or tobacco, or sulphur burns freely. If the former, such as is sold in cans, by wetting it, it burns slowly; place the articles infested by moths in a tight box, or in a small, close room. By fumigating with the buhach, or tobacco, or sulphur, all insect life will be destroyed. However, thorough application is necessary, as the larvæ of beetles are not so easily destroyed as insects of more delicate structure. Place a few pieces of burning charcoal in a pan, and on them dust the buhach, tobacco or sulphur as prepared, place in apartment to be fumigated, and close up tight. This is also applicable to rooms infested by mosquitoes, house flies, etc.

See Remedies Nos. 80, 81, 83, 84, 85 and 123.

REMEDY NO. 83.

Mix the powder in water, one ounce to each one gallon of water used, and use immediately; for caterpillars, plant-lice, etc., infesting trees, plants or flowers, apply in the evening, or in cloudy weather. In many cases the solution should be strained through a cloth, so as not to stain the flowers, etc.

See Remedies Nos. 80, 81, 82, 84, 85 and 123.

REMEDY NO. 84.

In a gallon of alcohol put six pounds of buhach, and cork up tightly; this can be diluted with from ten to twenty parts of water to one part of the mixture; apply with spraying nozzle. Strain if necessary. See No. 83.

See Remedies Nos. 80, 81, 82, 83, 85 and 123.

REMEDY NO. 85.

Professor Hilgard, of the State University of California, obtained satisfactory results from a tea, or decoction made by pouring boiling water upon the buhach flowers (not ground), and covering as in making tea, it being found that boiling is injurious to the strength of the liquid. For plant-lice, etc., use one pound of the tea to twenty-four gallons of water. For beetles, one pound to every ten gallons. Apply with spray nozzle. The powdered buhach can be used in this manner. Strain if necessary. See No. 83.

See Remedies Nos. 80, 81, 82, 83, 84 and 123.

REMEDY NO. 86.

Dig a trench or ditch about twelve inches wide and twelve to fifteen inches deep between the invading army of worms and the fields to be protected, the side of the ditch next to the fields to be perpendicular, or dug under, if possible, to prevent the worms from creeping up. Every fifteen or twenty feet a deep hole should be dug in the bottom of the ditch, where the worms can collect. In these holes they can easily be destroyed by pouring coal oil on them, or dusting them with buhach, or placing straw in the ditch and setting fire to it. A little coal oil sprinkled on the straw will make it burn more readily. This plan proved effectual in this vicinity (Sacramento) in 1879. Effectual results were derived from plowing a furrow eight inches deep, and kept soft by dragging brush in it. But two or three furrows, some two or three feet apart, would be better, and the vegetation on the space between them dusted with Paris green or London purple, one pound of either (the former is preferable) mixed in twenty pounds of coarse flour and dusted on the vegetation, or one tablespoonful of either

mixed in a pail of water. This solution sprayed or sprinkled
on vegetation will poison the caterpillars or worms eating it.
These substances should be only used where there is no danger
of poisoning stock, poultry, or other animals. On small patches
of corn, etc., infested by the army-worm, No. 64 can be used
effectually, by spraying.

REMEDY NO. 87.

Spread fresh-slacked lime, mixed with wood ashes, on the
floor and around the pots and plants infested by these pests—
slugs. They may be trapped by laying fresh cabbage and other
leaves around the roots of flowers and plants; the pests will
feed upon them, and remain under cover until examined early
in the morning and destroyed. Plants have been protected
from the ravages of snails and slugs, by spreading fresh wood-
ashes on the ground around the roots. No. 72 or 78, or
79 spread on the ground around the roots of plants, and also
near the hiding places of the pests, will destroy them. No. 73
is a sure remedy against slugs, etc.

REMEDY NO. 88.

Extract of buhach applied with a dropping tube or glass—
one or two drops will destroy a small colony of insects, also
the eggs. The extract should not be used on very tender
leaves. This will effectually destroy mealy bugs in conservatories.

REMEDY NO. 89.

In places where plants, vines or trees are seriously infested,
use one pound of arsenic perfectly dissolved by boiling in ten
gallons of water; then add fifteen gallons of water, making
twenty-five gallons of water to each pound: or, see No. 10.
Especially in relation to young grapevines, one pound of whale
oil soap dissolved in one gallon of water, and added to every
ten gallons of the above will make the latter more effective.

See Remedies Nos. 79, 103, 112 and 129.

REMEDY NO. 90.

As soon as the moths appear, or about the middle of April, the contents of wardrobes, closets, etc., should be removed, carpets taken up, and tapestry, etc., removed and exposed to the air and sun for several hours, and then be thoroughly brushed and fumigated as directed in No. 82. In infested apartments all cracks or crevices in the floors and walls, in the wainscoting and shelving of closets, etc., should be brushed over with spirits of turpentine. Powdered black or cayenne pepper, or Scotch snuff, strewn under the edges of carpets, etc., will repel the moths. Sheets of paper saturated with turpentine, camphor in coarse powder, tobacco leaves or snuff, are said to be an excellent preventive for placing among cloths laid away for the Summer, or articles not in use. Chests or closets made of cedar wood are said to afford a protection against clothes-moths, etc. Pieces of cedar wood placed among clothing, etc., is also effective. Carpets that will not be used during the Summer, when taken up and thoroughly cleaned, as directed, should be placed in boxes, and all seams or joints in the boxes pasted over with paper, to prevent the moths from entering. Cloth covering will do, if carefully folded, to prevent the moths from getting in through the folds. When carpets that cannot be taken up conveniently are infested, dampen it slightly with turpentine applied with a sponge. This will leave a disagreeable odor for a short time, but will destroy the larvæ, etc. Fumigating as recommended in No. 82, will give excellent results.

REMEDY NO. 91.

In forty gallons of water boil fifty pounds of potash until dissolved; then add one hundred pounds of arsenic, and boil until dissolved; when cool, put in barrel and cork. To one gallon of this mixture add eight gallons of water, and sprinkle the infested hides, etc., with a broom. This will effectually destroy all insect life upon hides, etc.

See Remedy No. 123.

REMEDY NO. 92.

To each sixty gallons of water used, add twenty pounds of sulphur and five pounds of lime, as follows: When the water is boiling, add the required proportion of sulphur; let it boil for twenty-five minutes, then add the lime, and boil a short time.

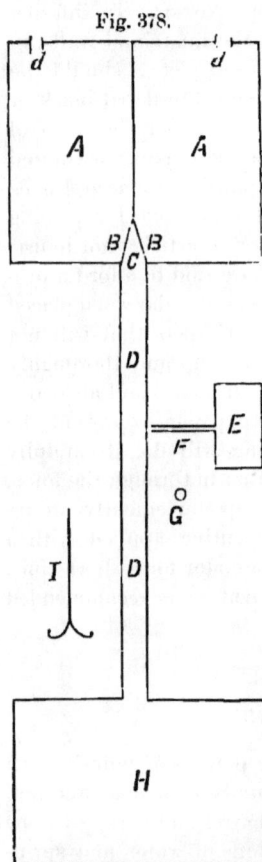

Fig. 378.

This water is emptied from the boiler or kettle (E, Fig. 378,) into the tank or trough, until sufficient is prepared to allow the sheep to float.

Fig. 378 is a plan of the arrangements used by Messrs. Green & Trainor, of Sacramento, at their farm in Placer County.

A A are two corrals, 10 by 12 feet, floored with tongue and grooved lumber, with fall to drain in center, allowing the wash dripping off of the sheep after dipping to drain into the dipping trough.

B B are gates to the corrals.

C. Bridge at end of dipping trough over which the sheep enter corrals.

D D. Dipping trough, 38 feet long by 2 feet wide and 3 feet deep.

E. Boiler, of a capacity of four hundred gallons.

F. Pipe with faucet leading to dipping trough from boiler.

G. Pump to supply boiler, etc., with water.

H. Corral, 24 by 20 feet, into which the sheep are gathered before dipping.

I. Pole, with outer-turned iron prongs, used for forcing the sheep under the surface of the liquid when passing through the trough, D D.

When the sheep are gathered in corral H, and the arrangements for dipping are completed, the sheep are forced into the trough, D D, (in which the liquid is twenty-seven inches in depth, and kept at a temperature of 80° Fahr.) ; with the pole I, the animal is forced under the surface of the liquid. One of the gates B is open; the animal reaches the bridge C and enters the corral through the open gate; when this corral is filled with sheep the gate is closed and the gate of the other opened. During the time the second corral is being filled the sheep placed in the first are sufficiently dripped (or clear of the liquid), and are allowed to go to the pasture; thus the corrals are filled and emptied alternately. A great many sheep can be dipped in a day in this way. The supply of liquid and temperature is kept up by the boiler E. Two hundred and seventy-five pounds of sulphur and 70 pounds of lime will make sufficient wash for 4,000 sheep after the tank is filled at first.

By covering the trough D D, the wash can be kept from one season to another.

If the above directions are complied with after each clipping, it is an effectual remedy for the scab.

It is not necessary to boil all the water required to make the solution.

This industry should be protected by legislation, by preventing infested flocks being driven from one section of the State to another, as beyond doubt the scab is spread to all sheep grazing on lands over which infested flocks are driven.

Such laws have been enacted in Australia, and have been of great benefit to sheep-raisers.

REMEDY NO. 93.

Spirits of turpentine injected into the nostrils with a syringe, will effectually destroy the maggot of the sheep bot-fly; also, maggot in the ears or any part of the body.

27

REMEDY NO. 94.

Placing Stockholm tar on the nose and inside the nostrils, will prevent the bot-fly from depositing her egg or larva in the nose of the sheep.

REMEDY NO. 95.

Place Stockholm tar in the bottom of a trough to the depth of one inch, and cover with salt. In this way the tar gets on the nose and into the nostrils while the sheep is gathering the salt.

REMEDY NO. 96.

Use No. 5 or 7, one pound of the mixture to each gallon of water used, and add one gallon of No. 9 to each gallon of the solution used. See Remedy No. 52.

REMEDY NO. 97.

In all cases where caterpillars congregate on a branch an effective remedy is to cut off the branch and destroy the caterpillars by burning or otherwise.

REMEDY NO. 98.

A.—Mounding. "In the Spring, before the moth emerges (April), a bank of dirt about one foot deep is thrown around the tree and pressed firmly about the trunk. Each subsequent Spring a little more earth is placed on the mound and pressed around the trunk as before. Mr. B. Pullen, of Centralia, Illinois, states that they should not be mounded till after they are four years old, but examine them in April and September of each year previous to that age, and with a knife destroy all borers that can be found. This has been found an excellent prevention, but where the trees are already suffering from them, the earth may be removed from the roots and a copious application of hot water made to the tree. This may be applied at any season, and will be very effectual in killing the larvæ or any eggs that may be present."—Professor G. H. French.

See Remedy No. 41.

B.—Grapevines attacked by *Ægeria polistiformis* may be treated as above, although the mounds need not be so high.

C.—Note.—In the latter end of April or early in May, the moth may be prevented from depositing her eggs by covering the stems of the squash plants with earth.

REMEDY NO. 99.

Hopvines, grapevines, and plants found withering or dying suddenly, should be carefully examined, and if the grub (**Fig.** 148) of any of the boring beetles or the larvæ of moths are found, they should be destroyed at once; or if found in hop-poles or rotten wood in the garden or field, they should in every instance be destroyed. (Hop-growers should examine their hop-poles carefully when taken from the ground for grubs, etc.) The beetle (Fig. 150) should be destroyed when-ever found. Grapevines, or hopvines, or fruit trees, should not be planted upon lands where decaying oak stumps are left in the ground, as there is danger of the trees or vines becom-ing infested by these grubs.

See Remedy No. 107.

REMEDY NO. 100.

Mr. R. B. Blowers, of Woodland, succeeded in preventing the spread of vine-moths on his premises by capturing the moths that came into his flower garden from the vineyards in the evening, his men capturing several hundred in one evening.

Also see Remedy No. 14.

REMEDY NO. 101.

To rid grounds infested by the pupæ of vine-moths (Figs. 158 and 164), white grubs (Figs. 109*a* and 181*a*), etc., early in the Spring inclose an area of ground by portable fence, say from forty square rods to one acre; in this inclosure place a number of hogs, and on the general principles of "root hog or die," the insects, in any state of their existence, within their reach, will be gathered clean; then move the corral on new ground.

REMEDY NO. 102.

Make a frame, say three feet square (or any suitable size), leaving an opening to the centre on one side, so that it will allow the stem or trunk of the vine or plant to enter, the frame to be covered with muslin cloth. On one side of the opening tack a piece of cloth wide enough to cover the space not already covered. Saturate the cloth with kerosene and place the frame on the ground under the vine or plant; cover the opening with a loose cloth, then jar the vine or plant; any insect falling upon the saturated cloth will die. This is used at night or in the daytime, according to the habits of the insect to be destroyed; or a frame in the form of a parallelogram, covered with cloth, may be used (of any length required), one placed on each side of the vine or vines, etc., provision being made for covering the open space. This mode of warfare may appear as a tedious operation, but for destroying the larvæ of saw-flies, grape flea-beetles, etc., it is a very effective remedy.

See Remedy No. 38.

REMEDY NO. 103.

Use Paris green, one tablespoonful to two gallons of water. The Paris green should be first mixed in two quarts of water, then pour in the balance of two gallons and apply by spraying; or one pound of Paris green mixed with fifteen pounds of flour.

The following precautions were suggested by Dr. Le Baron in relation to the application of Paris green :

1.—" Always dilute the poison with at least ten times its bulk of flour.

2.—" Apply it to the plants when wet with dew or rain.

3.—" Never entrust its use to young or careless persons.

4.—" Never use it near the house where young children resort.

5.—" Apply it with a gauze bag or other sifter attached to the end of a pole.

6.—" Let the operator always keep on the side from which the wind is blowing.

7.—" Do not apply it to a plant where it will come in contact with the fruit."

Professor Cyrus Thomas writes : " Recent experience appears to give decided preference to the application in liquid solution, especially since improved methods of applying it in this form have been invented. This not only avoids the danger of inhaling the powder, but it has the advantage of rendering it more certain that the application will reach the insects."

Great care should be taken in keeping such poisonous powders or mixtures where children cannot have access to them.

NOTE.—When used for cotton worms, mix one ounce to each gallon of water and spray thoroughly ; but mixing ten pounds of No. 4 or 3 to each barrel (say forty gallons) will make the solution more effective.

See Remedies Nos. 79, 112 and 123.

REMEDY NO. 104.

Where strawberries or other plants are infested by the grub or larva of the crown borer, peach moth, ægeria moths, etc., dig out all infested plants and burn them, and replace with new plants. Better, dig out the whole bed and replant for a sure remedy.

See Remedy No. 68.

REMEDY NO. 105.

Sprinkle the leaves with fresh slacked lime when they are wet with dew or rain; or No. 19 after sunset. These remedies should be applied as soon as the beetles appear.

REMEDY NO. 106.

A.—All grasses, weeds, debris of hopvines, etc., in the hop-fields and around the surrounding fences, roads, etc., should be carefully gathered and burned as soon as possible after the crop is gathered. There are many species of insects that feed upon the hopvine and pass the Winter (hibernate as perfect insects), taking shelter among grasses, weeds and hopvine debris that are allowed to remain on the grounds and around fences and roads that are located in or around the field. Therefore the necessity of cleaning and burning such shelter as they require to pass the Winter under.

B.—All poles used the previous year in the hopfields should be thoroughly scalded before using. (See Remedies Nos. 70 and 71, excepting that one pound of potash should be used for every ten gallons of water.)

C.—The necessity of scalding the poles arises from the fact that I have found the hop aphis hibernating in crevices and roughened portions of the poles. I have also found the Win-ter eggs of a species of plant-louse, but cannot say at present that they were the eggs of the hop aphis.

D.—All willow poles should be scalded before using. I have found the plant-lice infesting willows at the time the poles were cut, feeding upon the hopvine.

E.—I have also found the ova of red spiders on the hop-poles, after being taken from the ground in the Fall season.

F.—CAUTION.—I have found the hop aphis in several hop fields this season (1883). One grower remarked that they were harmless in this State. Do not depend on such state-ments; prevent them from spreading, by all means in your power; otherwise you may have the same experience of the hop-growers in some of the Eastern States, and also in Europe.

G.—The presence of the hop aphis, when in great numbers, is noticed by the *black smut* (*Fumago salicina*), which appears on the leaves, etc., of the vine. See black scale, Chapter LXXXIV.

H.—Clean cultivation and alkaline solutions properly applied will protect the hop crop; also,

See Remedy No. 99 (root borers).

See Remedy No. 55, A (wire worms).

See Remedy No. 55, A (cut worms).

See Remedy No. 73.

I.—Should plant-bugs or any species of beetles attack the vines or foliage, use No. 19. "Eternal vigilance is the price of"—a good hop crop. See Remedy No. 99.

REMEDY NO. 107.

Flowers, strawberries, plants, and grasses, often suddenly become withered. If the roots are examined it will, in at least a majority of cases, be found that the damage is caused by a grub, commonly called the white grub. It is difficult to capture or destroy these pests. Carefully examining the roots of strawberries, flowers, etc., and digging around them, will certainly bring some of the grubs to light; at such times poultry are excellent accessories. In fields they are raided on by crows, blackbirds, etc. The larvæ of the larger species of beetles that are leaf-eaters are known under the common name of white grubs. The only remedy that can be recommenced is hand-picking, as described above, with the assistance of poultry, birds, etc., and by capturing the beetles (perfect insects). See Remedy No. 38.

For excellent illustrations of the so-called white grubs, see Fig. 109*a*, Fig. 181*a*, and Fig. 290, *2*.

See Remedy No. 99.

REMEDY NO. 108.

To destroy plant-lice where a large area of corn is infested, in a financial point of view it is questionable if it can be done profitably. If a field is infested, a rotation of crops may afford temporary relief, but small garden lots of sweet corn and other varieties may be saved by spraying with No. 3 or 4; one gallon of No. 9 added to every ten gallons of the solution used will make the solution more efficient.

After the corn is harvested, the stalks and roots should be gathered in heaps and burned, and the grounds cleaned, as recommended in Nos. 20 and 106, A. Experiments should be made with Nos. 80 to 85.

REMEDY NO. 109.

Fortunately the fruit orchards of this State are free from the plum curculio at the present time. Professor C. H. Dwinelle has furnished the following remedy:

"BERKELEY, August 23, 1883.

"MR. MATTHEW COOKE—*Dear Sir:* In accordance with your request I will give you some notes on fighting the plum curculio or weevil.

"Some years ago I had charge of a number of plum trees at Rochester, N. Y. They were in their prime, and, as a rule, set full crops of fruit, which were as regularly destroyed by the plum curculio. The last season that I had to do with the trees I tried spraying them, when the fruit was about the size of small green peas, with a wash made after a recipe published in the *New York Observer*, as follows:

"To one pound of whale oil soap add four ounces of flour of sulphur; mix thoroughly and dissolve in twelve gallons of water. To one half peck of quick lime add four gallons of water, and stir well together. When fully settled, pour off the transparent lime water, and add to it the soap and sulphur mixture; add to the same also, say four gallons of tolerably strong tobacco water.

" Apply this mixture, when thus incorporated, with a garden syringe (or spray pump) to your plum or other fruit trees, so that the foliage shall be well drenched. If no rains succeed for three weeks, one application will be sufficient. Should frequent rains occur, the mixture should be again applied until the stone of the fruit becomes hardened, when the season of the curculio's ravages is past.

" The mixture is good to destroy the slug, caterpillar, green fly, thrips, and a host of the enemies of vegetation.

" It is my impression that I found the wash stronger than needed or desirable, and diluted it to a considerable extent, but how much I cannot remember. The application was made by means of a garden engine. As a result, the trees bore a full crop of perfect fruit, while that of my neighbors was destroyed as usual.

> " Respectfully yours,
> " C. H. DWINELLE.

" *University of California, August 23, 1883.*"

REMEDY NO. 110.

In relation to the Remedy No. 109, which has proved so effective, and is recommended by Professor C. H. Dwinelle, Remedy No. 66 will probably be equally effective should this pest appear in any orchard in this State.

REMEDY NO. 111.

As soon as the leaves of the cucumber vines appear above the ground, spray thoroughly with No. 5 or 7. Should plant-lice or the cucumber beetle appear on the foliage, spray thoroughly with No. 4, and cut out and destroy all stems infested by the grub of the beetle; or, use No. 65; but if the material is convenient, and the vines seriously infested, No. 51 is preferable.

NOTE.—For aphis, No. 5 or 7 is sufficient, and will not destroy the natural enemies of these insects.

REMEDY NO. 112.

In cases where beetles attack potato and other vines, good results have been obtained by placing a dish or pan under the vines, and brushing the beetles off of the vines into the pan, dish, or tray; if a little kerosene is placed in the bottom it will make short work of the pests.

See Remedies No. 79, 103, 112 and 123; also, No. 78.

REMEDY NO. 113.

Nos. 5 and 7 mixed, one pound to two gallons of water; spray thoroughly at least once each week, from the time the plants appear above the ground until the latter part of May. If thoroughly sprayed, digging a trench along the plants is unnecessary.

REMEDY NO. 114.

In many cases good results follow the capturing of butterflies and moths; also gathering the pupæ. Placing pieces of boards, cabbage leaves, etc., under plants, so that the night feeders take shelter under them, they can be easily captured in the morning. This will also apply to the squash bug and other species of plant bugs.

REMEDY NO. 115.

Great care should be taken in selecting peas, beans, and other seeds for planting that are liable to be attacked by weevils.

By careful examination, infested seeds can be detected by a small blotch on the skin of the pea or bean a little darker in color than the rest of the surface. If a pea or bean is broken open, the grub, pupa, or beetle will be found under the discolored place. It is claimed by some writers that dipping the seed before planting, in boiling water, and letting it remain from thirty to fifty-five seconds, that it will destroy any insect life in the pea or bean. Others claim that if the peas or beans are put in water the infested ones will float. This is not correct in all cases. A general remedy would be for the farmers in one locality not to plant any peas or beans for one season, but keep the seed from the previous year in closed vessels, so that any infested seed would mature the weevil, and the latter must perish, thus giving clean seed.

REMEDY NO. 116.

When this species of aphis appears, cut off the infested tops of stalks, and immediately destroy by burning or otherwise. In small garden patches, use No. 19, or spray with No. 64; or No. 4, 5, or 7 will give good results; but No. 19 or 64 are preferable; or use No. 83 or 85.

REMEDY NO. 117.

By scattering buhach with a small bellows (price twenty-five cents) in a room it will destroy insects, such as mosquitoes, house-flies, gnats, fleas, etc.; and by dusting in crevices of bedsteads, furniture, and cracks in floors, walls, or ceilings, it will destroy bedbugs and the larvæ of fleas, etc. Corrosive sublimate will also destroy insect life in joints and cracks in bedsteads, furniture, clothes-chests, and places where clothes, etc., are kept; it should be applied with a brush or feather. Great care should be taken if corrosive sublimate is used or kept around the house, especially where there are children, as it is very poisonous.

See Remedies Nos. 80, 81, 82, 83, 84, 85 and 118.

REMEDY NO. 118.

By rubbing the parts of the body exposed, such as the hands, face, etc., with the oil of pennyroyal, mosquitoes, gnats, etc., will not bite the parts to which it is applied; a few drops will be sufficient.

See Remedies Nos. 80 to 85, and 117.

REMEDY NO. 119.

To destroy weevils in coffee, rice, etc., fill into an air tight vessel, leaving a little space for a lamp or candle inside; when the candle or lamp is lighted, fasten on the cover tight. If perfectly air tight the lamp will burn up all the oxygen in the vessel and will then go out, thus creating a vacuum in the vessel; open in two days and all the beetles (weevils) will be found on top dead. I noticed this in an English authority some time ago, and have tried several experiments with wheat, rice, middlings, etc., and found it very effective.

NOTE.—A shipment of coffee was received in this city nearly two years ago seriously infested by weevil.

REMEDY NO. 120.

When flour or middlings, in sacks, are infested, place the sack in the sun, and as the beetles gather upon the outside sweep off and destroy.

REMEDY NO. 121.

Paris green dusted in and around infested places, etc., will effectually destroy cockroaches. No. 81 will also be effective if applied at intervals of twenty-four hours. Remember, great care should be taken in using Paris green where there are children, as it is a deadly poison. This caution also applies to the following: Arsenic mixed in corn meal and molasses, enough to form a dough; or arsenic mixed in mashed potatoes, laid in places that the roaches can reach, will effectually destroy them if repeated at intervals.

REMEDY NO. 122.

A.—Various designs of hives have been made to prevent the bee moth from entering, but, so far, all are more or less defective, and the best and surest remedy is to carefully examine the hives that are infested once every week, and collect all the nests, cocoons, etc., and destroy by burning; this should be done in the Spring and also in July and August.

B.—Professor C. V. Riley recommends that " a good way to entrap the worms would be to raise the front of the hive on two small wooden blocks, and put a piece of woolen cloth between the bottom board and the back of the hive. The worms (larvæ) find a cosy place of resort under the rag, where they may be sought and killed from time to time."

C.—Professor G. H. French writes : " With only moderately strong swarms of bees, vigilance, with properly constructed movable frame hives, is the only way to obtain immunity from their ravages."

I have witnessed the successful cleaning of hives seriously infested as described above, but with C and D the work would probably be still more effective.

D.—A loose band of cloth placed around the top of the hive outside, is an excellent trap for the moths to hide under in the daytime, where they can be easily captured.

REMEDY NO. 123.

1.—In using whale oil soap and sulphur, the opening in the nozzle should be made a little larger than the opening used for lye, etc.

2.—When buhach is used with any of the solutions recommended, it should be placed in a perfectly tight vessel and hot water poured on it and allowed to soak until the next day, or until ready for use, then strain and mix in solution and apply immediately. Calculation must be made so that only the original quantity of water recommended is used.

3.—In all cases where buhach is applied in liquid form to

flowers, it should be strained before using, so as not to stain the petals, etc.

4.—If possible, solutions for Summer use on trees, etc., should not be applied on very warm days, as it may injure the foliage; cloudy days and after sunset is the best time in hot weather.

5.—By straining through a coarse cloth all liquids or solutions to be used in spraying, will save time in application.

6.—Tobacco grown by fruit-growers is equal to so much money saved, as it is far superior to any refuse tobacco they can buy, and will make a solution fifty per cent. stronger; or in other words, one and a half gallons of water can be used to each pound, instead of one gallon, as recommended.

7.—Vineyardists should use one pound of buhach to every twenty pounds of sulphur (or one pound to every ten pounds if the vine-hoppers are numerous) when sulphuring their vines early in the Spring season; this will be death to the vine-hoppers, etc. The first sulphuring should be done when the shoots are from fifteen to twenty inches long.

8.—To those not acquainted with using the lye solutions, at first there will be some inconvenience experienced, but it must be remembered that serious diseases require serious treatment.

9.—If the owner of an orchard, vineyard, etc., cannot personally attend to the application of the remedies recommended, it is an imperative necessity that only a reliable person be intrusted with the charge of the work. Only thorough application will produce satisfactory results.

10.—Be careful in using the remedies that are poisonous, such as arsenic, Paris green, London purple, corrosive sublimate, etc., especially where children are around the premises. Under no consideration are they to be applied to fruit, etc., that is to be used as food, as serious results might follow.

11.—The soap and lye solutions recommended are excellent fertilizers, and produce smooth, healthy bark on the trees washed or sprayed.

REMEDY NO. 124.

It must be admitted that proper attention has not been given in the past to the proper treatment of fruit trees, grapevines, etc., as regards the fertilization of the grounds on which they are planted. Chemistry has demonstrated the elements of which the earth is composed, and a list of over sixty is given; yet strange to say, the vegetable kingdom selects its food from only a few of these elements. That the virgin soil of California abounded with the elements necessary for the food of plant life, there can be no question; but the growing of continuous crops has deprived the earth of such ingredients as are principally required for the food of plants, namely: potash, phosphorus, lime, ammonia, etc. When the supply of either of these elements falls below a fixed limit, the plants will not produce as heretofore; and if attacked by insect pests they will soon become worthless. Therefore the necessity of using the best known means of restoring the necessary plant-food to the impoverished ground. As fruit trees and grapevines require a large supply of potash and phosphorus, also some ammonia and lime, these elements can be supplied by using the remedies described in No. 35.

CONCLUSION.

The measurements throughout this work above one twentieth of an inch, are given in inches and lines—a line being the twelfth part of an inch. See illustrations in Chapters XXII, XXIII, XLVIII, etc.

The insects are usually figured the natural size, and when enlarged the natural size is generally indicated by a line or by a cross.

The remedies given are mostly the results of personal experience; in cases where remedies are recommended for insects not found in this State, they are given from analogy of insects belonging to closely allied species.

If new remedies are recommended, do not give up the use of those herein recommended, unless you find by repeated experiments that the new ones are more efficient. It is to be hoped that fruit-growers and others will carefully experiment with the remedies herein recommended, and give the results of their investigations in some horticultural publication, such as the *Pacific Rural Press, Cultivator's Guide*, etc., that others may be benefitted by their experience.

In regard to beneficial insects, the reader is referred to my forthcoming work entitled "INSECTS : INJURIOUS AND BENEFICIAL : THEIR NATURAL HISTORY AND CLASSIFICATION," which is shortly to issue from the press.

EXPLANATION OF PLATE 1.

Fig. 32.—Gray Bark-eating Weevil -color, gray.
Fig. 40.—Yellow Canker Worms—colors. yellow and black.
Fig. 41.—Female Yellow Canker Worm Moth—colors. white and black.
Fig. 42.—Male Yellow Canker Worm Moth—colors. yellowish and brown.
Fig. 51a.—Bracon Fly—colors, black and yellow.
Fig. 64.—Red-humped Caterpillar—colors, yellow. white and black.
Fig. 65.—Cocoon of Red-humped Caterpillar—color, whitish.
Fig. 66.—Red-humped Caterpillar Moth—colors. light and dark brown.
Fig. 74.—Ten-lined Leaf-eater (male)—colors. grayish-brown and white.
Fig. 75.—Ten-lined Leaf-eater (female)—colors, grayish-brown and white.

EXPLANATION OF PLATE 2.

Fig. 78.—Robust Leaf-beetle—color, brown.
Fig. 82.—Apple Maggot—color, white.
Fig. 83.—Pupa of Apple Maggot—color, brown.
Fig. 84.—Apple Maggot Fly—colors, black and white.
Fig. 92.—Branch bored by the Branch and Twig Burrower.
Fig. 93.—Grape cane bored by the Branch and Twig Burrower.
Fig. 94.—Branch and Twig Burrower entering a branch.
Fig. 95.—Branch and Twig Burrower, enlarged—color, brown.
Fig. 106.—Peach Moth and Larva; upper figure, the moth—colors, gray and black; lower figure, the larva—color, yellowish, tinged with pink.
Fig. 107.—Peach infested by the larva of the Peach Moth.
Fig. 110.—Red-bodied Saw-fly, enlarged—colors. reddish-brown and black.
Fig. 111.—Apricot Leaf-roller; upper figure, the moth—colors, yellowish and brown; lower figure, the larva—color, green.
Fig. 112.—Striped Bud-beetle—colors, pale yellow and black.
Fig. 115.—Cherry Worm—color. yellowish-white.
28

PLATE 1.

Fig. 32.

Fig. 41.

Fig. 42.

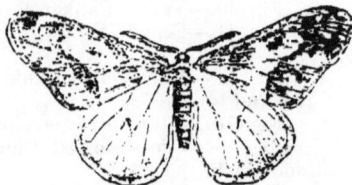

Fig. 51a.

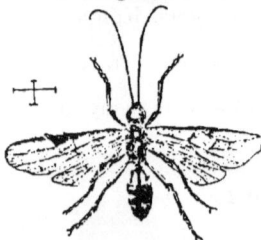

Fig. 40.

Fig. 74.

Fig. 65.

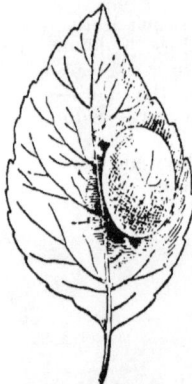

Fig. 64.

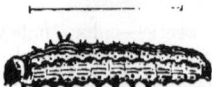

Fig. 66.

Fig. 75.

[For Figs. 78, 82, 83, 84, 92, 93, 94, 95, 106, 107, 110, 111, 112 and 115, see Plate 2.]

PLATE 2.

Fig. 78.

Fig. 82. Fig. 83.

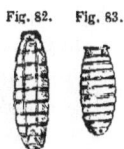

Fig. 84.

Fig. 92. Fig. 93.

Fig. 94.

Fig. 95.

Fig. 106.

Fig. 107.

Fig. 111.

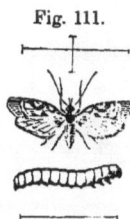

Fig. 112.

Fig. 115.

Fig. 110.

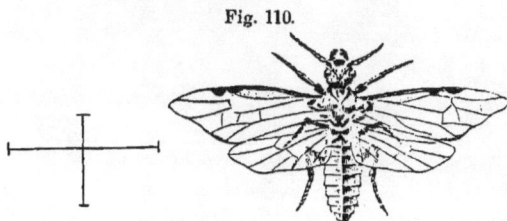

EXPLANATION OF PLATE 3.

Fig. 127.—Brown Strawberry Weevil—color, brown.

Fig. 146.—Male Cottony Cushion Scale-insect—color, brown.

Fig. 179.—Imported Grape Flea-beetle—color, black.

Fig. 182.—California Grape-vine Hopper—colors, pale yellow, orange and black.

Fig. 184.—Male Yellow Mite, highly magnified—color, yellowish.

Fig. 185.—Female Yellow Mite, highly magnified—color, yellowish.

Fig. 220.—Branch bored by the Brown Chestnut-bud Beetle.

Fig. 221.—Brown Chestnut-bud Beetle—color, brown.

Fig. 231.—Weeping Willow Borer Moth—colors, black, yellow and brown.

Fig. 230.—Weeping Willow Borer—color, whitish.

Fig. 246.—Horned Flower Beetle—colors, brownish-yellow and black.

Fig. 277.—Eggs of Clover-stem Borer—color, yellowish.

Fig. 278.—Larva of Clover-stem Borer—color, yellow.

Fig. 279.—Pupa of Clover-stem Borer—color, yellow.

Fig. 280.—Clover-stem Borer—colors, blue-black and dull red.

PLATE 3.

Fig. 146.

Fig. 179.

Fig. 182.

Fig. 184.

Fig. 185.

Fig. 221.

1

Fig. 230.

Fig. 127.

Fig. 231.

Fig. 220.

Fig. 246.

Fig. 278.

Fig. 279.

Fig. 280.

Fig. 277.

EXPLANATION OF PLATE 4.

Fig. 297.—Melon Worm; at the left, several of the worms on some leaves—color of worms, yellowish-green; at the right, a moth—colors, black and white.

Fig. 310.—Small Potato-beetle (No. 1)—color, black.

Fig. 341.—Asparagus Beetle—colors, blue-black, yellow and red.

Fig. 342.—Eggs of Asparagus Beetle on a stalk of asparagus —color of eggs, blackish.

Fig. 343.—Larva of Asparagus Beetle—color, grayish.

Fig. 353.—Dried-fruit Moth—colors, whitish and rusty-brown.

Fig. 354.—Dried-fruit Moth—colors, gray and brown.

Fig. 355.—Grain Weevil—color, blackish.

Fig. 358.—Bran and Flour Bug—color, brown.

Fig. 371.—Wasp— colors, black, the wings reddish.

Fig. 370.—California Lady-bird — colors, reddish-brown, black and white.

Fig. 372.—Mud nests of a wasp.

PLATE 4.

Fig. 297.

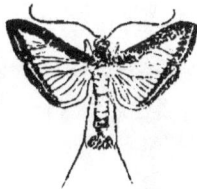

Fig. 358.

Fig. 310.   Fig. 341.

Fig. 370.

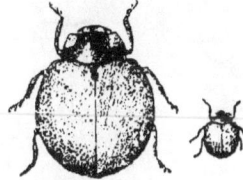

Fig. 371.

Fig. 342.   Fig. 372.

Fig. 353.

Fig. 343.

Fig. 355.

Fig. 354.

BEETLES.

Fig. 25.

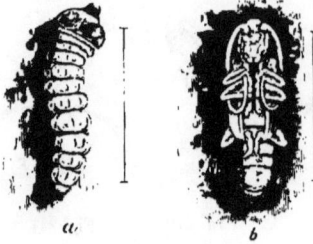

Fig. 26.

Fig. 30.

Fig. 31.

Fig. 77.

Fig. 86.

Fig. 108.

Fig. 89.

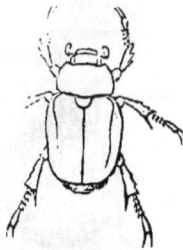

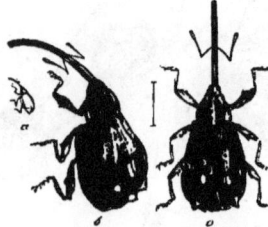

Fig. 113.

Fig. 109.

Fig. 119.

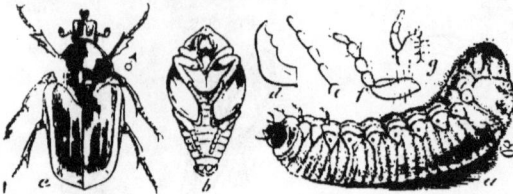

Fig. 103.

Fig. 120.

Fig. 197.

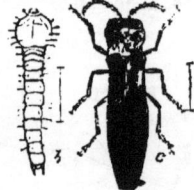

Fig. 156.

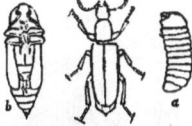

Fig. 201.

Fig. 180.

Fig. 150.

Fig. 178.

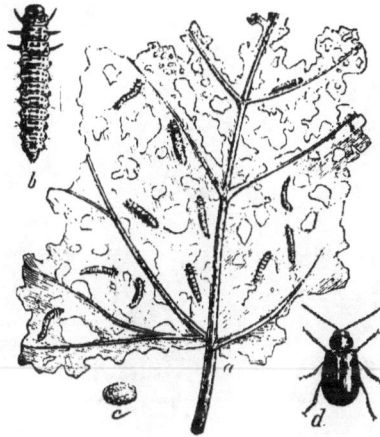

Fig. 181.

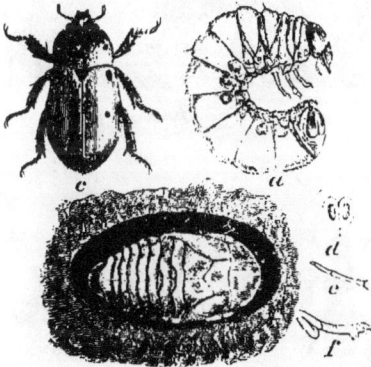

Fig. 204.

Fig. 229.

Fig. 258.

Fig. 195½.

Fig. 216.

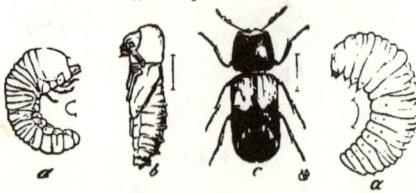

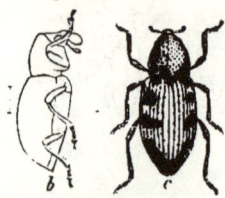

Fig. 191.

Fig. 312.

Fig.1 b

Fig.1

Fig.1a

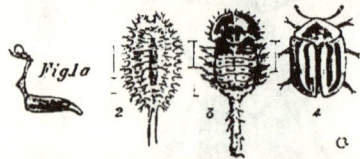

Fig. 234.

Fig. 290.

Fig. 303.

Fig. 306.

Fig. 335.

Fig. 308.

Fig. 314.

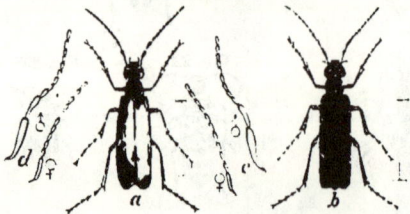

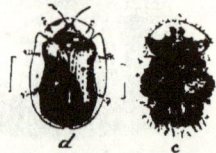

Fig. 282.

Fig. 359.

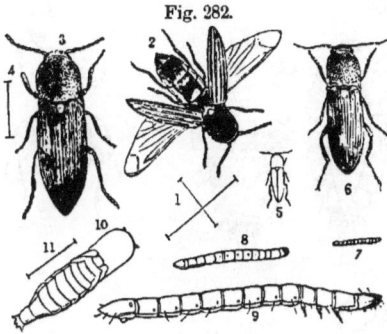

Fig. 305.

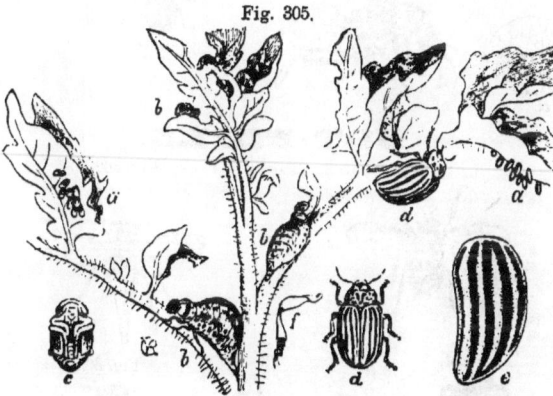

Fig. 309.

Fig. 313.

Fig. 316.

Fig. 339.

Fig. 350.

Fig. 374.

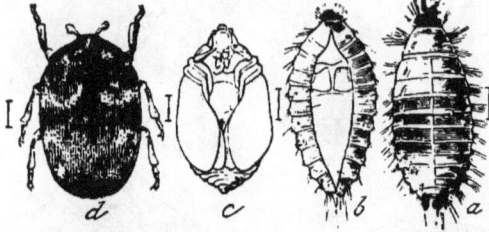

Fig. 338.

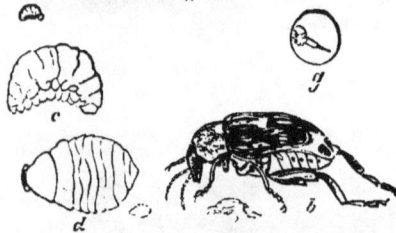

Fig. 300.

Fig. 373.

BOT-FLIES.

Fig. 361.

Fig. 362.

Fig. 364.

Fig. 365.

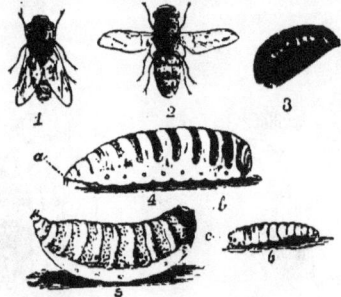

TRUE BUGS.

Fig. 36.

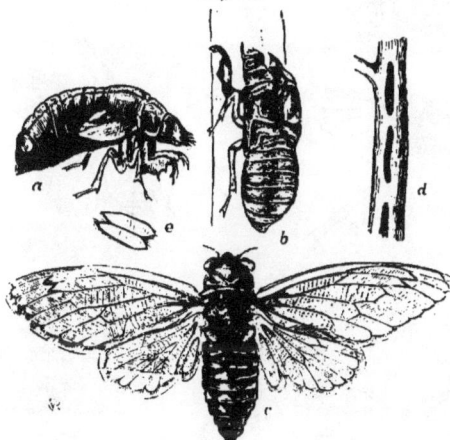

Fig. 266.

Fig. 183.

Fig. 296.

Fig. 200.

Fig. 333.

Fig. 33.

PLANT-LICE.

Fig. 11.

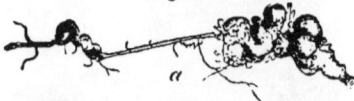

Fig. 74.

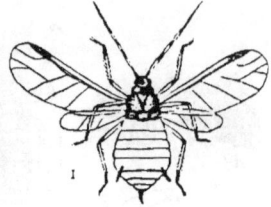

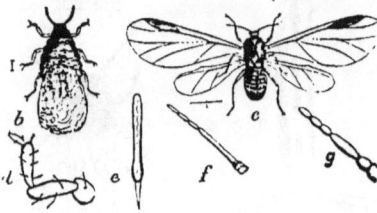

Fig. 73.   Fig. 267.

Fig. 212.

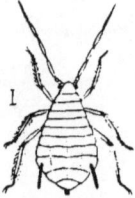

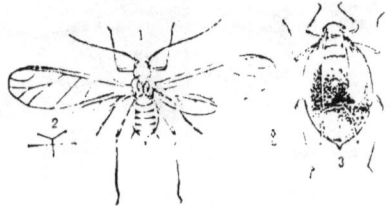

Fig. 268.

Fig. 243.

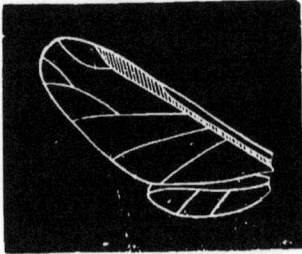

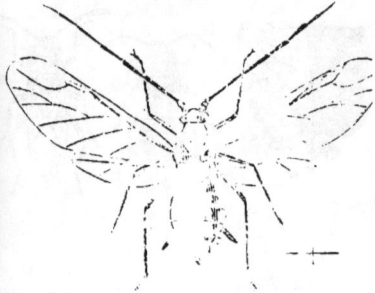

Fig. 256.

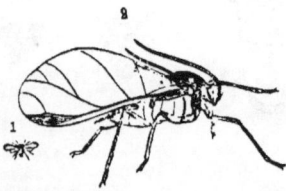

Fig. 152.

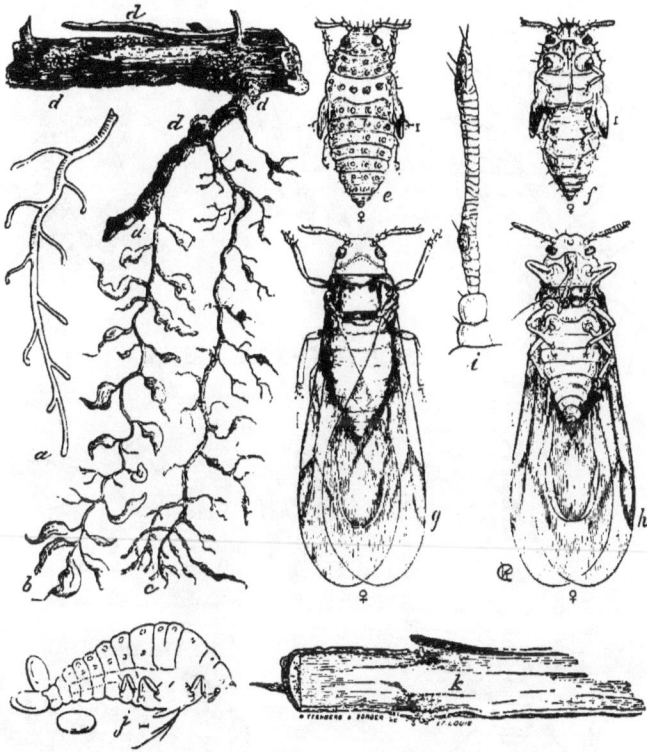

Fig. 225.

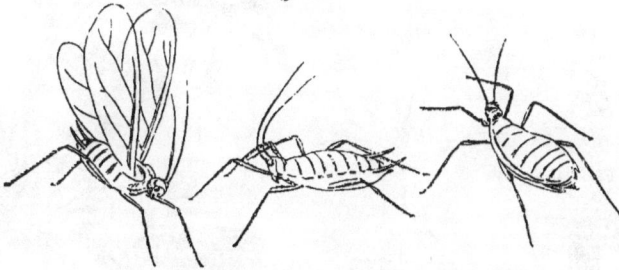

Fig. 334.

Fig. 340.

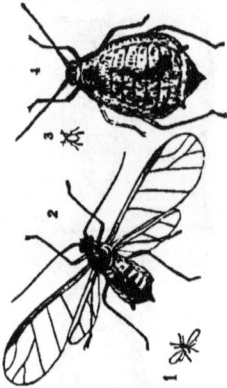

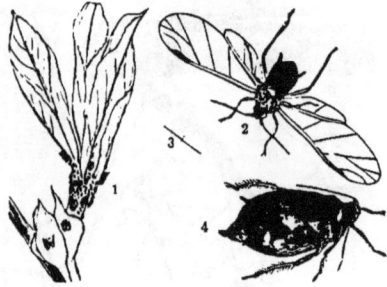

SAW-FLIES.

Fig. 207.

Fig. 209.

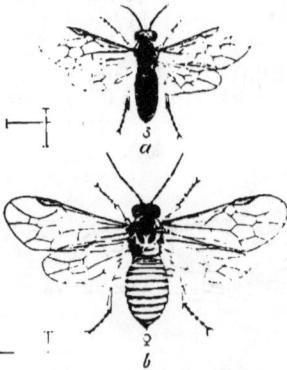

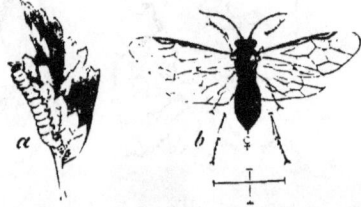

Fig. 218.

Fig. 97.

Fig. 99.

Fig. 245.

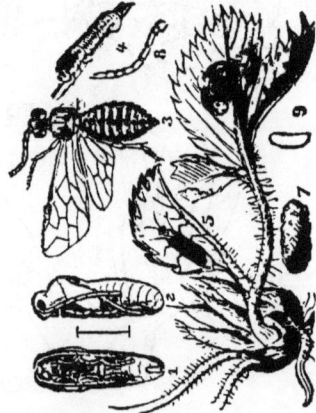

CATERPILLARS, GRUBS, Etc.

Fig. 52.

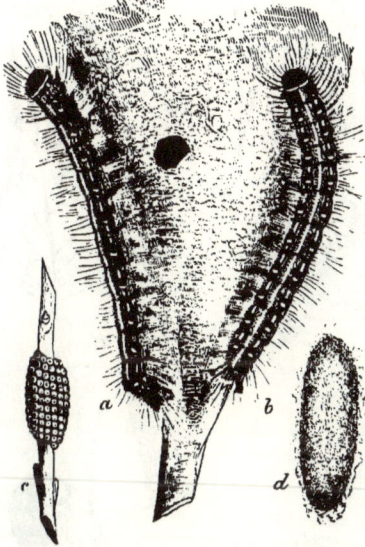

Fig. 55.

Fig. 27.

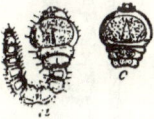

Fig. 58.

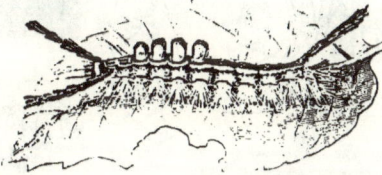

Fig. 37.

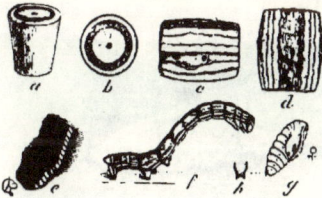

Fig. 43.

29

Fig. 96.

Fig. 98.

Fig. 147.

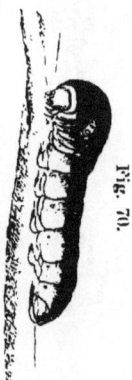

Fig. 70.

Fig. 85.

Fig. 104.

Fig. 117.

Fig. 148.

Fig. 157,

Fig. 166.

Fig. 163.

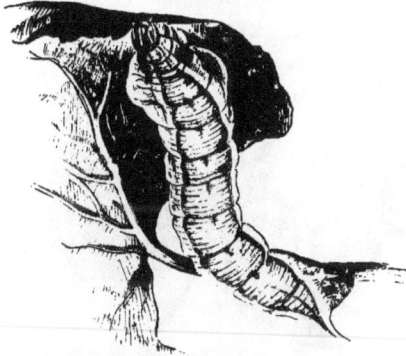

Fig. 160.

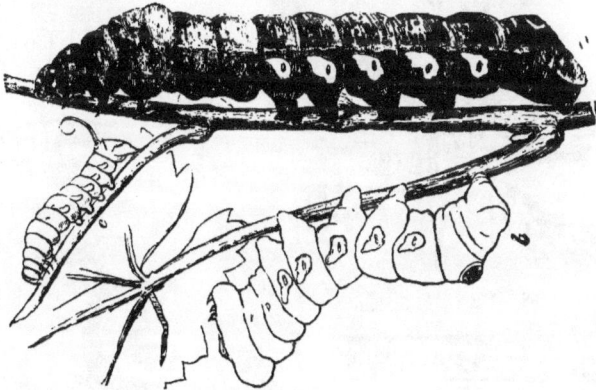

Fig. 167.

Fig. 174.

Fig. 170.

Fig. 193.

Fig. 206.

Fig. 202.

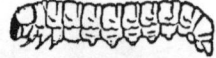

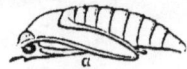

Fig. 192.

Fig. 172.

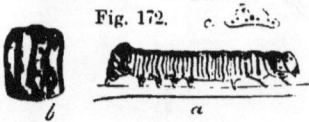

Fig. 210.

Fig. 214.

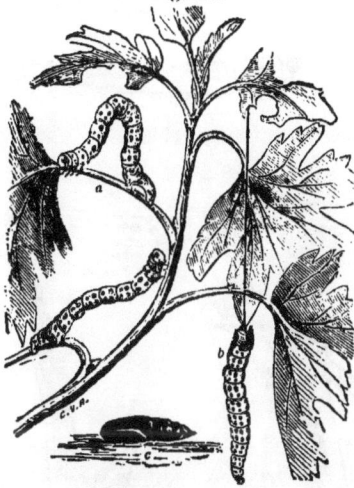

Fig. 224.

Fig. 285.

Fig. 227.

Fig. 269.

Fig. 283.

Fig. 284.

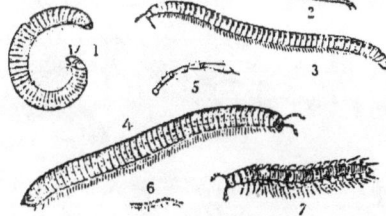

Fig. 286.

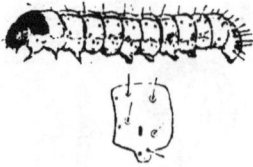

Fig. 299.

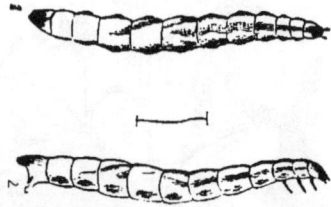

Fig. 292.

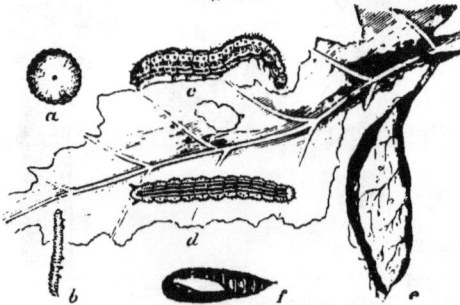

Fig. 327.

Fig. 302.

Fig. 321.

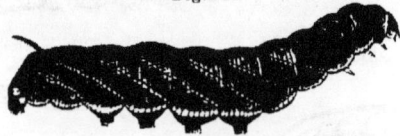

Fig. 324.

Fig. 294.

Fig. 336.

Fig. 348.

Fig. 351.

Fig. 347.

Fig. 363.

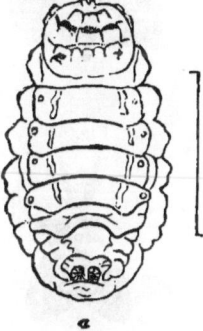

Fig. 344.

Fig. 369.

Fig. 357.

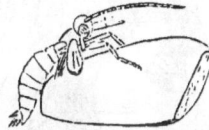

BUTTERFLIES AND MOTHS.

Fig. 39.

Fig. 48.

Fig. 49.

Fig. 45.

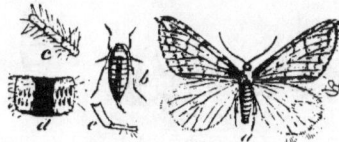

Fig. 68.

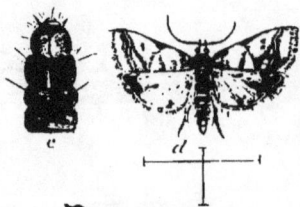

Fig. 56.

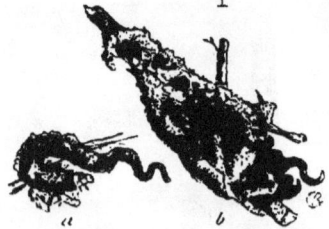

Fig. 69.

Fig. 67.

Fig. 63.

Fig. 60

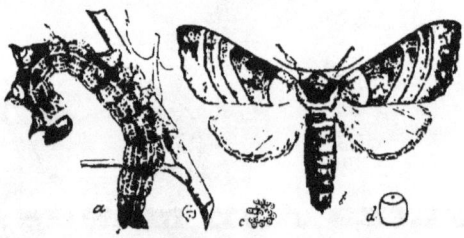

Fig. 57.

Fig. 53.

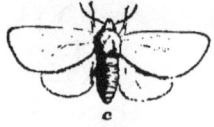

Fig. 71.

Fig. 72.

Fig. 88.

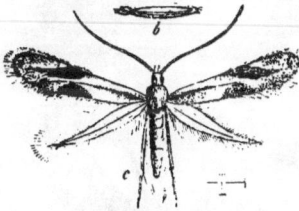

Fig. 114.

Fig. 105.

Fig. 79.

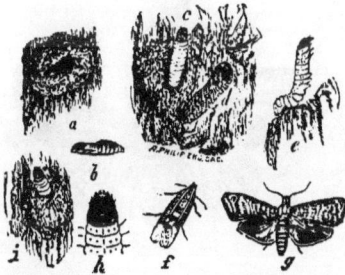

Fig. 169.

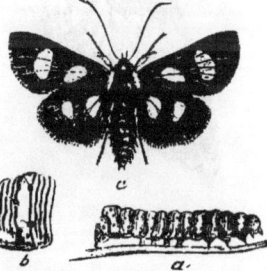

Fig. 161.

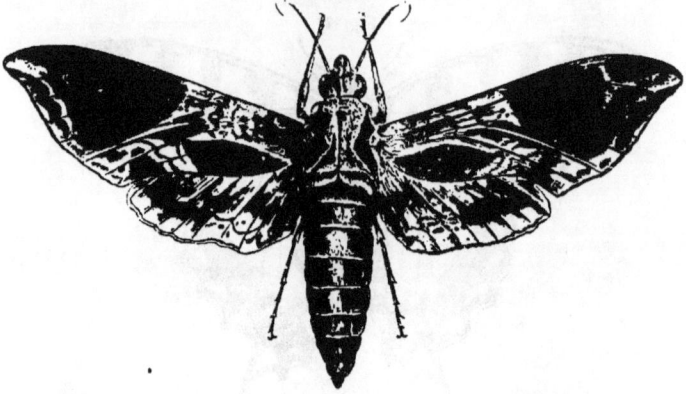

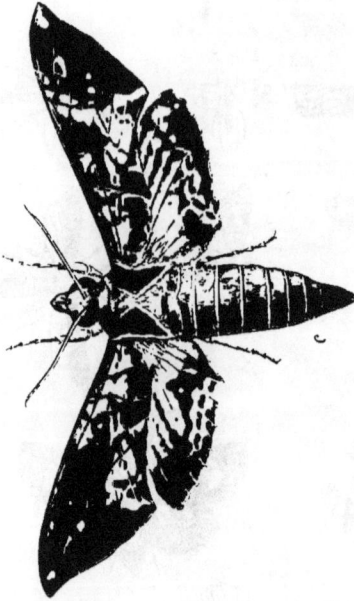

Fig. 159.

Fig. 165.

Fig. 171.

Fig. 162.

Fig. 175.

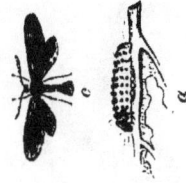

Fig. 168.

Fig. 173.

Fig. 203.

Fig. 176.

Fig. 215.

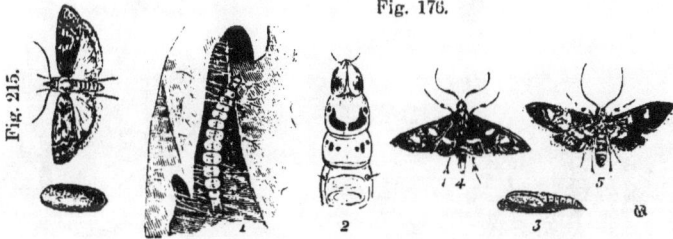

Fig. 177.

Fig. 196.

Fig. 199.

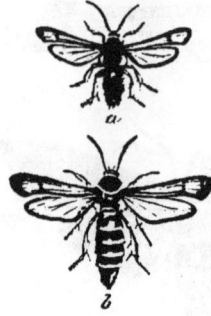

Fig. 211.

Fig. 219.

Fig. 228.

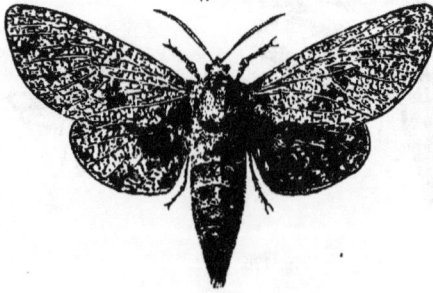

Fig. 232.

Fig. 254.

Fig. 291.

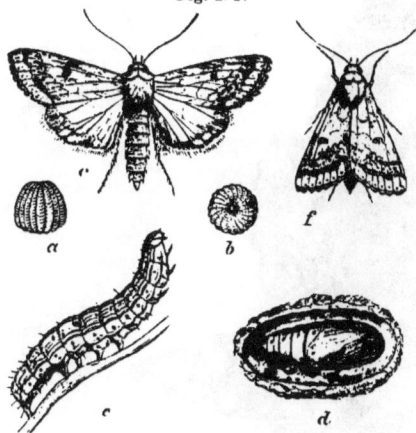

Fig. 271.

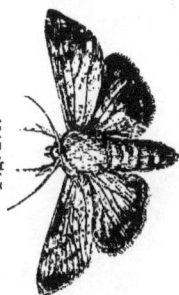

Fig. 295.

Fig. 287.

Fig. 289.

Fig. 283.

Fig. 255.

Fig. 333.

Fig. 341.

Fig. 346.

Fig. 349.

Fig. 329.

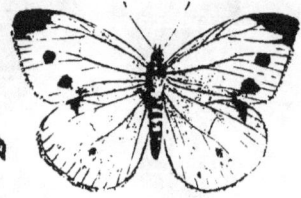

Fig. 304.

1

2

Fig. 301.

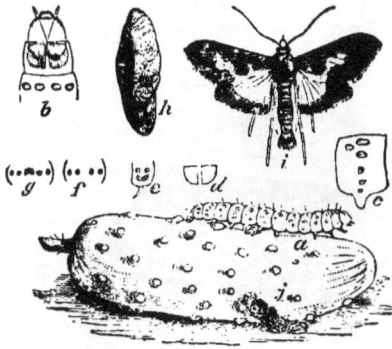

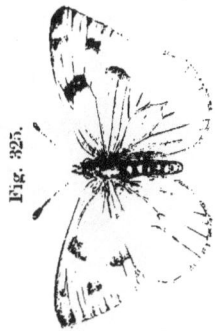

Fig. 325.

Fig. 326.

Fig. 328.

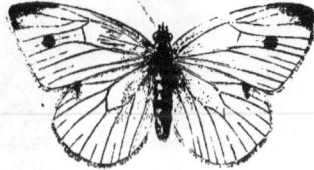

Fig. 330.

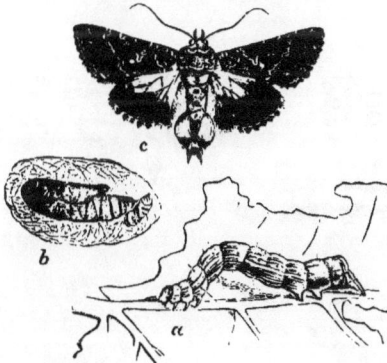

Fig. 356.

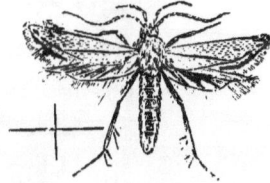

Fig. 331.

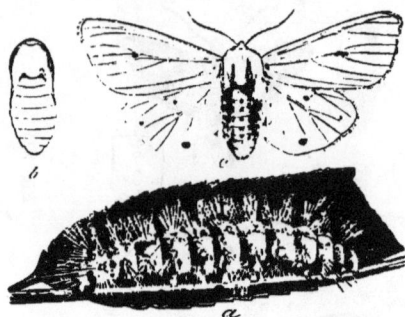

Fig. 345.

Fig. 368.

INDEX TO SCIENTIFIC NAMES.

INDEX TO COMMON NAMES.

www.ingramcontent.com/pod-product-compliance
Lightning Source LLC
Chambersburg PA
CBHW031817270326
41932CB00008B/457